T0092672

Quick Start Guide to VHDL

Quick Start Guide to VHDL

2nd Edition

Brock J. LaMeres

Brock J. LaMeres
Department of Electrical and Computer Engineering
Montana State University
Bozeman, MT, USA

ISBN 978-3-031-42542-4 ISBN 978-3-031-42543-1 (eBook)
https://doi.org/10.1007/978-3-031-42543-1

This Springer imprint is published by the registered company Springer Nature Switzerland AG
The registered company address is: Gewerbestrasse 11, 6330 Cham, Switzerland

Paper in this product is recyclable.

Preface

The classical digital design approach (i.e., manual synthesis and minimization of logic) quickly becomes impractical as systems become more complex. This is the motivation for the modern digital design flow, which uses hardware description languages (HDL) and computer-aided synthesis/minimization to create the final circuitry. The purpose of this book is to provide a quick start guide to the VHDL language, which is one of the two most common languages used to describe logic in the modern digital design flow. This book is intended for anyone that has already learned the classical digital design approach and is ready to begin learning HDL-based design. This book is also suitable for practicing engineers that already know VHDL and need quick reference for syntax and examples of common circuits. This book assumes that the reader already understands digital logic (i.e., binary numbers, combinational and sequential logic design, finite state machines, memory, and binary arithmetic basics).

Since this book is designed to accommodate a designer that is new to VHDL, the language is presented in a manner that builds foundational knowledge first before moving into more complex topics. As such, Chaps. 1, 2, 3, 4, and 5 only present functionality built into the VHDL standard package. Only after a comprehensive explanation of the most commonly used packages from the IEEE library is presented in Chap. 7 are examples presented that use data types from the widely adopted STD_LOGIC_1164 package. For a reader that is using the book as a reference guide, it may be more practical to pull examples from Chaps. 7, 8, 9, 10, and 11 as they use the types *std_logic* and *std_logic_vector*. For a VHDL novice, understanding the history and fundamentals of the VHDL base release will help form a comprehensive understanding of the language; thus, it is recommended that the early chapters are covered in the sequence they are written. The book culminates with a full computer design in Chap. 12 and a detailed look at floating-point systems in Chap. 13.

The second edition adds in a new chapter on floating-point systems. The detailed background of floating-point numbers is given before moving into VHDL modeling. A discussion of the packages in the IEEE and IEEE_Proposed libraries is presented in order to provide an understanding of what models are synthesizable and which are not.

Bozeman, MT, USA

Brock J. LaMeres

Acknowledgments

For my amazing daughter Alexis. You are the kindest person I have ever met. You have been blessed with intelligence, beauty, and a heart that cares more for others than yourself. Watching you become the person you are today has been one of the greatest joys of my life. You will make the world a better place just by being who you are. As you begin your journey into the world, know that you have everything you need to succeed and that you never need to change. Go forward with courage, learn from your mistakes, and know that your family will always be behind you every step of the way. With love, Dad.

Contents

Chapter 1: The Modern Digital Design Flow

The purpose of hardware description languages is to describe digital circuitry using a text-based language. HDLs provide a means to describe large digital systems without the need for schematics, which can become impractical in very large designs. HDLs have evolved to support logic simulation at different levels of abstraction. This provides designers the ability to begin designing and verifying functionality of large systems at a high level of abstraction and postpone the details of the circuit implementation until later in the design cycle. This enables a top-down design approach that is scalable across different logic families. HDLs have also evolved to support automated *synthesis*, which allows the CAD tools to take a functional description of a system (e.g., a truth table) and automatically create the gate-level circuitry to be implemented in real hardware. This allows designers to focus their attention on designing the behavior of a system and not spend as much time performing the formal logic synthesis steps as in the classical digital design approach.

There are two dominant hardware description languages in use today. They are VHDL and Verilog. VHDL stands for *very high speed integrated circuit hardware description language*. Verilog is not an acronym but rather a trade name. The use of these two HDLs is split nearly equally within the digital design industry. Once one language is learned it is simple to learn the other language, so the choice of the HDL to learn first is somewhat arbitrary. In this text we will use VHDL to learn the concepts of an HDL. VHDL is stricter in its syntax and typecasting than Verilog, so it is a good platform for beginners as it provides more of a scaffold for the description of circuits. This helps avoid some of the common pitfalls that beginners typically encounter. The goal of this chapter is to provide the background and context of the modern digital design flow using an HDL-based approach.

Learning Outcomes—After completing this chapter, you will be able to:

1.1 Describe the role of hardware description languages in modern digital design.
1.2 Describe the fundamentals of design abstraction in modern digital design.
1.3 Describe the modern digital design flow based on hardware description languages.

1.1 History of Hardware Description Languages

The invention of the integrated circuit is most commonly credited to two individuals who filed patents on different variations of the same basic concept within six months of each other in 1959. Jack Kilby filed the first patent on the integrated circuit in February of 1959 titled "Miniaturized Electronic Circuits" while working for *Texas Instruments*. Robert Noyce was the second to file a patent on the integrated circuit in July of 1959 titled "Semiconductor Device and Lead Structure" while at a company he cofounded called *Fairchild Semiconductor*. Kilby went on to win the Nobel Prize in Physics in 2000 for his invention, while Noyce went on to cofound *Intel Corporation* in 1968 with Gordon Moore. In 1971, Intel introduced the first single-chip microprocessor using integrated circuit technology, the *Intel 4004*. This microprocessor IC contained 2300 transistors. This series of inventions launched the semiconductor industry, which was the driving force behind the growth of Silicon Valley, and led to 40 years of unprecedented advancement in technology that has impacted every aspect of the modern world.

Gordon Moore, cofounder of Intel, predicted in 1965 that the number of transistors on an integrated circuit would double every two years. This prediction, now known as *Moore's Law*, has held true since

B. J. LaMeres, *Quick Start Guide to VHDL*, https://doi.org/10.1007/978-3-031-42543-1_1

the invention of the integrated circuit. As the number of transistors on an integrated circuit grew, so did the size of the design and the functionality that could be implemented. Once the first microprocessor was invented in 1971, the capability of CAD tools increased rapidly enabling larger designs to be accomplished. These larger designs, including newer microprocessors, enabled the CAD tools to become even more sophisticated and, in turn, yield even larger designs. The rapid expansion of electronic systems based on digital integrated circuits required that different manufacturers needed to produce designs that were compatible with each other. The adoption of logic family standards helped manufacturers ensure their parts would be compatible with other manufacturers at the physical layer (e.g., voltage and current); however, one challenge that was encountered by the industry was a way to document the complex behavior of larger systems. The use of schematics to document large digital designs became too cumbersome and difficult to understand by anyone besides the designer. Word descriptions of the behavior were easier to understand, but even this form of documentation became too voluminous to be effective for the size of designs that were emerging.

In 1983, the US Department of Defense (DoD) sponsored a program to create a means to document the behavior of digital systems that could be used across all of its suppliers. This program was motivated by a lack of adequate documentation for the functionality of application specific integrated circuits (ASICs) that were being supplied to the DoD. This lack of documentation was becoming a critical issue as ASICs would come to the end of their life cycle and need to be replaced. With the lack of a standardized documentation approach, suppliers had difficulty reproducing equivalent parts to those that had become obsolete. The DoD contracted three companies (Texas Instruments, IBM, and Intermetrics) to develop a standardized documentation tool that provided detailed information about both the interface (i.e., inputs and outputs) and the behavior of digital systems. The new tool was to be implemented in a format similar to a programming language. Due to the nature of this type of language-based tool, it was a natural extension of the original project scope to include the ability to *simulate* the behavior of a digital system. The simulation capability was desired to span multiple levels of abstraction to provide maximum flexibility. In 1985, the first version of this tool, called VHDL, was released. In order to gain widespread adoption and ensure consistency of use across the industry, VHDL was turned over to the *Institute of Electrical and Electronics Engineers* (IEEE) for standardization. The IEEE is a professional association that defines a broad range of open technology standards. In 1987, IEEE released the first industry standard version of VHDL. The release was titled IEEE 1076-1987. Feedback from the initial version resulted in a major revision of the standard in 1993 titled IEEE 1076-1993. While many minor revisions have been made to the 1993 release, the 1076-1993 standard contains the vast majority of VHDL functionality in use today. The most recent VHDL standard is IEEE 1076-2008.

Also in 1983, the Verilog HDL was developed by *Automated Integrated Design Systems* as a logic simulation language. The development of Verilog took place completely independent from the VHDL project. Automated Integrated Design Systems (renamed *Gateway Design Automation* in 1985) was acquired by CAD tool vendor *Cadence Design Systems* in 1990. In response to the rapid adoption of the open VHDL standard, Cadence made the Verilog HDL open to the public in order to stay competitive. The IEEE once again developed the open standard for this HDL and in 1995 released the Verilog standard titled IEEE 1364.

The development of CAD tools to accomplish automated logic synthesis can be dated back to the 1970s when IBM began developing a series of practical synthesis engines that were used in the design of their mainframe computers; however, the main advancement in logic synthesis came with the founding of a company called *Synopsis* in 1986. Synopsis was the first company to focus on logic synthesis directly from HDLs. This was a major contribution because designers were already using HDLs to describe and simulate their digital systems, and now logic synthesis became integrated in the same design flow. Due to the complexity of synthesizing highly abstract functional descriptions, only the lower levels of abstraction that were thoroughly elaborated were initially able to be synthesized. As CAD tool capability evolved, synthesis of the higher levels of abstraction became possible, but even today not all functionality that can be described in an HDL can be synthesized.

The history of HDLs, their standardization, and the creation of the associated logic synthesis tools are key to understanding the use and limitations of HDLs. HDLs were originally designed for documentation and behavioral simulation. Logic synthesis tools were developed independently and modified later to work with HDLs. This history provides some background into the most common pitfalls that beginning digital designers encounter, that being that most any type of behavior can be described and simulated in an HDL, but only a subset of well-described functionality can be synthesized. Beginning digital designers are often plagued by issues related to designs that simulate perfectly but that will not synthesize correctly. In this book, an effort is made to introduce VHDL at a level that provides a reasonable amount of abstraction while preserving the ability to be synthesized. Figure 1.1 shows a timeline of some of the major technology milestones that have occurred in the past 150 years in the field of digital logic and HDLs.

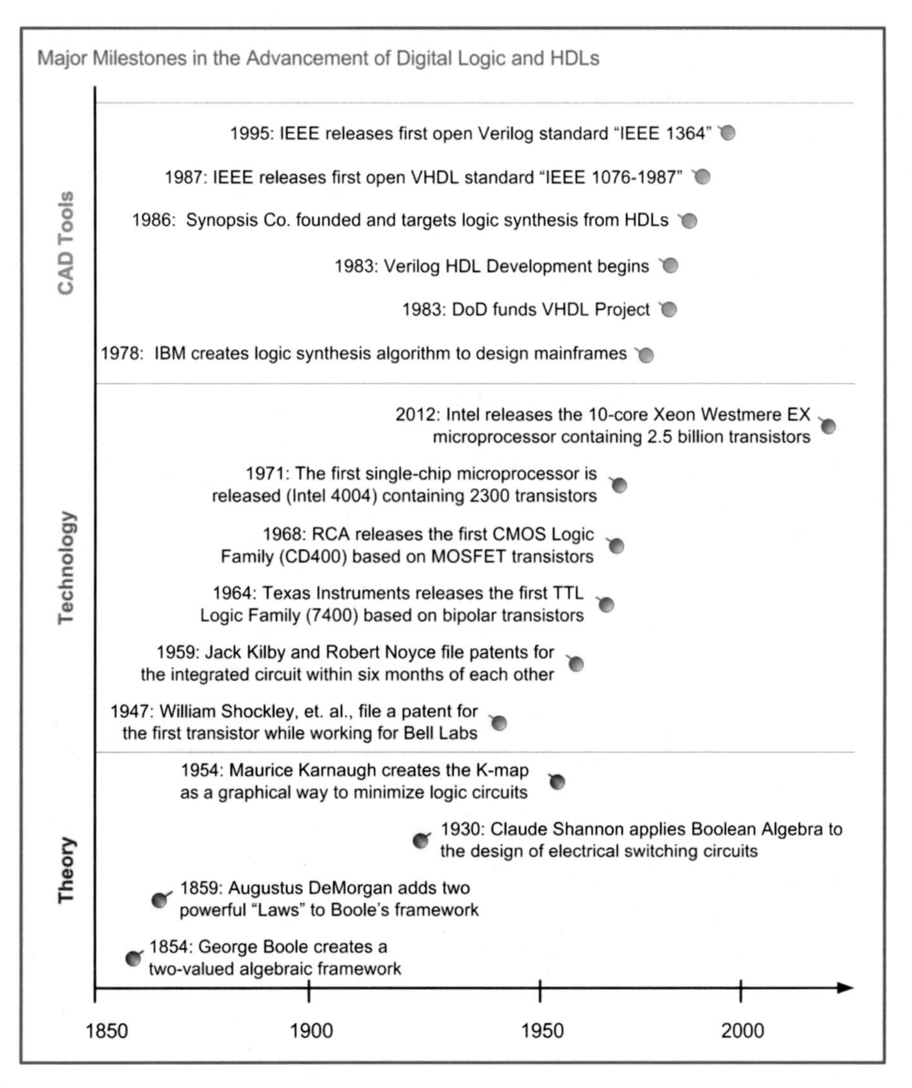

Fig. 1.1
The major milestones in the advancement of digital logic and HDLs

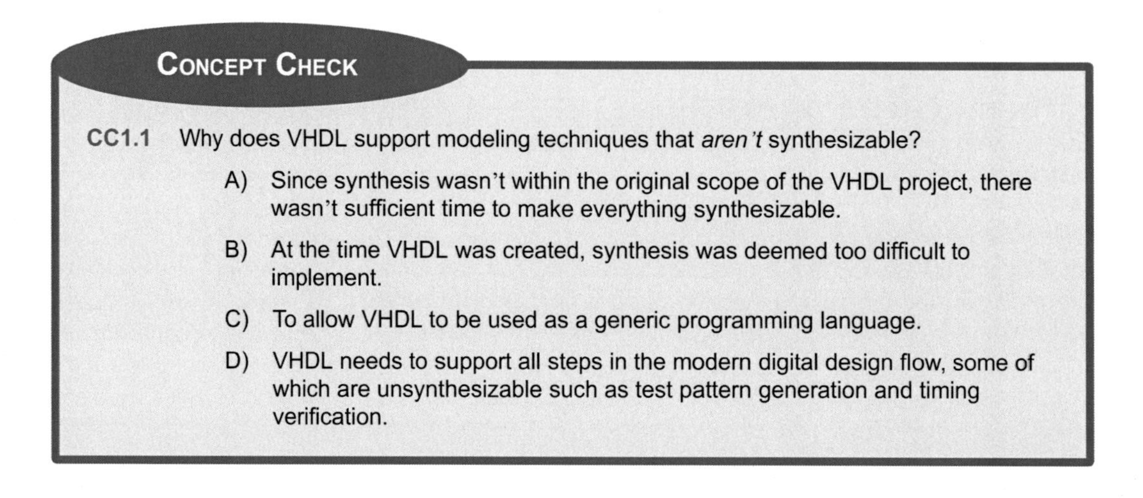

CONCEPT CHECK

CC1.1 Why does VHDL support modeling techniques that *aren't* synthesizable?

A) Since synthesis wasn't within the original scope of the VHDL project, there wasn't sufficient time to make everything synthesizable.

B) At the time VHDL was created, synthesis was deemed too difficult to implement.

C) To allow VHDL to be used as a generic programming language.

D) VHDL needs to support all steps in the modern digital design flow, some of which are unsynthesizable such as test pattern generation and timing verification.

1.2 HDL Abstraction

HDLs were originally defined to be able to model behavior at multiple levels of abstraction. Abstraction is an important concept in engineering design because it allows us to specify how systems will operate without getting consumed prematurely with implementation details. Also, by removing the details of the lower-level implementation, simulations can be conducted in reasonable amounts of time to model the higher-level functionality. If a full computer system was simulated using detailed models for every MOSFET, it would take an impracticable amount of time to complete. Figure 1.2 shows a graphical depiction of the different layers of abstraction in digital system design.

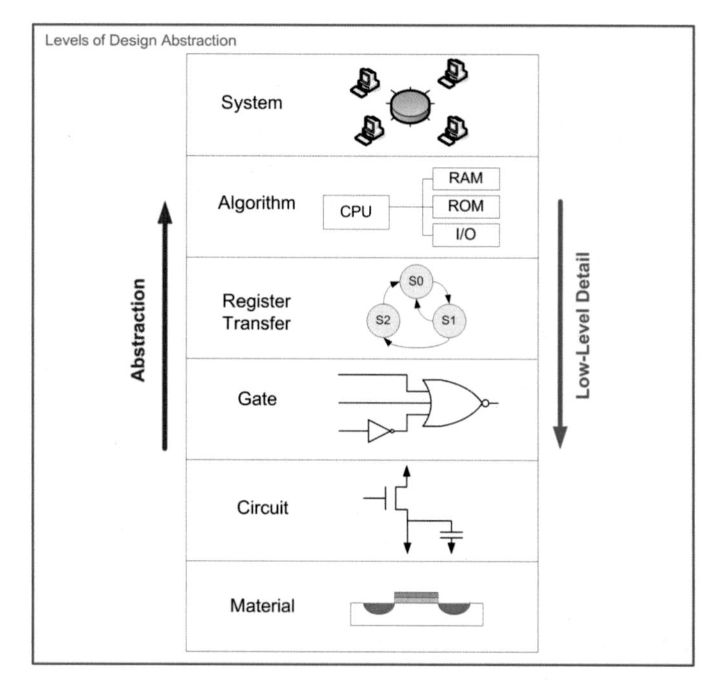

Fig. 1.2
Levels of design abstraction

The highest level of abstraction is the *system level*. At this level, the behavior of a system is described by stating a set of broad specifications. An example of a design at this level is a specification such as "the computer system will perform 10 Tera Floating-Point Operations per Second (10 TFLOPS) on double precision data and consume no more than 100 Watts of power." Notice that these specifications do not dictate the lower-level details such as the type of logic family or the type of computer architecture to use. One level down from the system level is the *algorithmic level*. At this level, the specifications begin to be broken down into sub-systems, each with an associated behavior that will accomplish a part of the primary task. At this level, the example computer specifications might be broken down into sub-systems such as a central processing unit (CPU) to perform the computation and random access memory (RAM) to hold the inputs and outputs of the computation. One level down from the algorithmic level is the *register transfer level (RTL)*. At this level, the details of how data is moved between and within sub-systems are described in addition to how the data is manipulated based on system inputs. One level down from the RTL level is the *gate level*. At this level, the design is described using basic gates and registers (or storage elements). The gate level is essentially a schematic (either graphically or text-based) that contains the components and connections that will implement the functionality from the above levels of abstraction. One level down from the gate level is the *circuit level*. The circuit level describes the operation of the basic gates and registers using transistors, wires, and other electrical components such as resistors and capacitors. Finally, the lowest level of design abstraction is the *material level*. This level describes how different materials are combined and shaped in order to implement the transistors, devices, and wires from the circuit level.

HDLs are designed to model behavior at all of these levels with the exception of the material level. While there is some capability to model circuit level behavior such as MOSFETs as ideal switches and pull-up/pull-down resistors, HDLs are not typically used at the circuit level. Another graphical depiction of design abstraction is known as the **Gajski and Kuhn's Y-chart**. A Y-chart depicts abstraction across three different design domains: behavioral, structural, and physical. Each of these design domains contains levels of abstraction (i.e., system, algorithm, RTL, gate, and circuit). An example Y-chart is shown in Fig. 1.3.

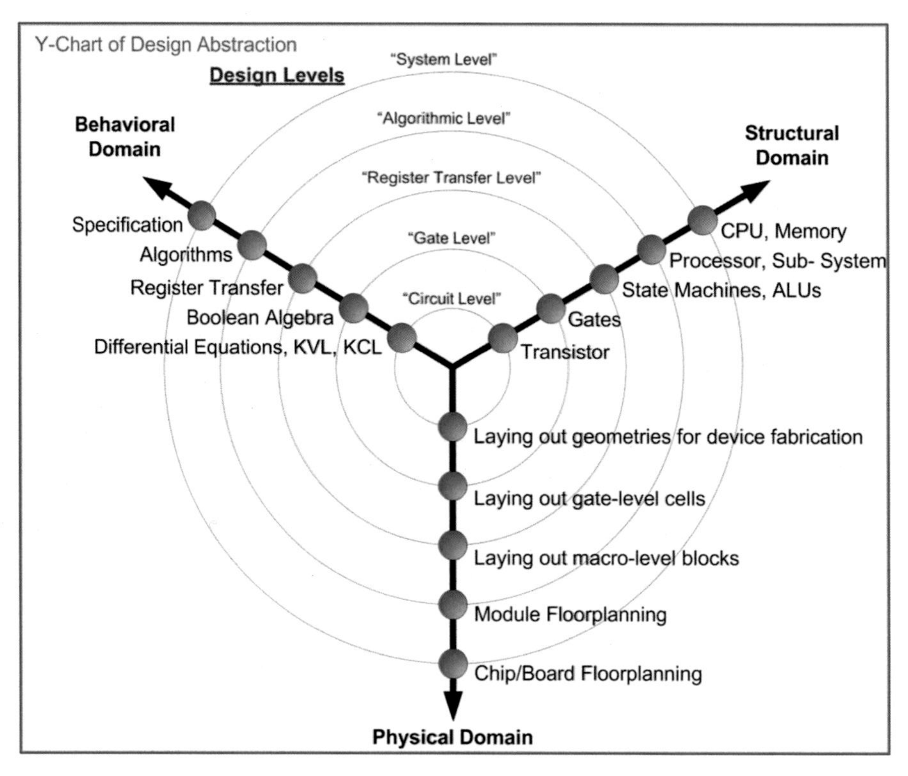

Fig. 1.3
Y-chart of design abstraction

A Y-chart also depicts how the abstraction levels of different design domains are related to each other. A top-down design flow can be visualized in a Y-chart by spiraling inward in a clockwise direction. Moving from the behavioral domain to the structural domain is the process of *synthesis*. Whenever synthesis is performed, the resulting system should be compared with the prior behavioral description. This checking is called *verification*. The process of creating the physical circuitry corresponding to the structural description is called *implementation*. The spiral continues down through the levels of abstraction until the design is implemented at a level that the geometries representing circuit elements (transistors, wires, etc.) are ready to be fabricated in silicon. Figure 1.4 shows the top-down design process depicted as an inward spiral on the Y-chart.

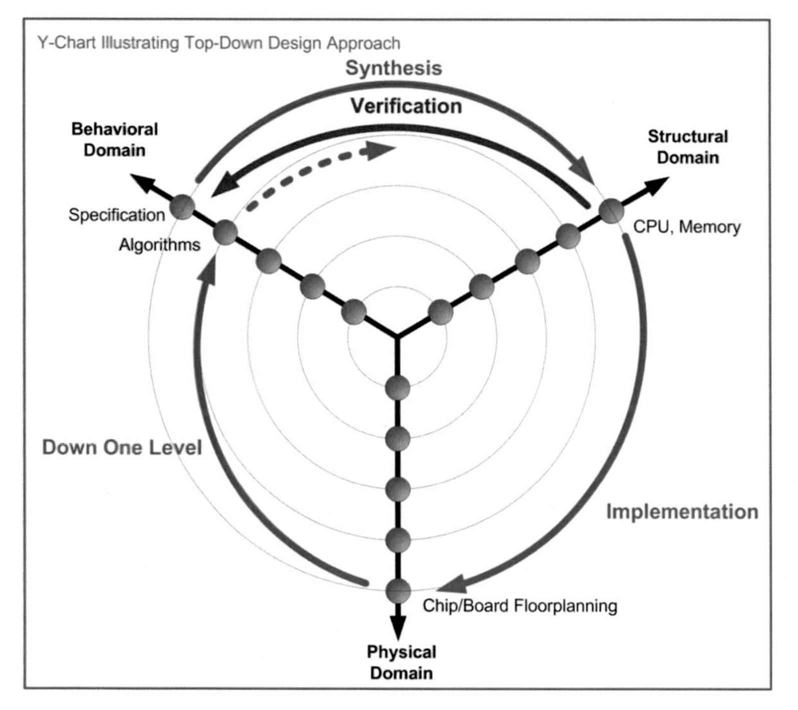

Fig. 1.4
Y-chart illustrating top-down design approach

The Y-chart represents a formal approach for large digital systems. For large systems that are designed by teams of engineers, it is critical that a formal, top-down design process is followed to eliminate potentially costly design errors as the implementation is carried out at the lower levels of abstraction.

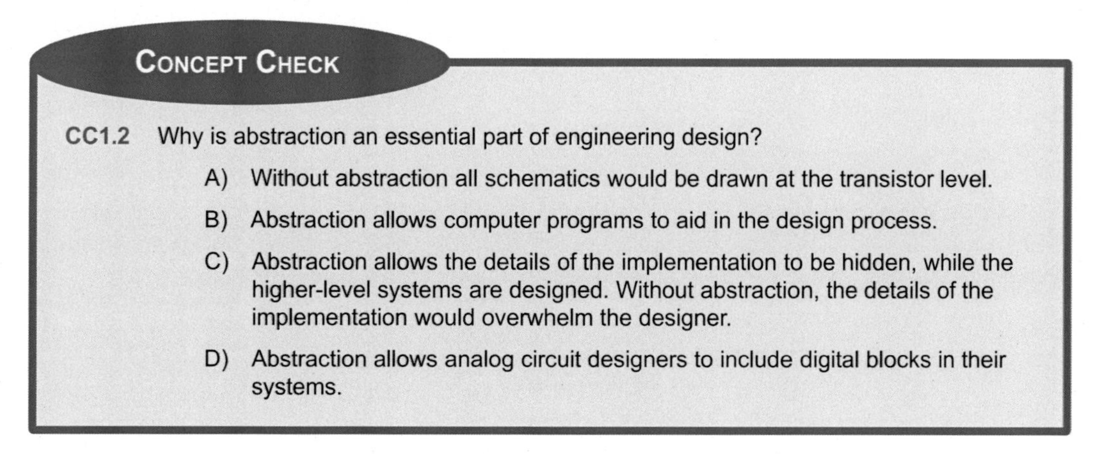

Concept Check

CC1.2 Why is abstraction an essential part of engineering design?

A) Without abstraction all schematics would be drawn at the transistor level.

B) Abstraction allows computer programs to aid in the design process.

C) Abstraction allows the details of the implementation to be hidden, while the higher-level systems are designed. Without abstraction, the details of the implementation would overwhelm the designer.

D) Abstraction allows analog circuit designers to include digital blocks in their systems.

1.3 The Modern Digital Design Flow

When performing a smaller design or the design of fully contained sub-systems, the process can be broken down into individual steps. These steps are shown in Fig. 1.5. This process is given generically and applies to both *classical* and *modern* digital design. The distinction between classical and modern is that modern digital design uses HDLs and automated CAD tools for simulation, synthesis, place and route, and verification.

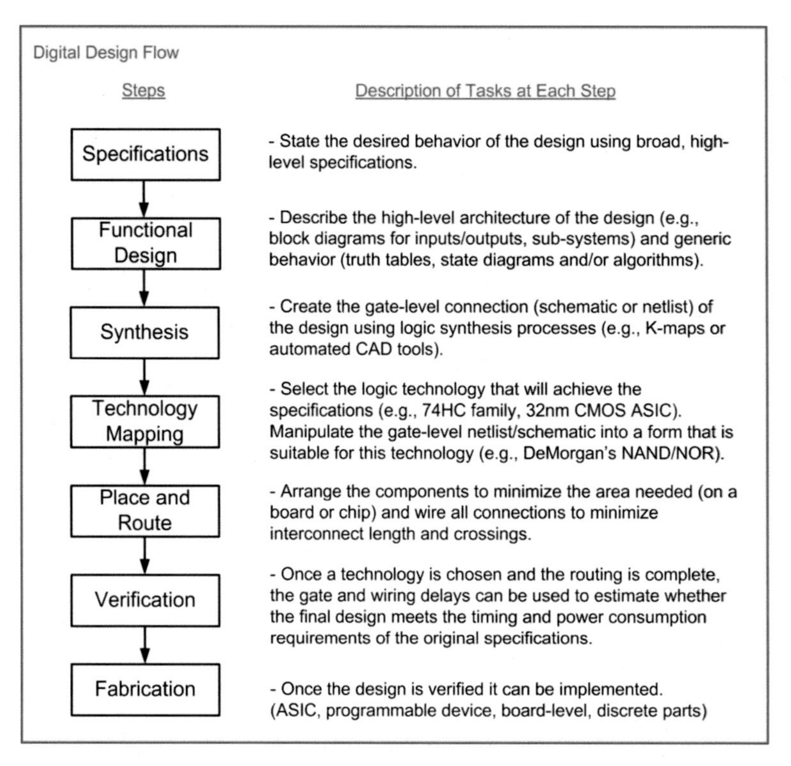

Fig. 1.5
Generic digital design flow

This generic design process flow can be used across classical and modern digital design, although modern digital design allows additional verification at each step using automated CAD tools. Figure 1.6 shows how this flow is used in the classical design approach of a combinational logic circuit.

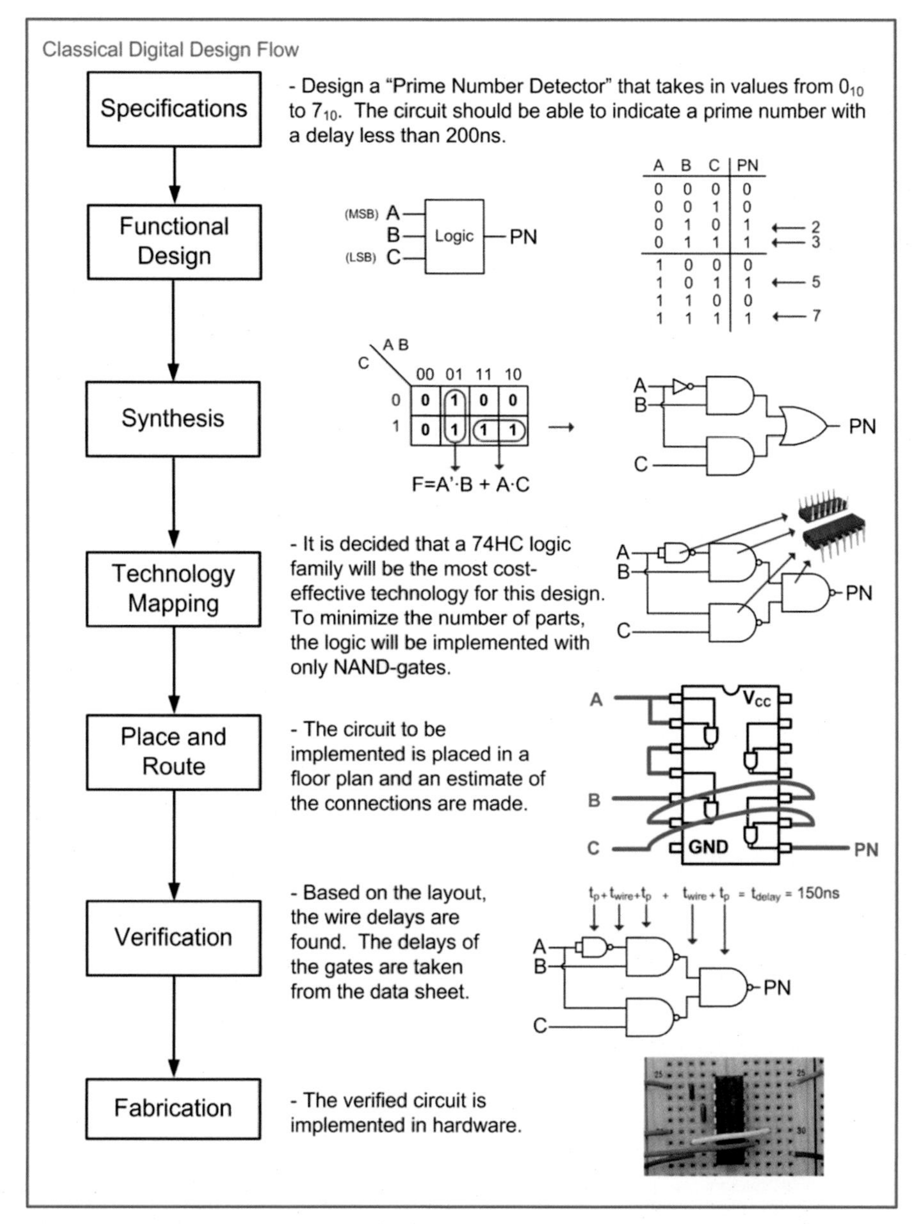

Fig. 1.6
Classical digital design flow

The modern design flow based on HDLs includes the ability to simulate functionality at each step of the process. Functional simulations can be performed on the initial behavioral description of the system. At each step of the design process, the functionality is described in more detail, ultimately moving toward the fabrication step. At each level, the detailed information can be included in the simulation to verify that the functionality is still correct and that the design is still meeting the original specifications. Figure 1.7 shows the modern digital design flow with the inclusion of simulation capability at each step.

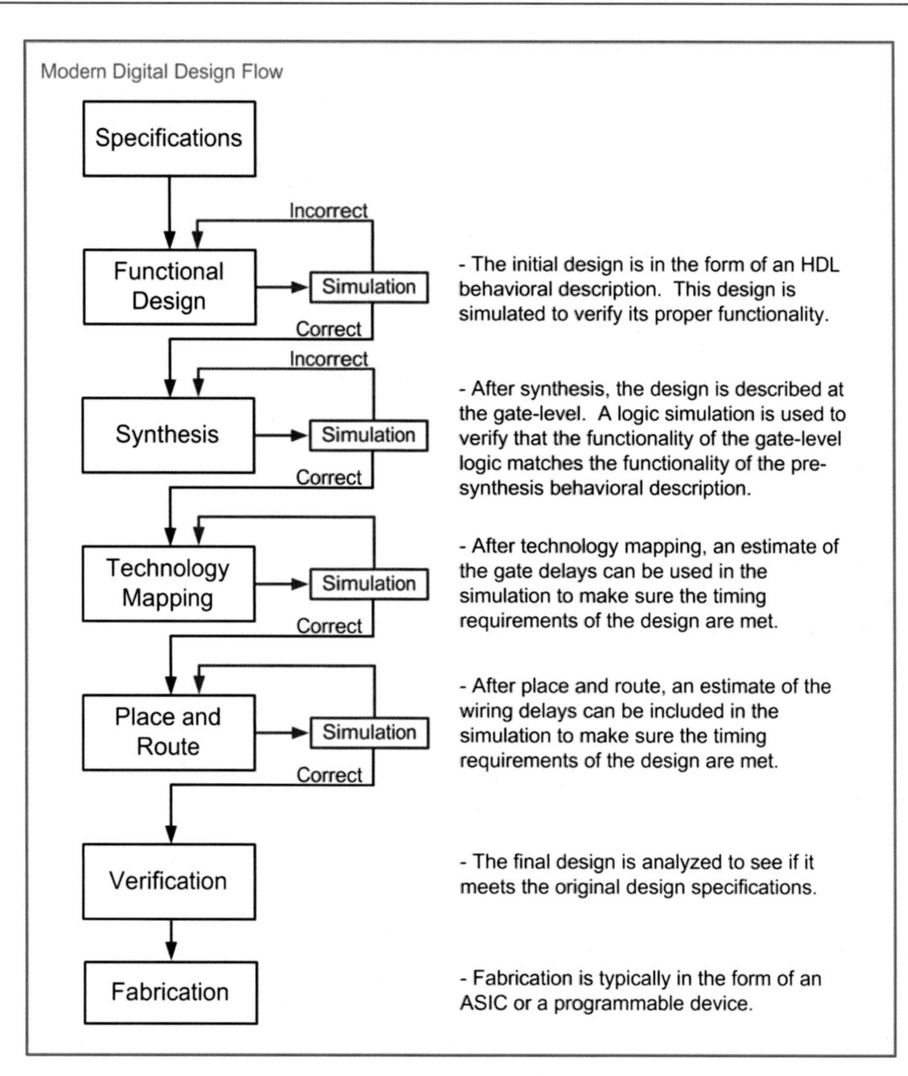

Fig. 1.7
Modern digital design flow

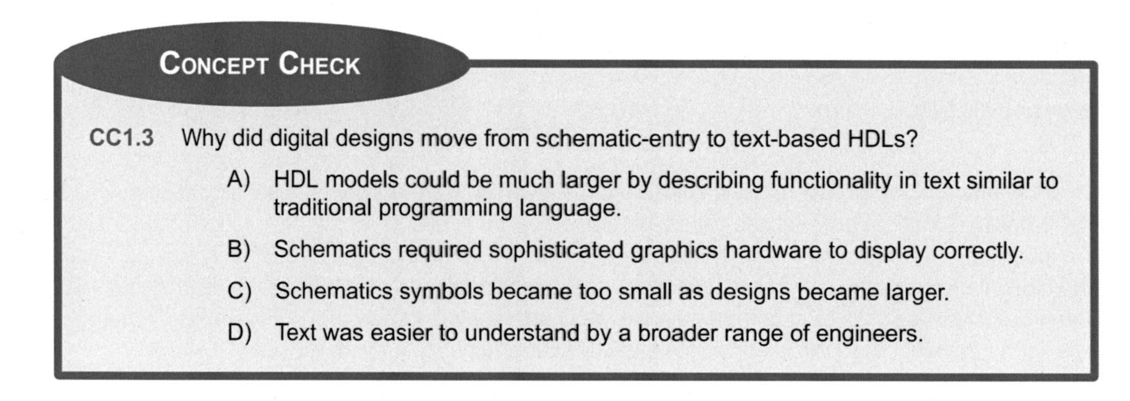

Concept Check

CC1.3 Why did digital designs move from schematic-entry to text-based HDLs?

A) HDL models could be much larger by describing functionality in text similar to traditional programming language.

B) Schematics required sophisticated graphics hardware to display correctly.

C) Schematics symbols became too small as designs became larger.

D) Text was easier to understand by a broader range of engineers.

Summary

- The modern digital design flow relies on computer-aided engineering (CAE) and computer-aided design (CAD) tools to manage the size and complexity of today's digital designs.
- Hardware description languages (HDLs) allow the functionality of digital systems to be entered using text. VHDL and Verilog are the two most common HDLs in use today.
- VHDL was originally created to document the behavior of large digital systems and support functional simulations.
- The ability to automatically synthesize a logic circuit from a VHDL behavioral description became possible approximately 10 years after the original definition of VHDL. As such, only a subset of the behavioral modeling techniques in VHDL can be automatically synthesized.
- HDLs can model digital systems at different levels of design abstraction. These include the *system*, *algorithmic*, *RTL*, *gate*, and *circuit* levels. Designing at a higher level of abstraction allows more complex systems to be modeled without worrying about the details of the implementation.

Exercise Problems

Section 1.1: History of HDLs

1.1.1 What was the original purpose of VHDL?

1.1.2 Can all of the functionality that can be described in VHDL be simulated?

1.1.3 Can all of the functionality that can be described in VHDL be synthesized?

Section 1.2: HDL Abstraction

1.2.1 Give the level of design abstraction that the following statement relates to: *if there is ever an error in the system, it should return to the reset state.*

1.2.2 Give the level of design abstraction that the following statement relates to: *once the design is implemented in a sum of products form, DeMorgan's theorem will be used to convert it to a NAND gate-only implementation.*

1.2.3 Give the level of design abstraction that the following statement relates to: *the design will be broken down into two sub-systems, one that will handle data collection and the other that will control data flow.*

1.2.4 Give the level of design abstraction that the following statement relates to: *the interconnect on the IC should be changed from aluminum to copper to achieve the performance needed in this design.*

1.2.5 Give the level of design abstraction that the following statement relates to: *the MOSFETs need to be able to drive at least eight other loads in this design.*

1.2.6 Give the level of design abstraction that the following statement relates to: *this system will contain 1 host computer and support up to 1000 client computers.*

1.2.7 Give the design domain that the following activity relates to: *drawing the physical layout of the CPU will require six months of engineering time.*

1.2.8 Give the design domain that the following activity relates to: *the CPU will be connected to four banks of memory.*

1.2.9 Give the design domain that the following activity relates to: *the fan-in specifications for this logic family require excessive logic circuitry to be used.*

1.2.10 Give the design domain that the following activity relates to: *the performance specifications for this system require 1 TFLOP at <5 W.*

Section 1.3: The Modern Digital Design Flow

1.3.1 Which step in the modern digital design flow does the following statement relate to: *a CAD tool will convert the behavioral model into a gate-level description of functionality.*

1.3.2 Which step in the modern digital design flow does the following statement relate to: *after realistic gate and wiring delays are determined, one last simulation should be performed to make sure the design meets the original timing requirements.*

1.3.3 Which step in the modern digital design flow does the following statement relate to: *if the memory is distributed around the perimeter of the CPU, the wiring density will be minimized.*

1.3.4 Which step in the modern digital design flow does the following statement relate to: *the design meets all requirements so now I'm building the hardware that will be shipped.*

1.3.5 Which step in the modern digital design flow does the following statement relate to: *the system will be broken down into three sub-systems with the following behaviors.*

1.3.6 Which step in the modern digital design flow does the following statement relate to: *this system needs to have 10 GB of memory.*

1.3.7 Which step in the modern digital design flow does the following statement relate to: *to meet the power requirements, the gates will be implemented in the 74HC logic family.*

Chapter 2: VHDL Constructs

This chapter begins looking at the basic construction of a VHDL model. This chapter begins by covering the built-in features of a VHDL model including the file structure, data types, operators, and declarations. This chapter provides a foundation of VHDL that will lead to modeling examples provided in Chap. 3. VHDL is not case sensitive. Each VHDL assignment, definition, or declaration is terminated with a semicolon (;). As such, line wraps are allowed and do not signify the end of an assignment, definition, or declaration. Line wraps can be used to make the VHDL more readable. Comments in VHDL are preceded with two dashes (i.e., --) and continue until the end of the line. All user-defined names in VHDL must start with an alphabetic letter, not a number. User-defined names are not allowed to be the same as any VHDL keyword. This chapter contains many definitions of syntax in VHDL. The following notations will be used throughout the chapter when introducing new constructs:

bold	= VHDL keyword, use as is
italics	= User-defined name
<>	= A required characteristic such as a data type, input/output, etc.

Learning Outcomes—After completing this chapter, you will be able to:

2.1 Describe the data types provided in the standard VHDL package.
2.2 Describe the basic construction of a VHDL model.

2.1 Data Types

In VHDL, every signal, constant, variable, and function must be assigned a *data type*. The IEEE standard package provides a variety of pre-defined data types. Some data types are synthesizable, while others are only for modeling abstract behavior. The following are the most commonly used data types in the VHDL standard package.

2.1.1 Enumerated Types

An *enumerated type* is one in which the exact values that the type can take on are defined.

Type	Values that the type can take on
bit	{0, 1}
boolean	{false, true}
character	{"any of the 256 ASCII characters defined in ISO 8859-1"}

The type bit is synthesizable, while boolean and character are not. The individual scalar values are indicated by putting them inside single quotes (e.g., '0,' 'a,' 'true').

B. J. LaMeres, *Quick Start Guide to VHDL*, https://doi.org/10.1007/978-3-031-42543-1_2

2.1.2 Range Types

A *range type* is one that can take on any value within a range.

Type	Values that the type can take on
integer	Whole numbers between –2,147,483,648 and +2,147,483,647
real	Fractional numbers between $-1.7e^{38}$ and $+1.7e^{38}$

The integer type is a 32-bit, signed, two's complement number and is synthesizable. If the full range of integer values is not desired, this type can be bounded by including *range <min> to <max>*. The real type is a 32-bit, floating-point value and is not directly synthesizable unless an additional package is included that defines the floating-point format. The values of these types are indicated by simply using the number without quotes (e.g., 33, 3.14).

2.1.3 Physical Types

A *physical type* is one that contains both a value and units. In VHDL, *time* is the primary supported physical type.

Type	Values that the type can take on	
time	Whole numbers between –2,147,483,648 and +2,147,483,647	
(unit relationships)	**fs**	(femtosecond, 10^{-15}), base unit
	ps = 1000 fs	(picosecond, 10^{-12})
	ns = 1000 ps	(nanosecond, 10^{-9})
	us = 1000 ns	(microsecond, 10^{-6})
	ms = 1000 us	(millisecond, 10^{-3})
	sec = 1000 ms	(second)
	min = 60 s	(minute)
	hr = 60 min	(hour)

The base unit for time is fs, meaning that if no units are provided, the value is assumed to be in femtoseconds. The value of time is held as a 32-bit, signed number and is not synthesizable.

2.1.4 Vector Types

A *vector type* is one that consists of a linear array of scalar types.

Type	Construction
bit_vector	A linear array of type bit
string	A linear array of type character

The size of a vector type is defined by including the maximum index, the keyword **downto**, and the minimum index. For example, if a signal called *BUS_A* was given the type bit_vector(7 downto 0), it would create a vector of 8 scalars, each of type bit. The leftmost scalar would have an index of 7, and the rightmost scalar would have an index of 0. Each of the individual scalars within the vector can be accessed by providing the index number in parentheses. For example, BUS_A(0) would access the scalar in the rightmost position. The indices do not always need to have a minimum value of 0, but this is the most common indexing approach in logic design. The type bit_vector is synthesizable, while string is not. The values of these types are indicated by enclosing them inside double quotes (e.g., "0011," "abcd").

2.1.5 User-Defined Enumerated Types

A *user-defined enumerated type* is one in which the name of the type is specified by the user in addition to all of the possible values that the type can assume. The creation of a user-defined enumerated type is shown below.

```
type name is (value1, value2, ...);
```

Example:

```
type traffic_light is (red, yellow, green);
```

In this example, a new type is created called *traffic_light*. If we declared a new signal called Sig1 and assigned it the type traffic_light, the signal could only take on values of red, yellow, and green. User-defined enumerated types are synthesizable in specific applications.

2.1.6 Array Type

An *array* contains multiple elements of the same type. Elements within an array can be scalar or vectors. In order to use an array, a new type must be declared that defines the configuration of the array. Once the new type is created, signals may be declared of that type. The *range* of the array must be defined in the array-type declaration. The range is specified with integers (min and max) and either the keywords *downto* or *to*. The creation of an array type is shown below.

```
type name is array (<range>) of <element_type>;
```

Example:

```
type block_8x16 is array (0 to 7) bit_vector(15 downto 0);
signal my_array : block_8x16;
```

In this example, the new array type is declared with eight elements. The beginning index of the array is 0, and the ending index is 7. Each element in the array is a 16-bit vector of type bit_vector.

2.1.7 Subtypes

A *subtype* is a constrained version or subset of another type. Subtypes are user-defined, although a few commonly used subtypes are pre-defined in the standard package. The following are the syntax for declaring a subtype and two examples of commonly used subtypes (NATURAL and POSTIVE) that are defined in the standard package:

```
subtype name is <type> range <min> to <max>;
```

Example:

```
subtype NATURAL is integer range 0 to 255;
subtype POSTIVE is integer range 1 to 256;
```

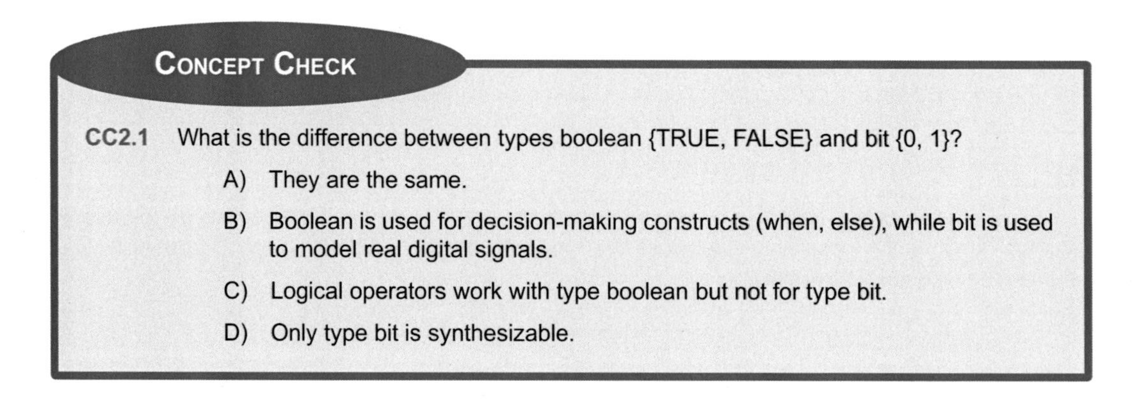

Concept Check

CC2.1 What is the difference between types boolean {TRUE, FALSE} and bit {0, 1}?

A) They are the same.

B) Boolean is used for decision-making constructs (when, else), while bit is used to model real digital signals.

C) Logical operators work with type boolean but not for type bit.

D) Only type bit is synthesizable.

2.2 VHDL Model Construction

A VHDL design describes a single system in a single file. The file has the suffix *.vhd. Within the file, there are two parts that describe the system: the **entity** and the **architecture**. The entity describes the interface to the system (i.e., the inputs and outputs), and the architecture describes the behavior. The functionality of VHDL (e.g., operators, signal types, functions, etc.) is defined in the **package**. Packages are grouped within a **library**. The IEEE defines the base set of functionality for VHDL in the *standard* package. This package is contained within a library called *IEEE*. The library and package inclusion is stated at the beginning of a VHDL file before the entity and architecture. Additional functionality can be added to VHDL by including other packages, but all packages are based on the core functionality defined in the standard package. As a result, it is not necessary to explicitly state that a design is using the IEEE standard package because it is inherent in the use of VHDL. All functionality described in this chapter is for the IEEE standard package, while other common packages are covered in subsequent chapters. Figure 2.1 shows a graphical depiction of a VHDL file.

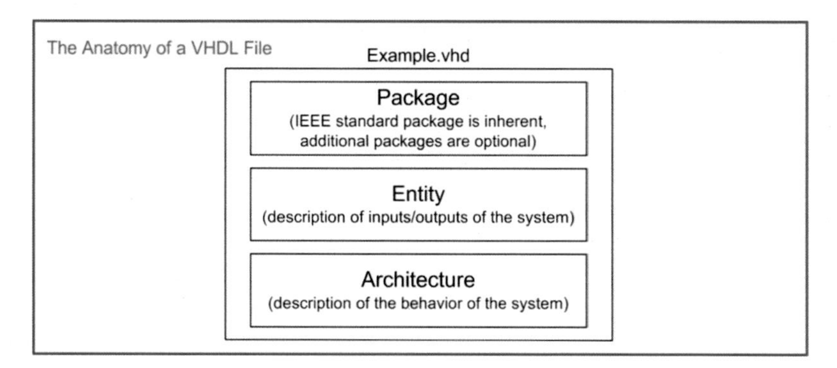

Fig. 2.1
The anatomy of a VHDL file

2.2.1 Libraries and Packages

As mentioned earlier, the IEEE standard package is implied when using VHDL; however, we can use it as an example of how to include packages in VHDL. The keyword **library** is used to signify that packages are going to be added to the VHDL design from the specified library. The name of the library follows this keyword. To include a specific package from the library, a new line is used with the keyword **use** followed by the package details. The package syntax has three fields separated with a period. The first field is the library name. The second field is the package name. The third field is the specific functionality of the package to be included. If all functionality of a package is to be used, then the keyword **all** is used in the third field. Examples of how to include some of the commonly used packages from the IEEE library are shown below.

```
library IEEE;
use IEEE.std_logic_1164.all;
use IEEE.numeric_std.all;
use IEEE.std_logic_textio.all;
```

2.2.2 The Entity

The entity in VHDL describes the inputs and outputs of the system. These are called **ports**. Each port needs to have its name, mode, and type specified. The name is user-defined. The mode describes the direction data is transferred through the port and can take on values of **in**, **out**, **inout**, and **buffer**. The type is one of the legal data types described above. Port names with the same mode and type can be listed on the same line separated by commas. The definition of an entity is given below.

```
entity entity_name is
  port (port_name  : <mode> <type>;
        port_name  : <mode> <type>);
end entity;
```

Example 2.1 shows multiple approaches for defining an entity.

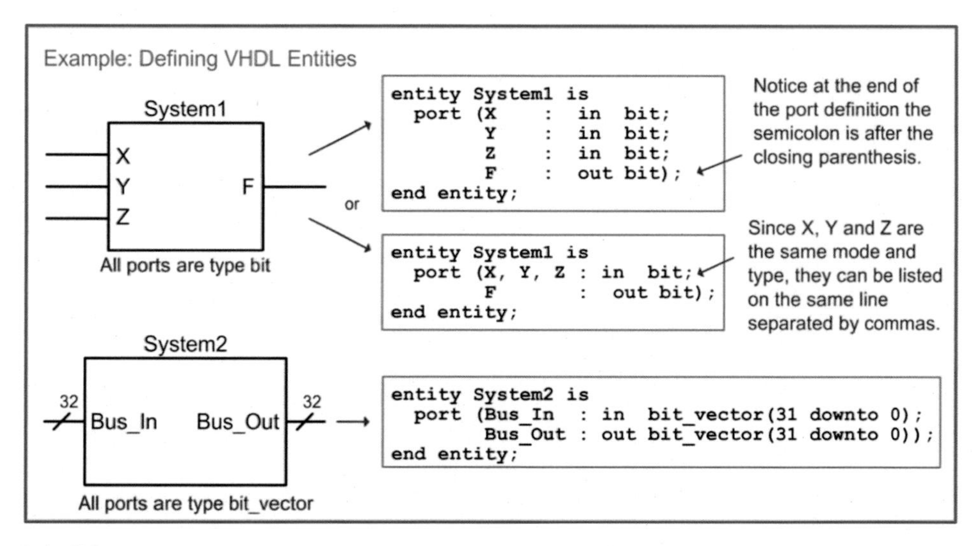

Example 2.1
Defining VHDL entities

2.2.3 The Architecture

The architecture in VHDL describes the behavior of a system. There are numerous techniques to describe behavior in VHDL that span multiple levels of abstraction. The architecture is where the majority of the design work is conducted. The form of a generic architecture is given below.

```
architecture architecture_name of <entity associated with> is

  -- user-defined enumerated type declarations   (optional)
  -- signal declarations                          (optional)
  -- constant declarations                        (optional)
  -- component declarations                       (optional)

  begin

  -- behavioral description of the system goes here

end architecture;
```

2.2.3.1 Signal Declarations

A signal that is used for internal connections within a system is declared in the architecture. Each signal must be declared with a type. The signal can only be used to make connections of like types. A signal is declared with the keyword **signal** followed by a user-defined name, colon, and the type. Signals of like type can be declared on the same line separated with a comma. All of the legal data types described above can be used for signals. Signals represent wires within the system, so they do not have a direction or mode. Signals cannot have the same name as a port in the system in which they reside. The syntax for a signal declaration is as follows:

```
signal name : <type>;
```

Example:

```
signal node1  : bit;
signal a1, b1 : integer;
signal Bus3   : bit_vector (15 downto 0);
signal C_int  : integer range 0 to 255;
```

VHDL supports a hierarchical design approach. Signal names can be the same within a sub-system as those at a higher level without conflict. Figure 2.2 shows an example of legal signal naming in a hierarchical design.

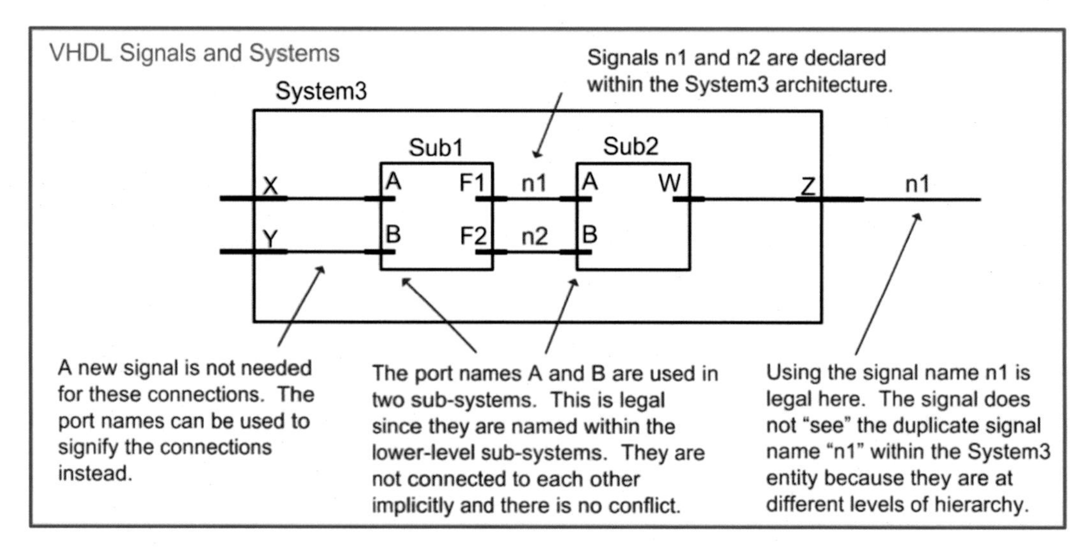

Fig. 2.2
VHDL signals and systems

2.2.3.2 Constant Declarations

A constant is useful for representing a quantity that will be used multiple times in the architecture. The syntax for declaring a constant is as follows:

```
constant constant_name : <type> := <value>;
```

Example:

```
constant BUS_WIDTH : integer := 32;
```

Once declared, the constant name can now be used throughout the architecture. The following example illustrates how we can use a constant to define the size of a vector. Notice that since we defined the constant to be the actual width of the vector (i.e., 32 bits), we need to subtract one from its value when defining the indices (i.e., 31 downto 0).

Example:

```
signal BUS_A : bit_vector (BUS_WIDTH-1 downto 0);
```

2.2.3.3 Component Declarations

A **component** is the term used for a VHDL sub-system that is instantiated within a higher-level system. If a component is going to be used within a system, it must be declared in the architecture before the begin statement. The syntax for a component declaration is as follows:

```
component component_name
  port (port_name  : <mode> <type>;
        port_name  : <mode> <type>);
end component;
```

The port definitions of the component must match the port definitions of the sub-system's entity exactly. As such, these lines are typically copied directly from the lower-level systems VHDL entity description. Once declared, a component can be instantiated after the begin statement in the architecture as many times as needed.

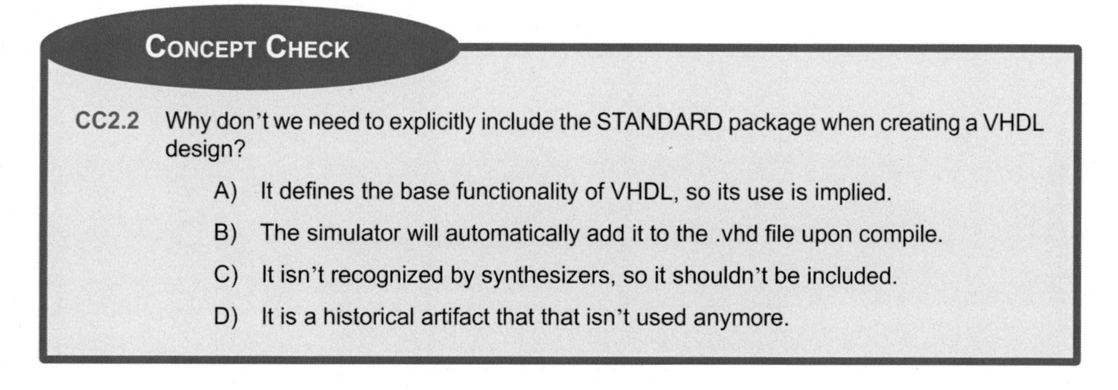

CONCEPT CHECK

CC2.2 Why don't we need to explicitly include the STANDARD package when creating a VHDL design?

A) It defines the base functionality of VHDL, so its use is implied.

B) The simulator will automatically add it to the .vhd file upon compile.

C) It isn't recognized by synthesizers, so it shouldn't be included.

D) It is a historical artifact that that isn't used anymore.

Summary

- Every signal and port in VHDL needs to be associated with a data type.
- A data type defines the values that can be taken on by a signal or port.
- In a VHDL source file, there are three main sections. These are the package, the entity, and the architecture. Including a package allows additional functionality to be included in VHDL. The entity is where the inputs and outputs of the system are declared. The architecture is where the behavior of the system is described.
- A *port* is an input or output to a system that is declared in the entity. A *signal* is an internal connection within the system that is declared in the architecture. A signal is not visible outside of the system.
- A *component* is how a VHDL system uses another sub-system. A component is first *declared*, which defines the name and entity of the sub-system to be used. The component can then be *instantiated* one or more times.

Exercise Problems

Section 2.1: Data Types

2.1.1 What are all the possible values that the type *bit* can take on in VHDL?

2.1.2 What are all the possible values that the type *boolean* can take on in VHDL?

2.1.3 What is the range of decimal numbers that can be represented using the type *integer* in VHDL?

2.1.4 What is the width of the vector defined using the type *bit_vector(63 downto 0)*?

2.1.5 What is the syntax for indexing the most significant bit in the type *bit_vector(31 downto 0)*? Assume the vector is named *example*.

2.1.6 What is the syntax for indexing the least significant bit in the type *bit_vector(31 downto 0)*? Assume the vector is named *example*.

2.1.7 What is the difference between an *enumerated* type and a *range* type?

2.1.8 What scalar type does a *bit_vector* consist?

Section 2.2: VHDL Model Construction

2.2.1 In which construct of VHDL are the inputs and outputs of the system defined?

2.2.2 In which construct of VHDL is the behavior of the system described?

2.2.3 Which construct is used to add additional functionality such as data types to VHDL?

Chapter 3: Modeling Concurrent Functionality in VHDL

This chapter presents a set of built-in operators that will allow logic to be modeled within the VHDL architecture. This chapter then presents a series of combinational logic model examples.

Learning Outcomes—After completing this chapter, you will be able to:

3.1 Describe the various built-in operators within VHDL.
3.2 Design a VHDL model for a combinational logic circuit using concurrent signal assignments and logical operators.
3.3 Design a VHDL model for a combinational logic circuit using conditional signal assignments.
3.4 Design a VHDL model for a combinational logic circuit using selected signal assignments.
3.5 Design a VHDL model for a combinational logic circuit that contains delay.

3.1 VHDL Operators

There are a variety of pre-defined operators in the IEEE standard package. It is important to note that operators are defined to work on specific data types and that not all operators are synthesizable. It is also important to remember that VHDL is a hardware description language, not a programming language. In a programming language, the lines of code are executed sequentially as they appear in the source file. In VHDL, the lines of code represent the behavior of real hardware. As a result, all signal assignments are by default executed concurrently unless specifically noted otherwise. All operations in VHDL must be on like types, and the result must be assigned to the same type as the operation inputs.

3.1.1 Assignment Operator

VHDL uses **<=** for all signal assignments and **:=** for all variable and initialization assignments. These assignment operators work on all data types. The target of the assignment goes on the left of these operators, and the input arguments go on the right.

Example:

```
F1 <= A;          -- F1 and A must be the same size and type
F2 <= '0';        -- F2 is type bit in this example
F3 <= "0000";     -- F3 is type bit_vector(3 downto 0) in this example
F4 <= "hello";    -- F4 is type string in this example
F5 <= 3.14;       -- F5 is type real in this example
F6 <= x"1A";      -- F6 is type bit_vector(7 downto 0), x"1A" is in HEX
```


B. J. LaMeres, *Quick Start Guide to VHDL*, https://doi.org/10.1007/978-3-031-42543-1_3

3.1.2 Logical Operators

VHDL contains the following logical operators:

Operator	Operation
not	Logical negation
and	Logical AND
nand	Logical NAND
or	Logical OR
nor	Logical NOR
xor	Logical Exclusive-OR
xnor	Logical Exclusive-NOR

These operators work on types bit, bit_vector, and boolean. For operations on the type bit_vector, the input vectors must be the same size and will take place in a bit-wise fashion. For example, if two 8-bit buses called BusA and BusB were AND'd together, BusA(0) would be individually AND'd with BusB(0), BusA(1) would be individually AND'd with BusB(1), etc. The not operator is a unary operation (i.e., it operates on a single input), and the keyword is put before the signal being operated on. All other operators have two or more inputs and are placed in-between the input names.

Example:

```
F1 <= not A;
F2 <= B and C;
```

The order of precedence in VHDL is different from in Boolean algebra. The NOT operator is a higher priority than all other operators. All other logical operators have the same priority and have no inherent precedence. This means that in VHDL, the AND operator will *not* precede the OR operation as it does in Boolean algebra. Parentheses are used to explicitly describe precedence. If operators that have the same priority are used and parentheses are not provided, then the operations will take place on the signals listed first moving left to right in the signal assignment. The following are examples on how to use these operators:

Example:

```
F3 <= not D nand E;     -- D will be complemented first, the result
                        -- will then be NAND'd with E, then the
                        -- result will be assigned to F3
F4 <= not (F or G);     -- the parentheses take precedence so
                        -- F will be OR'd with G first, then
                        -- complemented, and then assigned to F4.

F5 <= H nor I nor J;    -- logic operations can have any number of
                        -- inputs.

F6 <= K xor L xnor M;   -- XOR and XNOR have the same priority so with
                        -- no parentheses given, the logic operations
                        -- will take place on the signals from
                        -- left to right. K will be XOR'd with L first,
                        -- then the result will be XNOR'd with M.
```

3.1.3 Numerical Operators

VHDL contains the following numerical operators:

Operator	Operation
+	Addition
-	Subtraction
*	Multiplication
/	Division
mod	Modulus
rem	Remainder
abs	Absolute value
**	Exponential

These operators work on types integer and real. Note that the default VHDL standard does not support numerical operators on types bit and bit_vector.

3.1.4 Relational Operators

VHDL contains the following relational operators. These operators compare two inputs of the same type and return the type boolean (i.e., true or false).

Operator	Returns true if the comparison is:
=	Equal
/=	Not equal
<	Less than
<=	Less than or equal
>	Greater than
>=	Greater than or equal

3.1.5 Shift Operators

VHDL contains the following shift operators. These operators work on vector types bit_vector and string.

Operator	Operation
sll	Shift left logical
srl	Shift right logical
sla	Shift left arithmetic
sra	Shift right arithmetic
rol	Rotate left
ror	Rotate right

The syntax for using a shift operation is to provide the name of the vector followed by the desired shift operator, followed by an integer indicating how many shift operations to perform. The target of the assignment must be the same type and size as the input.

Example:

```
A <= B srl 3;        -- A is assigned the result of a logical shift
                     -- right 3 times on B.
```

3.1.6 Concatenation Operator

In VHDL the **&** is used to concatenate multiple signals. The target of this operation must be the same size of the sum of the sizes of the input arguments.

Example:

```
Bus1 <= "11" & "00";   -- Bus1 must be 4-bits and will be assigned
                       -- the value "1100"

Bus2 <= BusA & BusB;   -- If BusA and BusB are 4-bits, then Bus2
                       -- must be 8-bits.

Bus3 <= '0' & BusA;    -- This attaches a leading '0' to BusA. Bus3
                       -- must be 5-bits
```

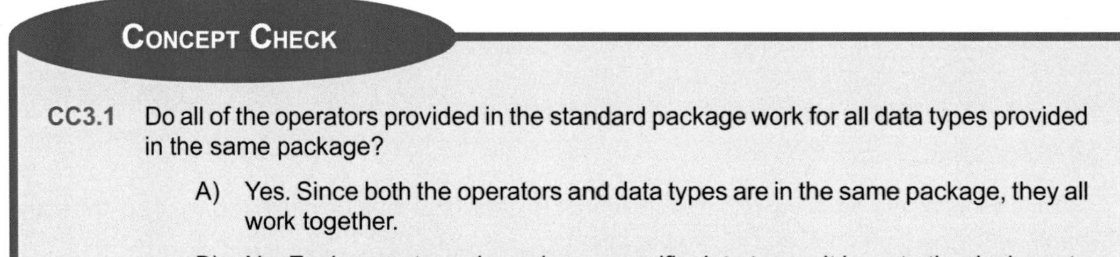

Concept Check

CC3.1 Do all of the operators provided in the standard package work for all data types provided in the same package?

A) Yes. Since both the operators and data types are in the same package, they all work together.

B) No. Each operator only works on specific data types. It is up to the designer to know what types the operator work with.

3.2 Concurrent Signal Assignments with Logical Operators

Concurrent signal assignments are accomplished by simply using the <= operator after the begin statement in the architecture. Each individual assignment will be executed concurrently and synthesized as separate logic circuits. Consider the following example:

Example:

```
X <= A;
Y <= B;
Z <= C;
```

When simulated, these three lines of VHDL will make three separate signal assignments at the exact same time. This is different from a programming language that will first assign A to X, then B to Y, and finally C to Z. In VHDL this functionality is identical to three separate wires. This description will be directly synthesized into three separate wires.

Below is another example of how concurrent signal assignments in VHDL differ from a sequentially executed programming language.

Example:

```
A <= B;
B <= C;
```

In a VHDL simulation, the signal assignments of C to B and B to A will take place at the same time; however, during synthesis, the signal B will be eliminated from the design since this functionality describes two wires in series. Automated synthesis tools will eliminate this unnecessary signal name. This is not the same functionality that would result if this example was implemented as a sequentially executed computer program. A computer program would execute the assignment of B to A first and then assign the value of C to B second. In this way, B represents a storage element that is passed to A before it is updated with C.

Each of the logical operators described in Sect. 3.1.2 can be used in conjunction with concurrent signal assignments to create individual combinational logic circuits.

3.2.1 Logical Operator Example: SOP Circuit

Example 3.1 shows how to design a VHDL model of a standard sum of products combinational logic circuit using concurrent signal assignments with logical operators.

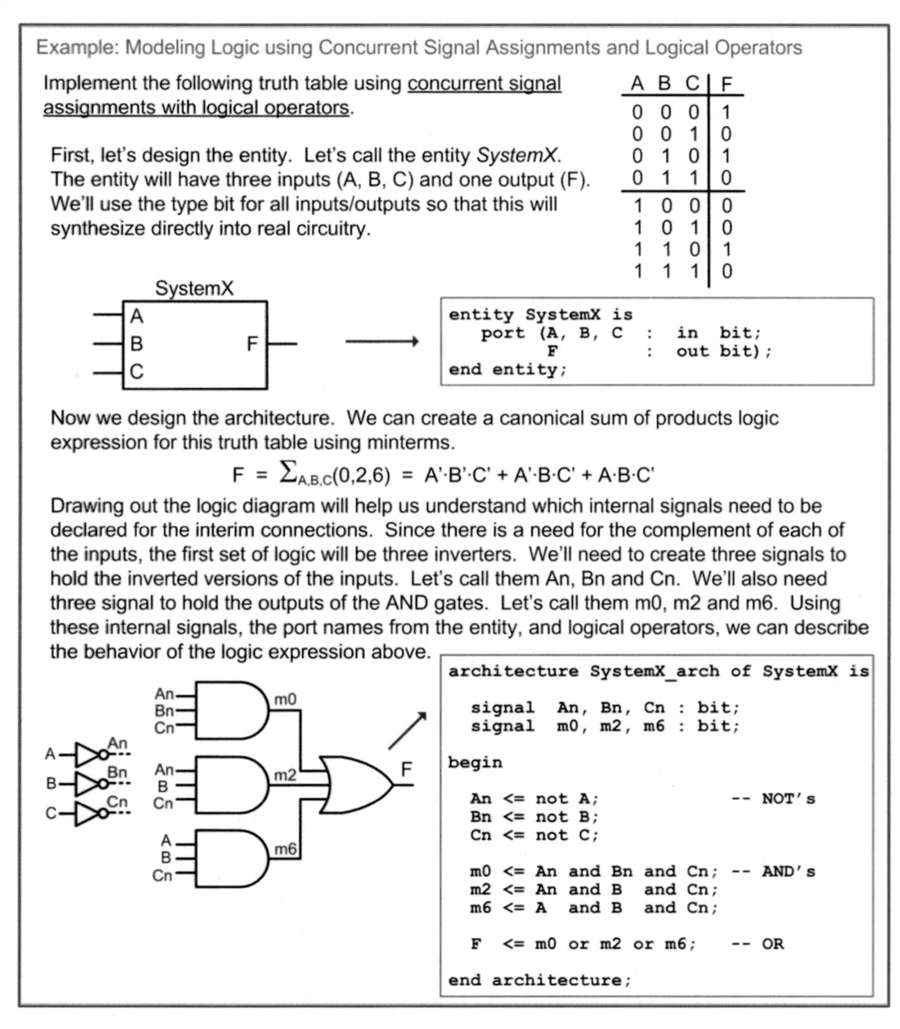

Example 3.1
SOP logic circuit – VHDL modeling using logical operators

3.2.2 Logical Operator Example: One-Hot Decoder

A one-hot decoder is a circuit that has n inputs and 2^n outputs. Each output will assert for one and only one input code. Since there are 2^n outputs, there will always be one and only one output asserted at any given time. Example 3.2 shows how to model a 3-to-8 one-hot decoder in VHDL with concurrent signal assignments and logic operators.

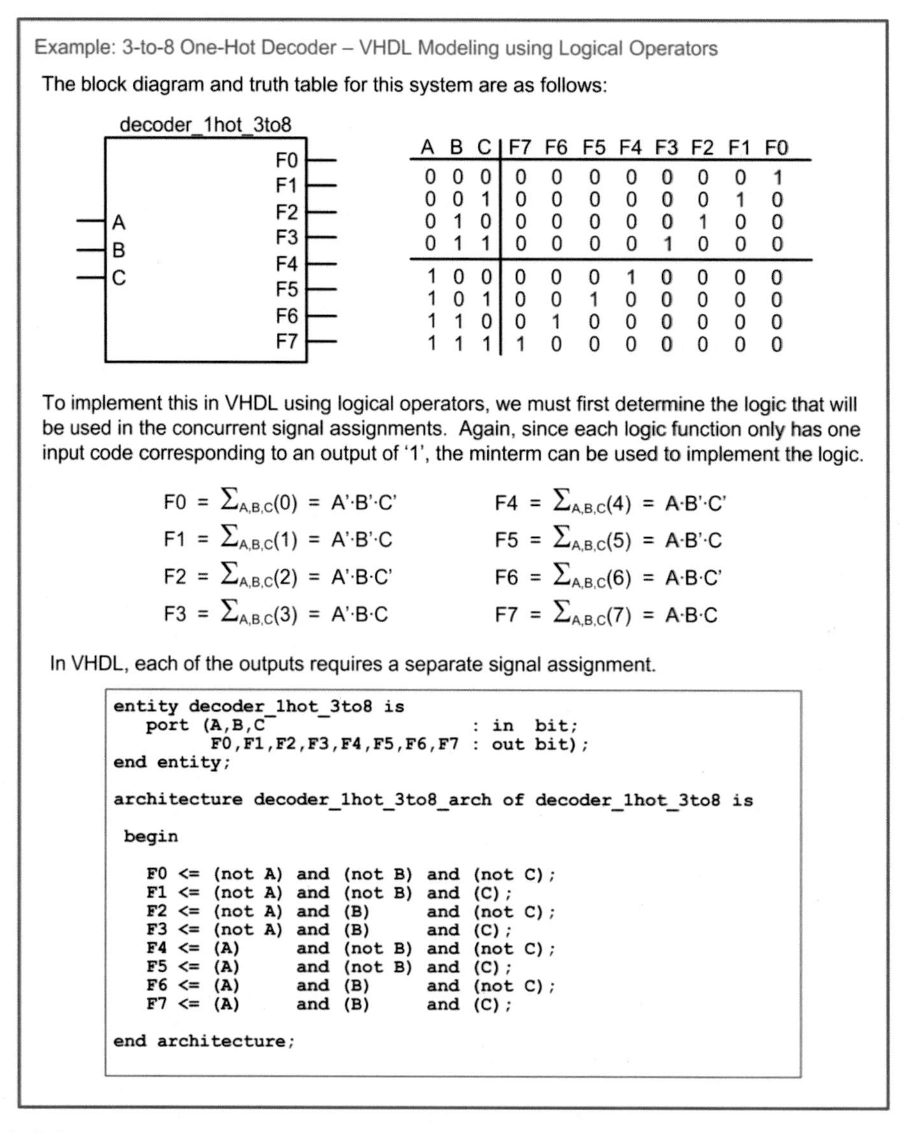

Example: 3-to-8 One-Hot Decoder – VHDL Modeling using Logical Operators

The block diagram and truth table for this system are as follows:

A	B	C	F7	F6	F5	F4	F3	F2	F1	F0
0	0	0	0	0	0	0	0	0	0	1
0	0	1	0	0	0	0	0	0	1	0
0	1	0	0	0	0	0	0	1	0	0
0	1	1	0	0	0	0	1	0	0	0
1	0	0	0	0	0	1	0	0	0	0
1	0	1	0	0	1	0	0	0	0	0
1	1	0	0	1	0	0	0	0	0	0
1	1	1	1	0	0	0	0	0	0	0

To implement this in VHDL using logical operators, we must first determine the logic that will be used in the concurrent signal assignments. Again, since each logic function only has one input code corresponding to an output of '1', the minterm can be used to implement the logic.

$$F0 = \Sigma_{A,B,C}(0) = A' \cdot B' \cdot C' \qquad F4 = \Sigma_{A,B,C}(4) = A \cdot B' \cdot C'$$

$$F1 = \Sigma_{A,B,C}(1) = A' \cdot B' \cdot C \qquad F5 = \Sigma_{A,B,C}(5) = A \cdot B' \cdot C$$

$$F2 = \Sigma_{A,B,C}(2) = A' \cdot B \cdot C' \qquad F6 = \Sigma_{A,B,C}(6) = A \cdot B \cdot C'$$

$$F3 = \Sigma_{A,B,C}(3) = A' \cdot B \cdot C \qquad F7 = \Sigma_{A,B,C}(7) = A \cdot B \cdot C$$

In VHDL, each of the outputs requires a separate signal assignment.

```
entity decoder_1hot_3to8 is
   port (A,B,C                       : in  bit;
         F0,F1,F2,F3,F4,F5,F6,F7 : out bit);
end entity;

architecture decoder_1hot_3to8_arch of decoder_1hot_3to8 is

 begin

   F0 <= (not A) and (not B) and (not C);
   F1 <= (not A) and (not B) and (C);
   F2 <= (not A) and (B)     and (not C);
   F3 <= (not A) and (B)     and (C);
   F4 <= (A)     and (not B) and (not C);
   F5 <= (A)     and (not B) and (C);
   F6 <= (A)     and (B)     and (not C);
   F7 <= (A)     and (B)     and (C);

end architecture;
```

Example 3.2
3-to-8 one-hot decoder – VHDL modeling using logical operators

3.2.3 Logical Operator Example: 7-Segment Display Decoder

A 7-segment display decoder is a circuit used to drive character displays that are commonly found in applications such as digital clocks and household appliances. A character display is made up of 7 individual LEDs, typically labeled a–g. The input to the decoder is the binary equivalent of the decimal or Hex character that is to be displayed. The output of the decoder is the arrangement of LEDs that will form the character. Decoders with 2-inputs can drive characters "0" to "3." Decoders with 3-inputs can drive characters "0" to "7." Decoders with 4-inputs can drive characters "0" to "F" with the case of the Hex characters being "A, b, c or C, d, E and F."

Let's look at an example of how to design a 3-input, 7-segment decoder by hand. The first step in the process is to create the truth table for the outputs that will drive the LEDs in the display. We'll call these outputs F_a, F_b, . . ., F_g. Example 3.3 shows how to construct the truth table for the 7-segment display decoder. In this table, a logic 1 corresponds to the LED being ON.

Example: 7-Segment Display Decoder - Truth Table

LED Labels: a, b, c, d, e, f, g

A	B	C	F_a	F_b	F_c	F_d	F_e	F_f	F_g
0	0	0	1	1	1	1	1	1	0
0	0	1	0	1	1	0	0	0	0
0	1	0	1	1	0	1	1	0	1
0	1	1	1	1	1	1	0	0	1
1	0	0	0	1	1	0	0	1	1
1	0	1	1	0	1	1	0	1	1
1	1	0	1	0	1	1	1	1	1
1	1	1	1	1	1	0	0	0	0

Example 3.3
7-segment display decoder – truth table

If we wish to design this decoder by hand, we need to create seven separate combinational logic circuits. Each of the outputs (F_a–F_g) can be put into a 3-input K-map to find the minimized logic expression. Example 3.4 shows the design of the decoder from the truth table in Example 3.3 by hand.

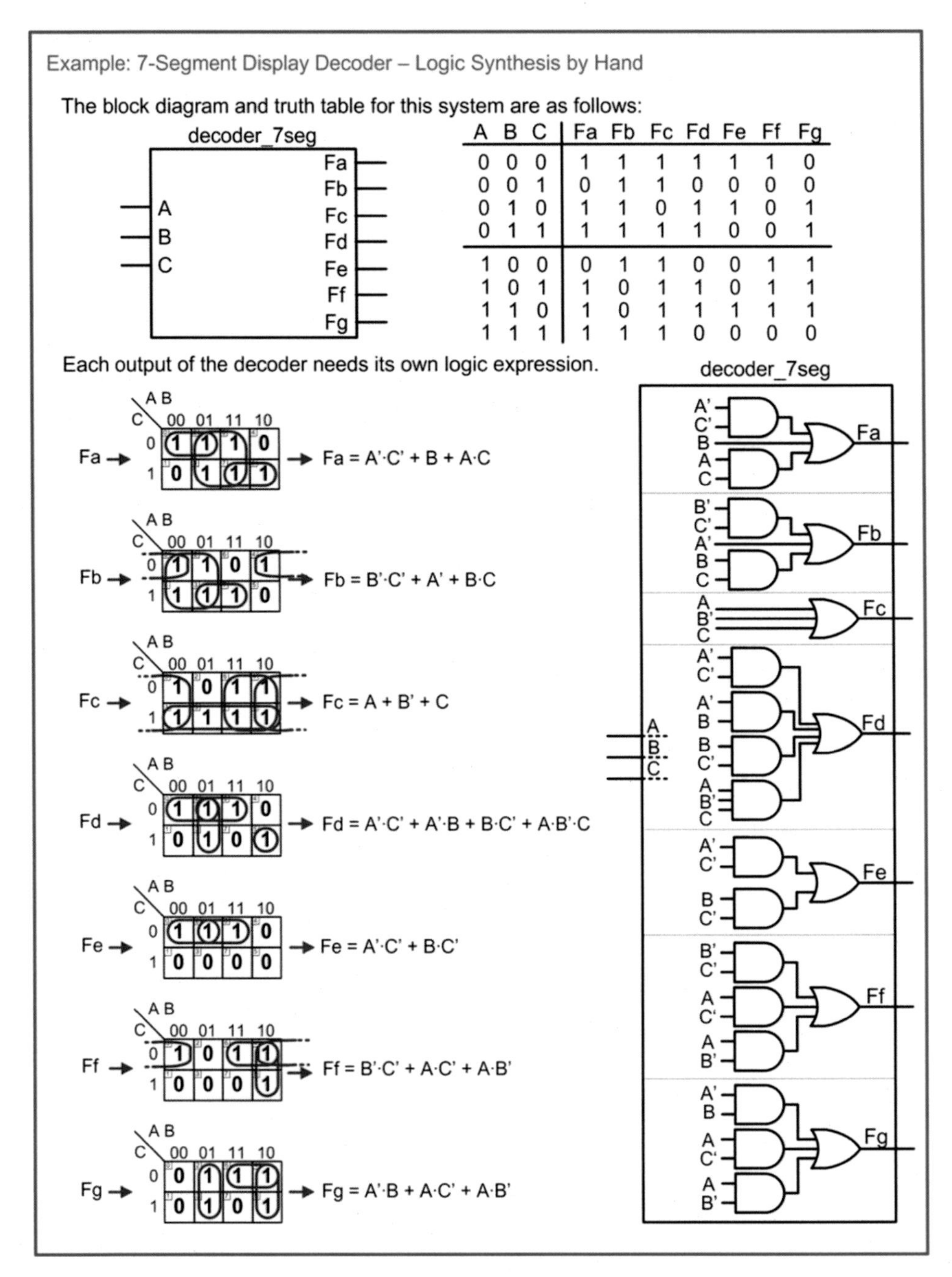

A	B	C	Fa	Fb	Fc	Fd	Fe	Ff	Fg
0	0	0	1	1	1	1	1	1	0
0	0	1	0	1	1	0	0	0	0
0	1	0	1	1	0	1	1	0	1
0	1	1	1	1	1	1	0	0	1
1	0	0	0	1	1	0	0	1	1
1	0	1	1	0	1	1	0	1	1
1	1	0	1	0	1	1	1	1	1
1	1	1	1	1	1	0	0	0	0

Example 3.4
7-segment display decoder – logic synthesis by hand

This same functionality can be modeled in VHDL using concurrent signal assignments with logical operators. Example 3.5 shows how to model the 7-segment decoder in VHDL using concurrent signal assignments with logic operators. It should be noted that this is example is somewhat artificial because a design would typically not be minimized before modeling in VHDL. Instead, model would be entered at the behavioral level, and then the CAD tool would be allowed to synthesize and minimize the final logic.

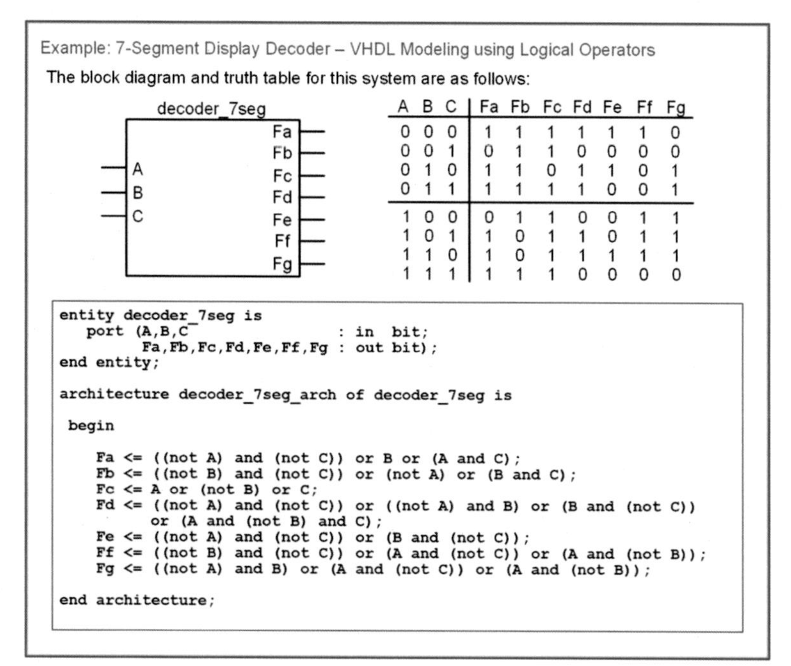
Example: 7-Segment Display Decoder – VHDL Modeling using Logical Operators

The block diagram and truth table for this system are as follows:

A	B	C	Fa	Fb	Fc	Fd	Fe	Ff	Fg
0	0	0	1	1	1	1	1	1	0
0	0	1	0	1	1	0	0	0	0
0	1	0	1	1	0	1	1	0	1
0	1	1	1	1	1	1	0	0	1
1	0	0	0	1	1	0	0	1	1
1	0	1	1	0	1	1	0	1	1
1	1	0	1	0	1	1	1	1	1
1	1	1	1	1	1	0	0	0	0

```
entity decoder_7seg is
   port (A,B,C                  : in  bit;
         Fa,Fb,Fc,Fd,Fe,Ff,Fg : out bit);
end entity;

architecture decoder_7seg_arch of decoder_7seg is

 begin

    Fa <= ((not A) and (not C)) or B or (A and C);
    Fb <= ((not B) and (not C)) or (not A) or (B and C);
    Fc <= A or (not B) or C;
    Fd <= ((not A) and (not C)) or ((not A) and B) or (B and (not C))
          or (A and (not B) and C);
    Fe <= ((not A) and (not C)) or (B and (not C));
    Ff <= ((not B) and (not C)) or (A and (not C)) or (A and (not B));
    Fg <= ((not A) and B) or (A and (not C)) or (A and (not B));

end architecture;
```

Example 3.5
7-segment display decoder – VHDL modeling using logical operators

3.2.4 Logical Operator Example: One-Hot Encoder

A one-hot binary encoder has *n* outputs and 2^n inputs. The output will be an *n*-bit, binary code which corresponds to an assertion on one and only one of the inputs. Example 3.6 shows the process of designing a 4-to-2 binary encoder by hand (i.e., using the classical digital design approach).

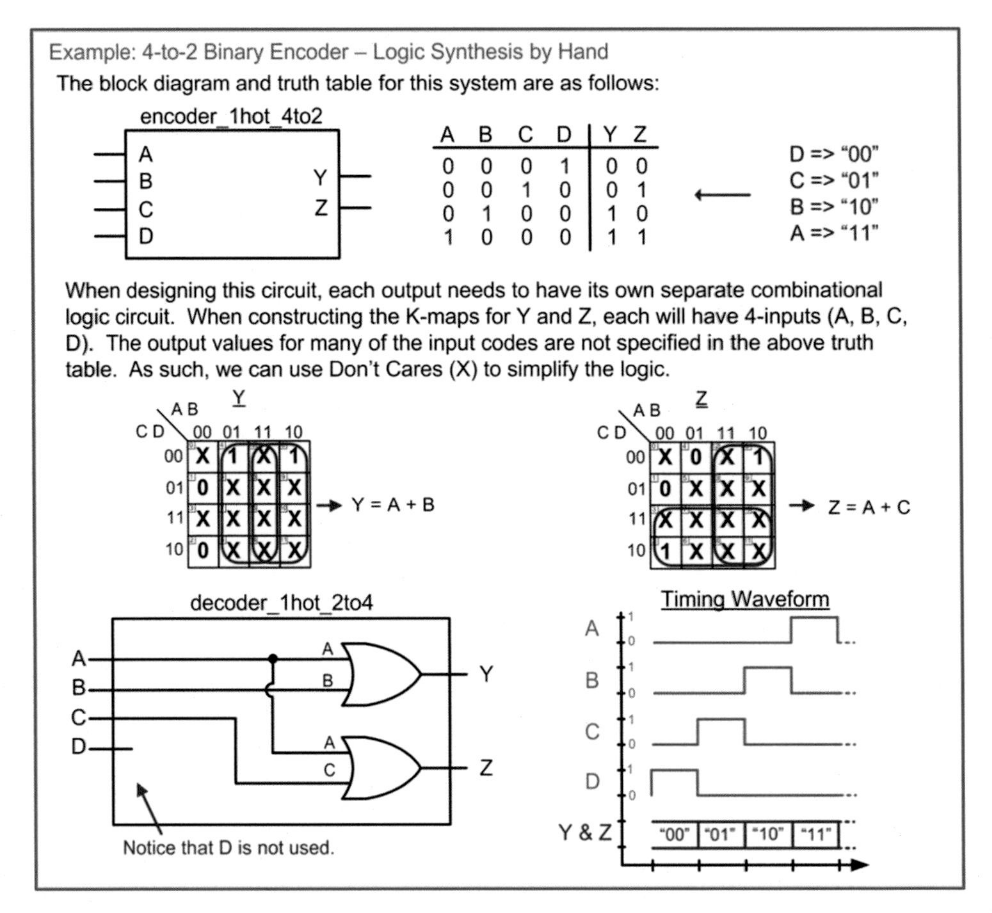

Example 3.6
4-to-2 binary encoder – logic synthesis by hand

In VHDL this can be implemented directly using logical operators. Example 3.7 shows how to model the encoder in VHDL using concurrent signal assignments with logical operators.

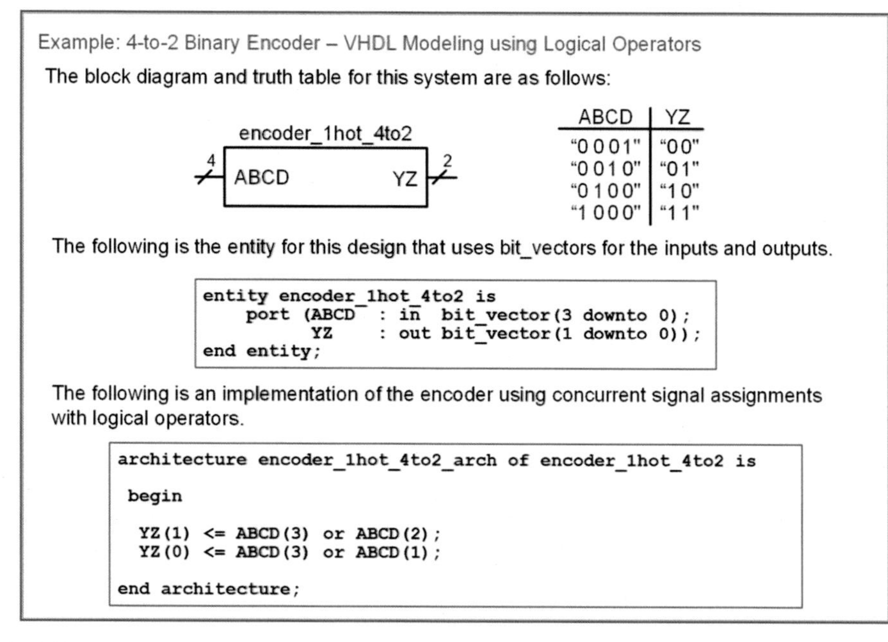

Example: 4-to-2 Binary Encoder – VHDL Modeling using Logical Operators

The block diagram and truth table for this system are as follows:

ABCD	YZ
"0001"	"00"
"0010"	"01"
"0100"	"10"
"1000"	"11"

The following is the entity for this design that uses bit_vectors for the inputs and outputs.

```
entity encoder_1hot_4to2 is
    port (ABCD  : in  bit_vector(3 downto 0);
          YZ    : out bit_vector(1 downto 0));
end entity;
```

The following is an implementation of the encoder using concurrent signal assignments with logical operators.

```
architecture encoder_1hot_4to2_arch of encoder_1hot_4to2 is

 begin

  YZ(1) <= ABCD(3) or ABCD(2);
  YZ(0) <= ABCD(3) or ABCD(1);

end architecture;
```

Example 3.7
4-to-2 binary encoder – VHDL modeling using logical operators

3.2.5 Logical Operator Example: Multiplexer

A multiplexer is a circuit that passes one of its multiple inputs to a single output based on a select input. This can be thought of as a digital routing switch. The multiplexer has *n* select lines, 2^n inputs, and one output. Example 3.8 shows the process of designing a 4-to-1 multiplexer using concurrent signal assignments and logical operators.

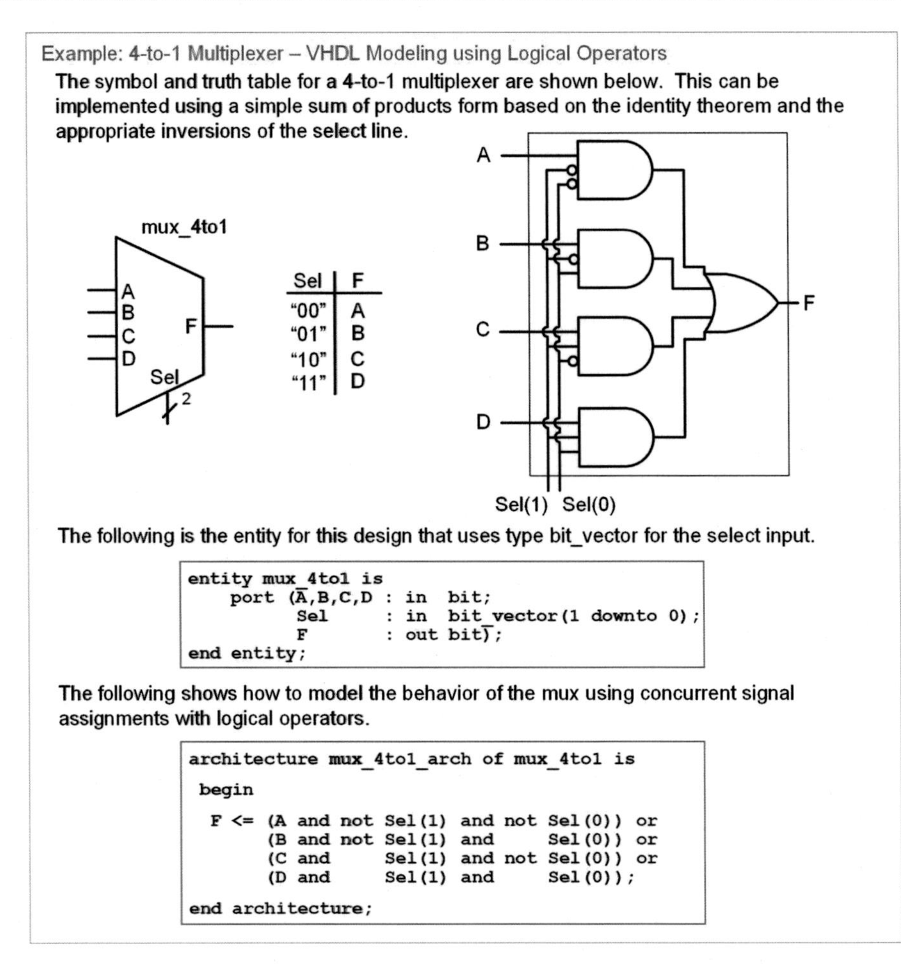

Sel	F
"00"	A
"01"	B
"10"	C
"11"	D

```
entity mux_4to1 is
    port (A,B,C,D : in  bit;
          Sel     : in  bit_vector(1 downto 0);
          F       : out bit);
end entity;
```

```
architecture mux_4to1_arch of mux_4to1 is
 begin
  F <= (A and not Sel(1) and not Sel(0)) or
       (B and not Sel(1) and     Sel(0)) or
       (C and     Sel(1) and not Sel(0)) or
       (D and     Sel(1) and     Sel(0));
end architecture;
```

Example 3.8
4-to-1 multiplexer – VHDL modeling using logical operators

3.2.6 Logical Operator Example: Demultiplexer

A demultiplexer works in a complementary fashion to a multiplexer. A demultiplexer has one input that is routed to one of its multiple outputs. The output that is active is dictated by a select input. A demux has n select lines that chooses to route the input to one of its 2^n outputs. When an output is not selected, it outputs a logic 0. Example 3.9 shows the process of designing a 1-to-4 demultiplexer using concurrent signal assignments and logical operators.

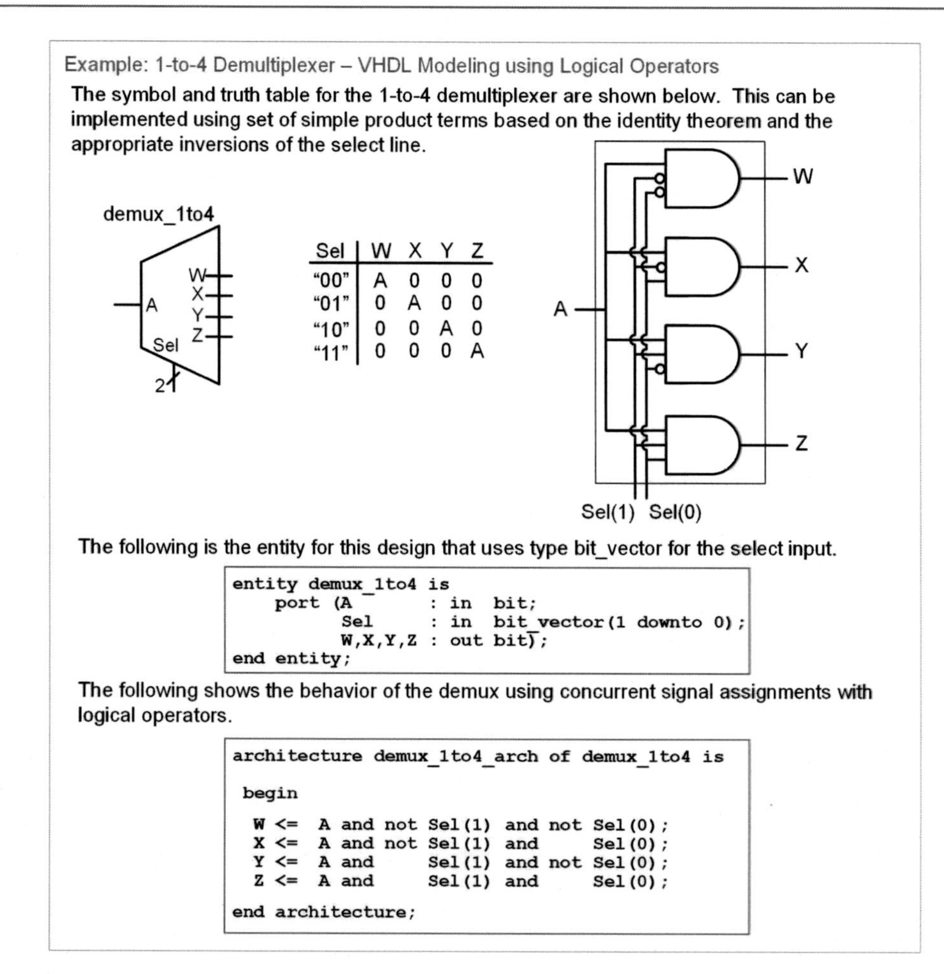

Example: 1-to-4 Demultiplexer – VHDL Modeling using Logical Operators

The symbol and truth table for the 1-to-4 demultiplexer are shown below. This can be implemented using set of simple product terms based on the identity theorem and the appropriate inversions of the select line.

Sel	W	X	Y	Z
"00"	A	0	0	0
"01"	0	A	0	0
"10"	0	0	A	0
"11"	0	0	0	A

The following is the entity for this design that uses type bit_vector for the select input.

```
entity demux_1to4 is
    port (A       : in  bit;
          Sel     : in  bit_vector(1 downto 0);
          W,X,Y,Z : out bit);
end entity;
```

The following shows the behavior of the demux using concurrent signal assignments with logical operators.

```
architecture demux_1to4_arch of demux_1to4 is

 begin

  W <=  A and not Sel(1) and not Sel(0);
  X <=  A and not Sel(1) and     Sel(0);
  Y <=  A and     Sel(1) and not Sel(0);
  Z <=  A and     Sel(1) and     Sel(0);

end architecture;
```

Example 3.9
1-to-4 demultiplexer – VHDL modeling using logical operators

Concept Check

CC3.2 Why does modeling combinational logic in its canonical form with concurrent signal assignments with logical operators defeat the purpose of the modern digital design flow?

A) It requires the designer to first create the circuit using the classical digital design approach and then enter it into the HDL in a form that is essentially a text-based netlist. This doesn't take advantage of the abstraction capabilities and automated synthesis in the modern flow.

B) It cannot be synthesized because the order of precedence of the logical operators in VHDL doesn't match the precedence defined in Boolean algebra.

C) The circuit is in its simplest form, so there is no work for the synthesizer to do.

D) It doesn't allow an *else* clause to cover the outputs for any remaining input codes not explicitly listed.

3.3 Conditional Signal Assignments

Logical operators are good for describing the behavior of small circuits; however, in the prior example, we still needed to create the canonical or minimal sum of products logic expression by hand before describing the functionality in VHDL. The true power of an HDL is when the behavior of the system can be described fully without requiring any hand design. A conditional signal assignment allows us to describe a concurrent signal assignment using Boolean conditions that affect the values of the result. In a conditional signal assignment, the keyword **when** is used to describe the signal assignment for a particular Boolean condition. The keyword **else** is used to describe the signal assignments for any other conditions. Multiple Boolean conditions can be used to fully describe the output of the circuit under all input conditions. Logical operators can also be used in the Boolean conditions to create more sophisticated conditions. The Boolean conditions can be encompassed within parentheses for readability. The syntax for a conditional signal assignment is shown below.

```
signal_name <= expression_1 when condition_1 else
               expression_2 when condition_2 else
                                  :
               expression_n;
```

Example:

```
F1 <= '0' when A='0' else '1';
F2 <= '1' when (A='0' and B='1') else '0';
F3 <= A when (C = D) else B;
```

An important consideration of conditional signal assignments is that they are still executed concurrently. Each assignment represents a separate, combinational logic circuit. In the above example, F1, F2, and F3 will be implemented as three separate, parallel circuits.

3.3.1 Conditional Signal Assignment Example: SOP Circuit

Example 3.10 shows how to design a VHDL model of a combinational logic circuit using conditional signal assignments. Note that this example uses the same truth table as in Example 3.1 to illustrate a comparison between approaches. This approach provides a model that can be created directly from the truth table without needing to do any synthesis or minimization by hand.

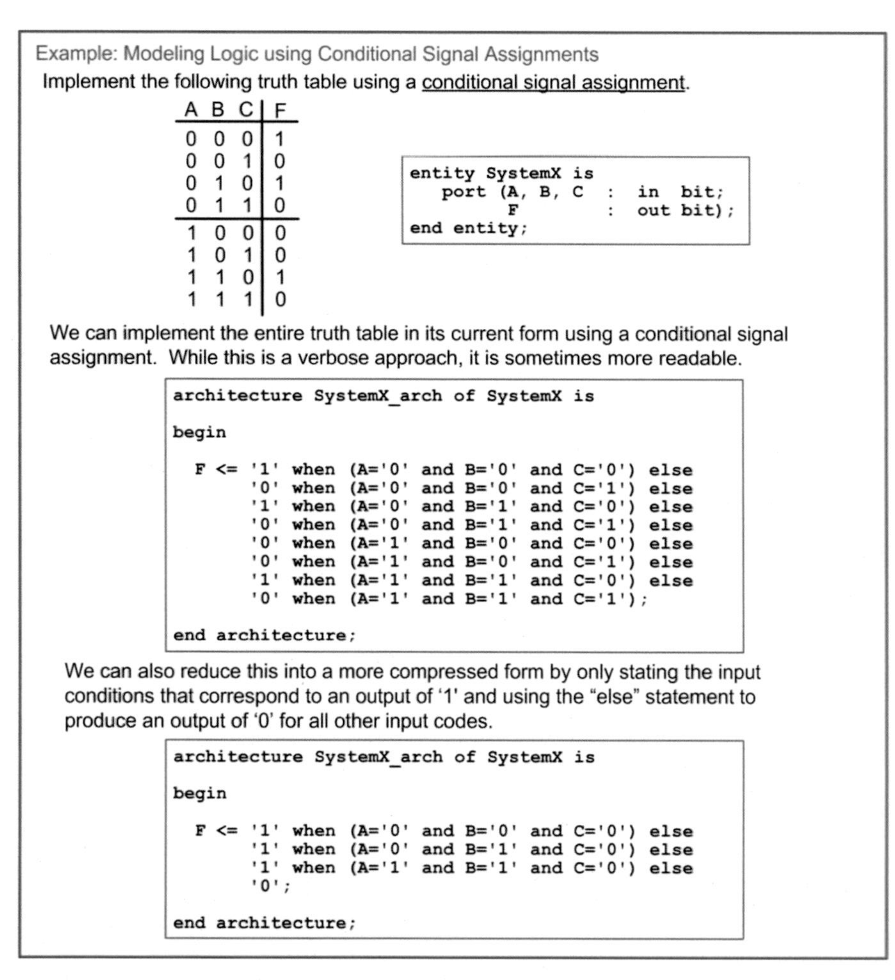

Example: Modeling Logic using Conditional Signal Assignments

Implement the following truth table using a conditional signal assignment.

A	B	C	F
0	0	0	1
0	0	1	0
0	1	0	1
0	1	1	0
1	0	0	0
1	0	1	0
1	1	0	1
1	1	1	0

```
entity SystemX is
   port (A, B, C  :  in  bit;
         F        :  out bit);
end entity;
```

We can implement the entire truth table in its current form using a conditional signal assignment. While this is a verbose approach, it is sometimes more readable.

```
architecture SystemX_arch of SystemX is

begin

  F <= '1' when (A='0' and B='0' and C='0') else
       '0' when (A='0' and B='0' and C='1') else
       '1' when (A='0' and B='1' and C='0') else
       '0' when (A='0' and B='1' and C='1') else
       '0' when (A='1' and B='0' and C='0') else
       '0' when (A='1' and B='0' and C='1') else
       '1' when (A='1' and B='1' and C='0') else
       '0' when (A='1' and B='1' and C='1');

end architecture;
```

We can also reduce this into a more compressed form by only stating the input conditions that correspond to an output of '1' and using the "else" statement to produce an output of '0' for all other input codes.

```
architecture SystemX_arch of SystemX is

begin

  F <= '1' when (A='0' and B='0' and C='0') else
       '1' when (A='0' and B='1' and C='0') else
       '1' when (A='1' and B='1' and C='0') else
       '0';

end architecture;
```

Example 3.10
SOP logic circuit – VHDL modeling using conditional signal assignments

3.3.2 Conditional Signal Assignment Example: One-Hot Decoder

Example 3.11 shows how to model a 3-to-8 one-hot decoder in VHDL with conditional signal assignments. Again, this approach allows the logic to be modeled directly from its functional description rather than having to perform any synthesis by hand.

Example: 3-to-8 One-Hot Decoder – VHDL Modeling – Conditional Signal Assignments

The block diagram and truth table for this system are as follows. Notice that the input and output ports now use type bit_vector in order to create a more compact description.

decoder_1hot_3to8 (input ABC, 3 bits; output F, 8 bits)

ABC	F(7)	F(6)	F(5)	F(4)	F(3)	F(2)	F(1)	F(0)
"000"	0	0	0	0	0	0	0	1
"001"	0	0	0	0	0	0	1	0
"010"	0	0	0	0	0	1	0	0
"011"	0	0	0	0	1	0	0	0
"100"	0	0	0	1	0	0	0	0
"101"	0	0	1	0	0	0	0	0
"110"	0	1	0	0	0	0	0	0
"111"	1	0	0	0	0	0	0	0

The following is the entity for this design that uses bit_vectors for the inputs and outputs.

```
entity decoder_1hot_3to8 is
    port (ABC  : in  bit_vector(2 downto 0);
          F    : out bit_vector(7 downto 0));
end entity;
```

The following is an implementation of the decoder using a conditional signal assignment. In this technique, the signal assignment can be made to the entire F vector instead of to the individual scalar outputs. This creates a compact VHDL model that will synthesis into eight separate combinational logic circuits.

```
architecture decoder_1hot_3to8_arch of decoder_1hot_3to8 is

 begin

  F <= "00000001" when (ABC = "000") else
       "00000010" when (ABC = "001") else
       "00000100" when (ABC = "010") else
       "00001000" when (ABC = "011") else
       "00010000" when (ABC = "100") else
       "00100000" when (ABC = "101") else
       "01000000" when (ABC = "110") else
       "10000000" when (ABC = "111");

end architecture;
```

Example 3.11
3-to-8 one-hot decoder – VHDL modeling using conditional signal assignments

3.3.3 Conditional Signal Assignment Example: 7-Segment Display Decoder

Back in Example 3.3 the truth table for a 7-segment display decoder was given along with the subsequent steps to create its logic using the classical digital design approach and model it in VHDL using logical operators. With a conditional signal assignment, this decoder can be modeled directly from the truth table without needing to do any design by hand. Example 3.12 shows how to model the logic for a 7-segment display decoder using a conditional signal assignment.

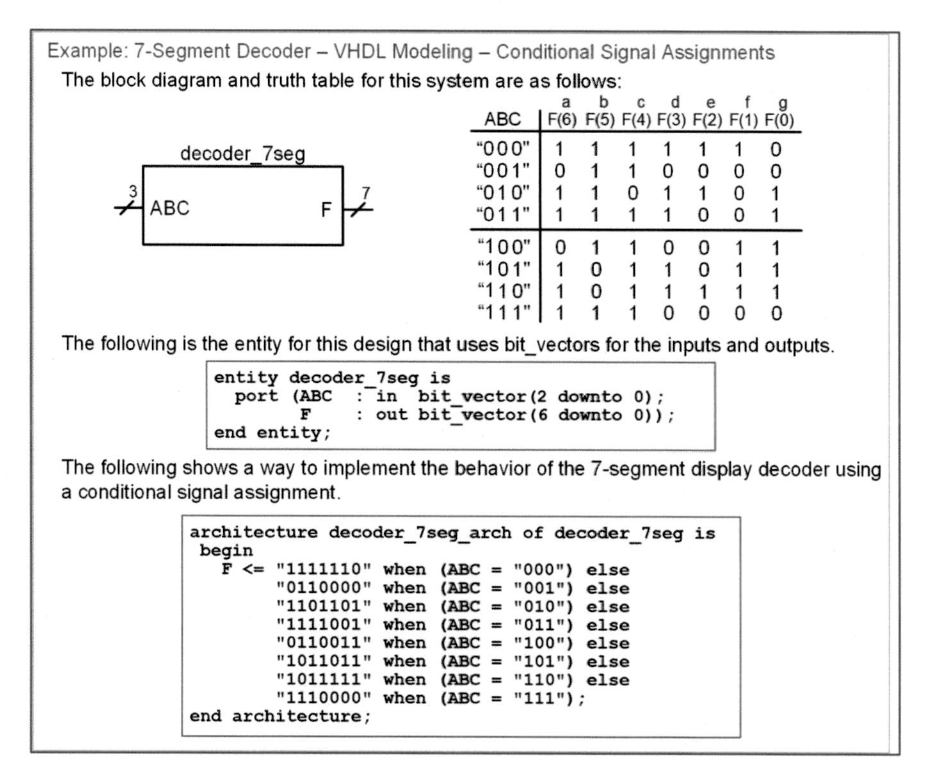

Example: 7-Segment Decoder – VHDL Modeling – Conditional Signal Assignments

The block diagram and truth table for this system are as follows:

ABC	a F(6)	b F(5)	c F(4)	d F(3)	e F(2)	f F(1)	g F(0)
"000"	1	1	1	1	1	1	0
"001"	0	1	1	0	0	0	0
"010"	1	1	0	1	1	0	1
"011"	1	1	1	1	0	0	1
"100"	0	1	1	0	0	1	1
"101"	1	0	1	1	0	1	1
"110"	1	0	1	1	1	1	1
"111"	1	1	1	0	0	0	0

The following is the entity for this design that uses bit_vectors for the inputs and outputs.

```
entity decoder_7seg is
  port (ABC   : in  bit_vector(2 downto 0);
        F     : out bit_vector(6 downto 0));
end entity;
```

The following shows a way to implement the behavior of the 7-segment display decoder using a conditional signal assignment.

```
architecture decoder_7seg_arch of decoder_7seg is
 begin
   F <= "1111110" when (ABC = "000") else
        "0110000" when (ABC = "001") else
        "1101101" when (ABC = "010") else
        "1111001" when (ABC = "011") else
        "0110011" when (ABC = "100") else
        "1011011" when (ABC = "101") else
        "1011111" when (ABC = "110") else
        "1110000" when (ABC = "111");
end architecture;
```

Example 3.12
7-segment display decoder – VHDL modeling using conditional signal assignments

3.3.4 Conditional Signal Assignment Example: One-Hot Encoder

Example 3.13 shows how to model a one-hot encoder in VHDL with conditional signal assignments. Again, this approach allows the logic to be modeled directly from its functional description rather than having to perform any synthesis by hand.

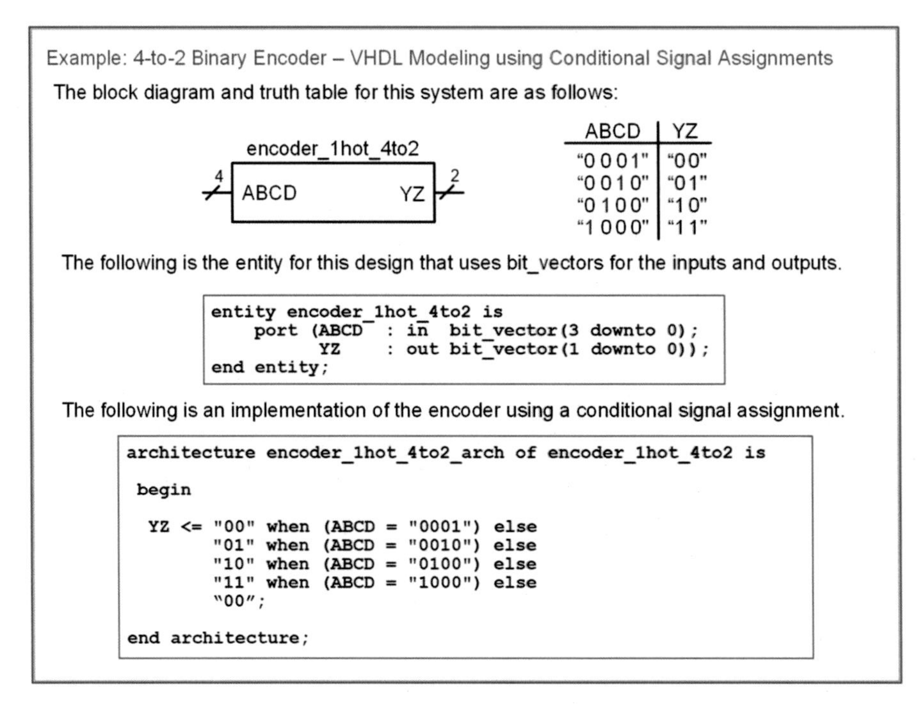

Example: 4-to-2 Binary Encoder – VHDL Modeling using Conditional Signal Assignments

The block diagram and truth table for this system are as follows:

ABCD	YZ
"0001"	"00"
"0010"	"01"
"0100"	"10"
"1000"	"11"

The following is the entity for this design that uses bit_vectors for the inputs and outputs.

```
entity encoder_1hot_4to2 is
    port (ABCD  : in  bit_vector(3 downto 0);
          YZ    : out bit_vector(1 downto 0));
end entity;
```

The following is an implementation of the encoder using a conditional signal assignment.

```
architecture encoder_1hot_4to2_arch of encoder_1hot_4to2 is

 begin

  YZ <= "00" when (ABCD = "0001") else
        "01" when (ABCD = "0010") else
        "10" when (ABCD = "0100") else
        "11" when (ABCD = "1000") else
        "00";

end architecture;
```

Example 3.13
4-to-2 binary encoder – VHDL modeling using conditional signal assignments

3.3.5 Conditional Signal Assignment Example: Multiplexer

Example 3.14 shows the process of designing a 4-to-1 multiplexer using conditional signal assignments.

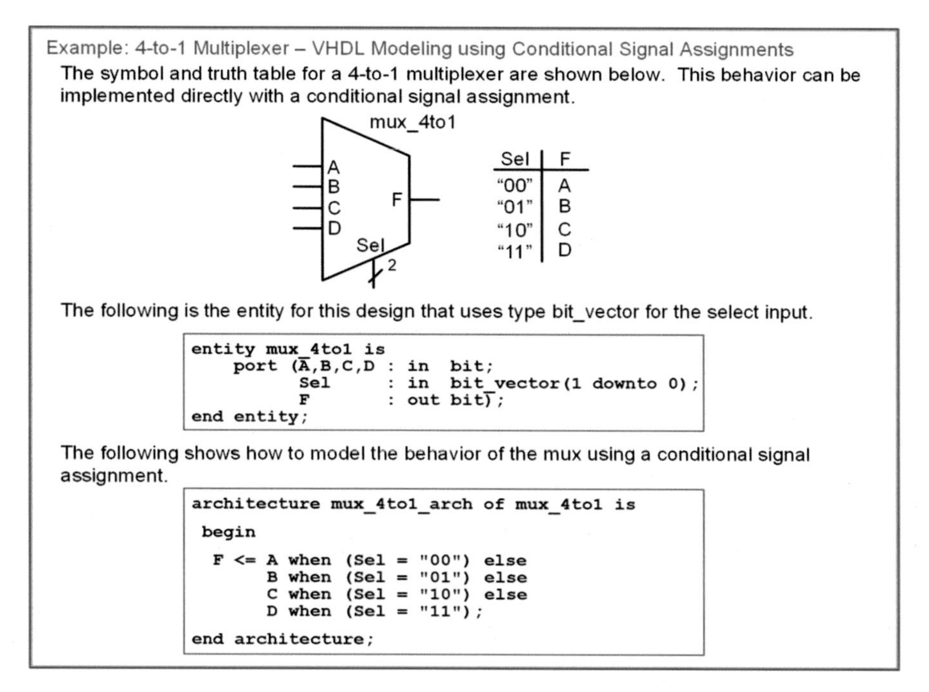

Example: 4-to-1 Multiplexer – VHDL Modeling using Conditional Signal Assignments

The symbol and truth table for a 4-to-1 multiplexer are shown below. This behavior can be implemented directly with a conditional signal assignment.

Sel	F
"00"	A
"01"	B
"10"	C
"11"	D

The following is the entity for this design that uses type bit_vector for the select input.

```
entity mux_4to1 is
    port (A,B,C,D : in  bit;
          Sel     : in  bit_vector(1 downto 0);
          F       : out bit);
end entity;
```

The following shows how to model the behavior of the mux using a conditional signal assignment.

```
architecture mux_4to1_arch of mux_4to1 is

 begin

  F <= A when (Sel = "00") else
       B when (Sel = "01") else
       C when (Sel = "10") else
       D when (Sel = "11");

end architecture;
```

Example 3.14
4-to-1 multiplexer – VHDL modeling using conditional signal assignments

3.3.6 Conditional Signal Assignment Example: Demultiplexer

Example 3.15 shows the process of designing a 1-to-4 demultiplexer using conditional signal assignments.

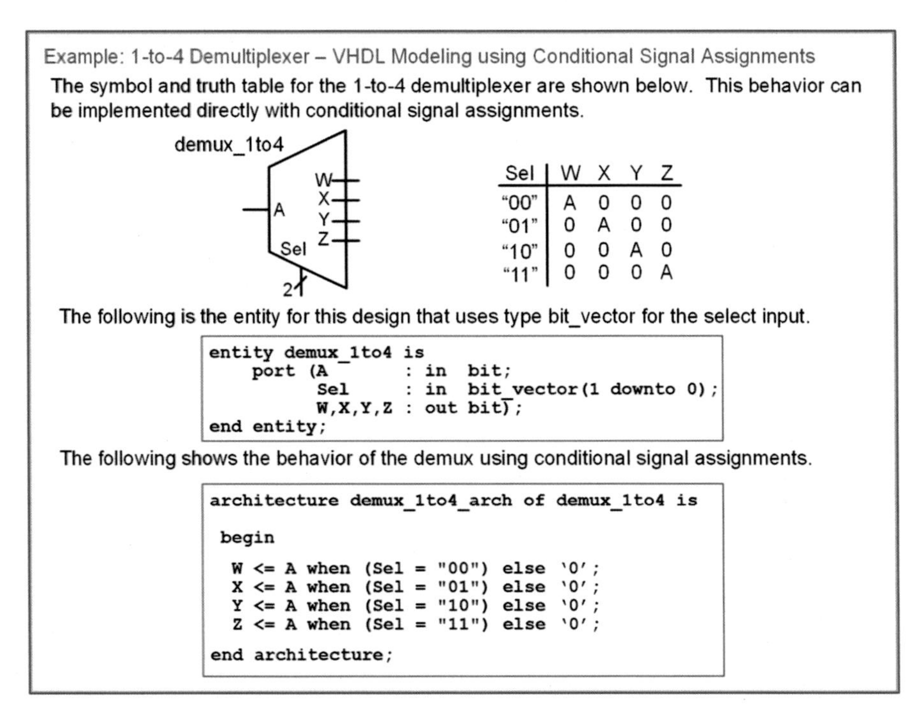

Example: 1-to-4 Demultiplexer – VHDL Modeling using Conditional Signal Assignments

The symbol and truth table for the 1-to-4 demultiplexer are shown below. This behavior can be implemented directly with conditional signal assignments.

Sel	W	X	Y	Z
"00"	A	0	0	0
"01"	0	A	0	0
"10"	0	0	A	0
"11"	0	0	0	A

The following is the entity for this design that uses type bit_vector for the select input.

```
entity demux_1to4 is
    port (A       : in  bit;
          Sel     : in  bit_vector(1 downto 0);
          W,X,Y,Z : out bit);
end entity;
```

The following shows the behavior of the demux using conditional signal assignments.

```
architecture demux_1to4_arch of demux_1to4 is

 begin

  W <= A when (Sel = "00") else '0';
  X <= A when (Sel = "01") else '0';
  Y <= A when (Sel = "10") else '0';
  Z <= A when (Sel = "11") else '0';

end architecture;
```

Example 3.15
1-to-4 demultiplexer – VHDL modeling using conditional signal assignments

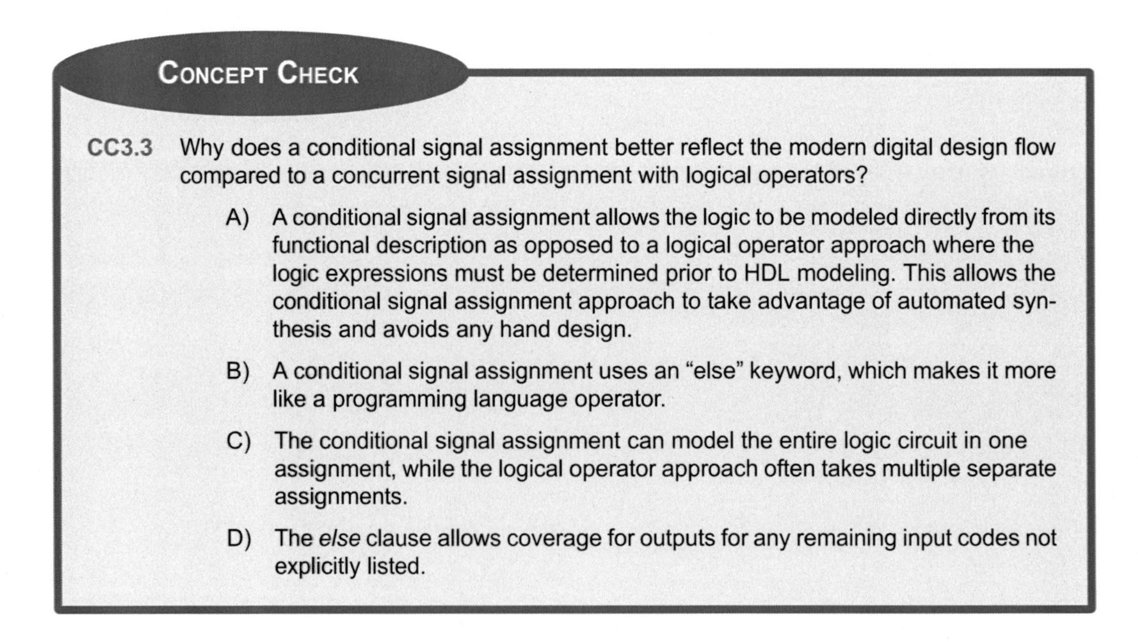

Concept Check

CC3.3 Why does a conditional signal assignment better reflect the modern digital design flow compared to a concurrent signal assignment with logical operators?

A) A conditional signal assignment allows the logic to be modeled directly from its functional description as opposed to a logical operator approach where the logic expressions must be determined prior to HDL modeling. This allows the conditional signal assignment approach to take advantage of automated synthesis and avoids any hand design.

B) A conditional signal assignment uses an "else" keyword, which makes it more like a programming language operator.

C) The conditional signal assignment can model the entire logic circuit in one assignment, while the logical operator approach often takes multiple separate assignments.

D) The *else* clause allows coverage for outputs for any remaining input codes not explicitly listed.

3.4 Selected Signal Assignments

A selected signal assignment provides another technique to implement concurrent signal assignments. In this approach, the signal assignment is based on a specific value on the input signal. The keyword **with** is used to begin the selected signal assignment. It is then followed by the name of the input that will be used to dictate the value of the output. Only a single variable name can be listed as the input. This means that if the assignment is going to be based on multiple variables, they must first be concatenated into a single vector name before starting the selected signal assignment. After the input is listed, the keyword **select** signifies the beginning of the signal assignments. An assignment is made to a signal based on a list of possible input values that follow the keyword **when**. Multiple values of the input codes can be used and are separated by commas. The keyword **others** is used to cover any input values that are not explicitly stated. The syntax for a selected signal assignment is as follows:

```
with input_name select
     signal_name <= expression_1 when condition_1,
                    expression_2 when condition_2,
                                  :
                    expression_n when others;
```

Example:

```
with A select
  F1 <= '1' when '0',        -- F1 will be assigned '1' when A='0'
        '0' when '1';        -- F1 will be assigned '0' when A='1'

AB <= A&B;                   -- concatenate A and B so that they can be used as a vector
with AB select
  F2 <= '0' when "00",       -- F2 will be assigned '0' when AB="00"
        '1' when "01",
        '1' when "10",
        '0' when "11";

with AB select
  F3 <= '1' when "01",
        '1' when "10",
        '0' when others;
```

One feature of selected signal assignments that makes its form even more compact than other techniques is that multiple input codes that correspond to the same output assignment can be listed on the same line pipe (|)-delimited. The example for F3 can be equivalently described as:

```
with AB select
  F3 <= '1' when "01" | "10",
        '0' when others;
```

3.4.1 Selected Signal Assignment Example: SOP Circuit

Example 3.16 shows how to design a VHDL model of a combinational logic circuit using selected signal assignments. Note that this example uses the same truth table as in Example 3.1 to illustrate a comparison between approaches. This approach provides a model that can be created directly from the truth table without needing to do any synthesis or minimization by hand.

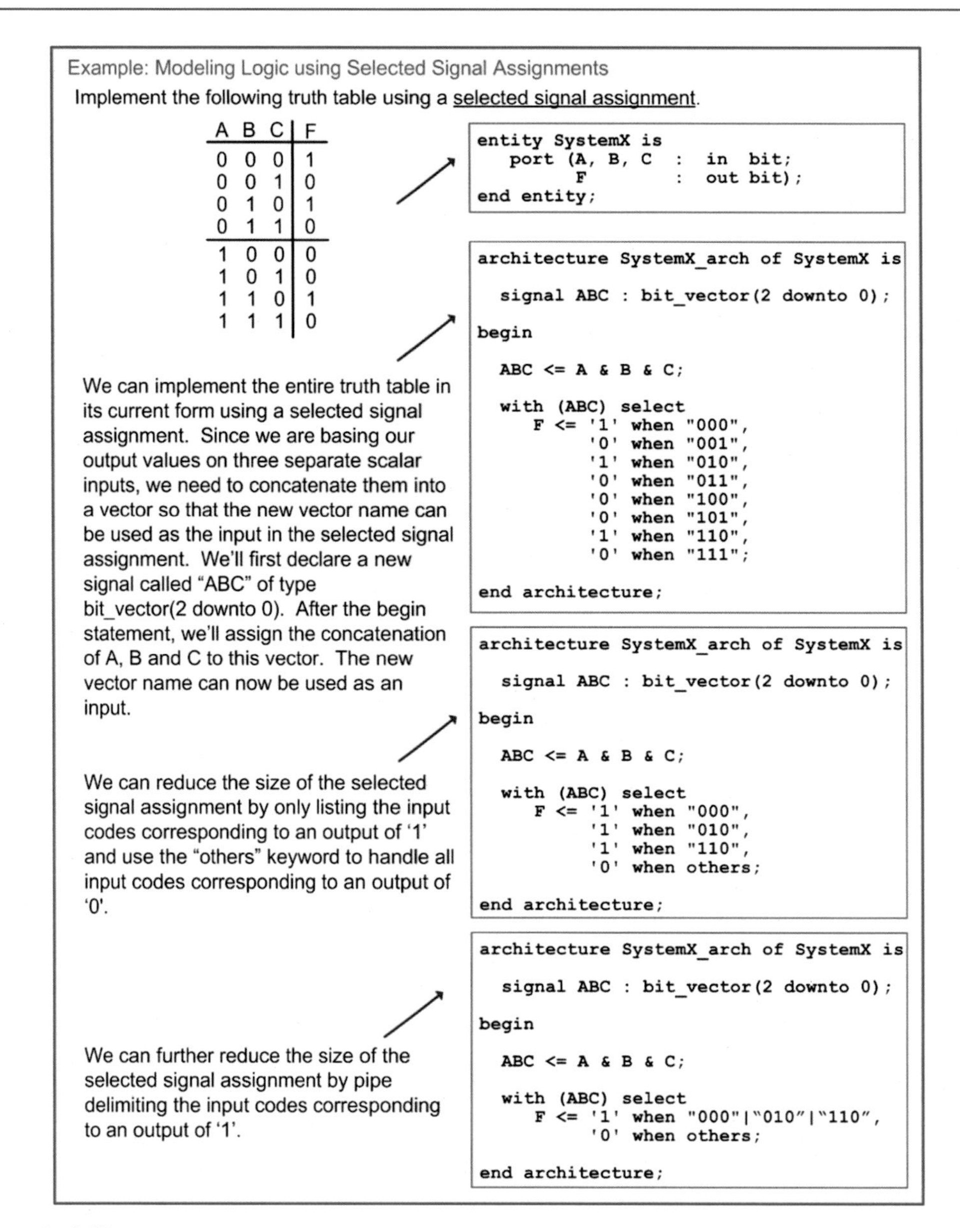

Example 3.16
SOP logic circuit – VHDL modeling using selected signal assignments

3.4.2 Selected Signal Assignment Example: One-Hot Decoder

Example 3.17 shows how to model a 3-to-8 one-hot decoder in VHDL with selected signal assignments. Again, this approach allows the logic to be modeled directly from its functional description rather than having to perform any synthesis by hand.

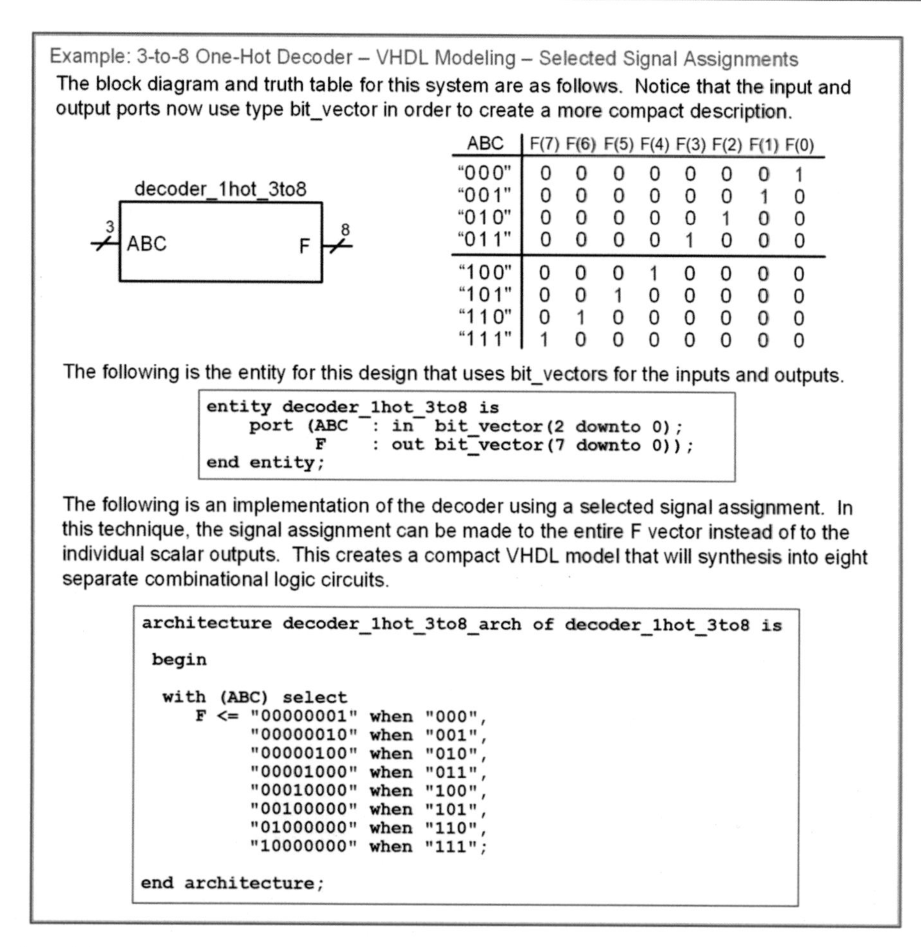

Example: 3-to-8 One-Hot Decoder – VHDL Modeling – Selected Signal Assignments

The block diagram and truth table for this system are as follows. Notice that the input and output ports now use type bit_vector in order to create a more compact description.

ABC	F(7)	F(6)	F(5)	F(4)	F(3)	F(2)	F(1)	F(0)
"000"	0	0	0	0	0	0	0	1
"001"	0	0	0	0	0	0	1	0
"010"	0	0	0	0	0	1	0	0
"011"	0	0	0	0	1	0	0	0
"100"	0	0	0	1	0	0	0	0
"101"	0	0	1	0	0	0	0	0
"110"	0	1	0	0	0	0	0	0
"111"	1	0	0	0	0	0	0	0

The following is the entity for this design that uses bit_vectors for the inputs and outputs.

```
entity decoder_1hot_3to8 is
    port (ABC  : in  bit_vector(2 downto 0);
          F    : out bit_vector(7 downto 0));
end entity;
```

The following is an implementation of the decoder using a selected signal assignment. In this technique, the signal assignment can be made to the entire F vector instead of to the individual scalar outputs. This creates a compact VHDL model that will synthesis into eight separate combinational logic circuits.

```
architecture decoder_1hot_3to8_arch of decoder_1hot_3to8 is

 begin

  with (ABC) select
     F <= "00000001" when "000",
          "00000010" when "001",
          "00000100" when "010",
          "00001000" when "011",
          "00010000" when "100",
          "00100000" when "101",
          "01000000" when "110",
          "10000000" when "111";

end architecture;
```

Example 3.17
3-to-8 one-hot decoder – VHDL modeling using selected signal assignments

3.4.3 Selected Signal Assignment Example: 7-Segment Display Decoder

Back in Example 3.3 the truth table for a 7-segment display decoder was given along with the subsequent steps to create its logic using the classical digital design approach and model it in VHDL using logical operators. With a selected signal assignment, this decoder can be modeled directly from the truth table without needing to do any design by hand. Example 3.18 shows how to model the logic for a 7-segment display decoder using a selected signal assignment.

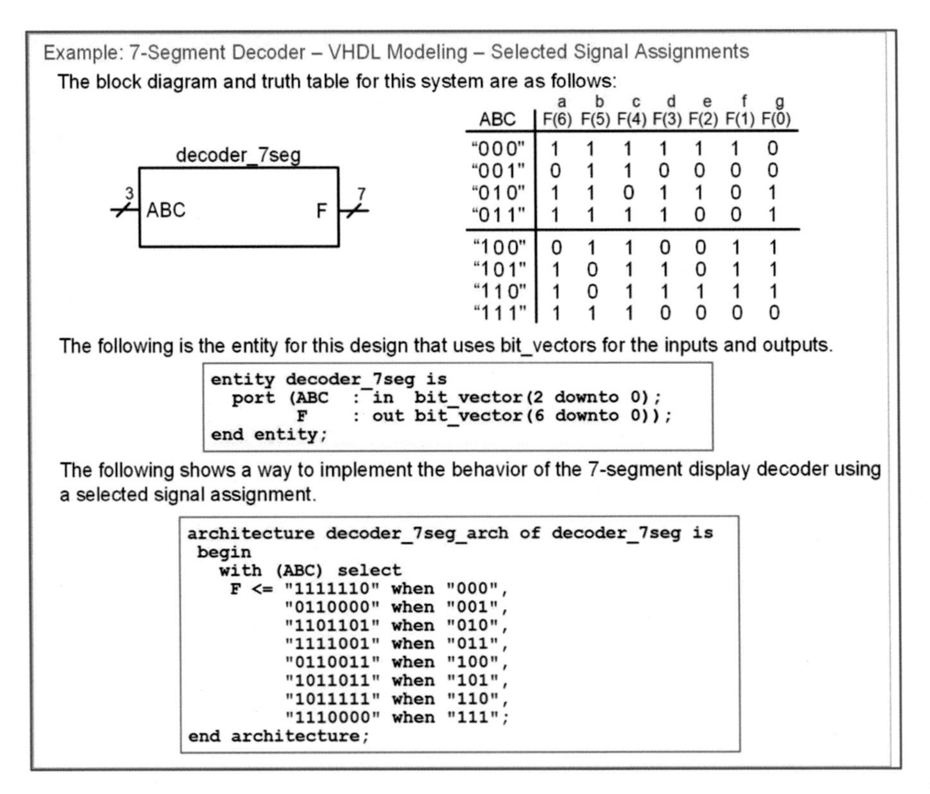

Example: 7-Segment Decoder – VHDL Modeling – Selected Signal Assignments

The block diagram and truth table for this system are as follows:

ABC	a F(6)	b F(5)	c F(4)	d F(3)	e F(2)	f F(1)	g F(0)
"000"	1	1	1	1	1	1	0
"001"	0	1	1	0	0	0	0
"010"	1	1	0	1	1	0	1
"011"	1	1	1	1	0	0	1
"100"	0	1	1	0	0	1	1
"101"	1	0	1	1	0	1	1
"110"	1	0	1	1	1	1	1
"111"	1	1	1	0	0	0	0

The following is the entity for this design that uses bit_vectors for the inputs and outputs.

```
entity decoder_7seg is
  port (ABC  : in  bit_vector(2 downto 0);
        F    : out bit_vector(6 downto 0));
end entity;
```

The following shows a way to implement the behavior of the 7-segment display decoder using a selected signal assignment.

```
architecture decoder_7seg_arch of decoder_7seg is
 begin
   with (ABC) select
    F <= "1111110" when "000",
         "0110000" when "001",
         "1101101" when "010",
         "1111001" when "011",
         "0110011" when "100",
         "1011011" when "101",
         "1011111" when "110",
         "1110000" when "111";
end architecture;
```

Example 3.18
7-segment display decoder– VHDL modeling using selected signal assignments

3.4.4 Selected Signal Assignment Example: One-Hot Encoder

Example 3.19 shows how to model a one-hot encoder in VHDL with selected signal assignments. Again, this approach allows the logic to be modeled directly from its functional description rather than having to perform any synthesis by hand.

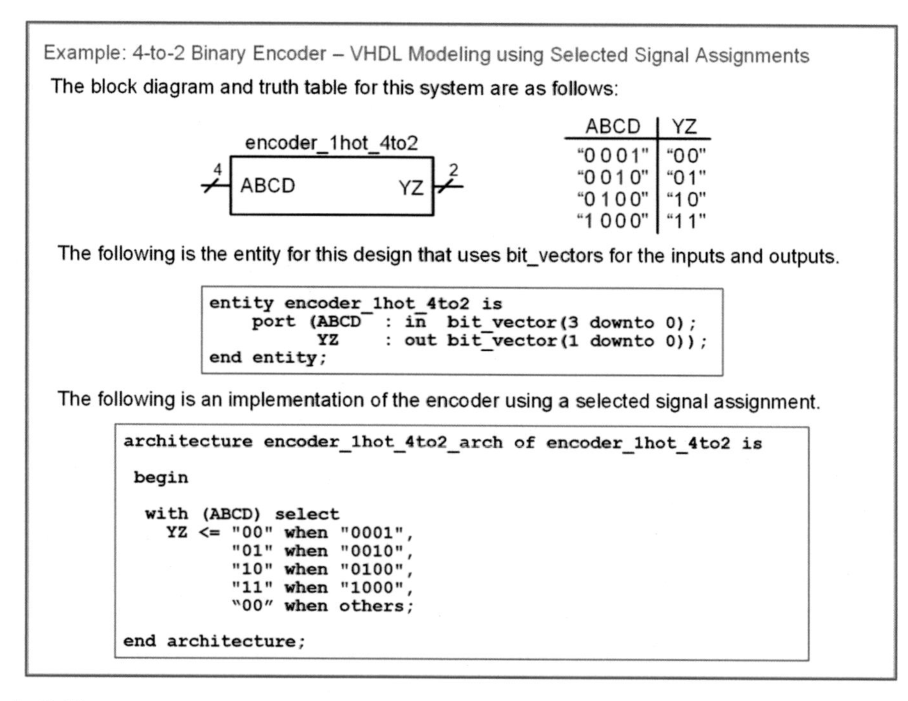

Example: 4-to-2 Binary Encoder – VHDL Modeling using Selected Signal Assignments

The block diagram and truth table for this system are as follows:

ABCD	YZ
"0 0 0 1"	"0 0"
"0 0 1 0"	"0 1"
"0 1 0 0"	"1 0"
"1 0 0 0"	"1 1"

The following is the entity for this design that uses bit_vectors for the inputs and outputs.

```
entity encoder_1hot_4to2 is
    port (ABCD  : in  bit_vector(3 downto 0);
          YZ    : out bit_vector(1 downto 0));
end entity;
```

The following is an implementation of the encoder using a selected signal assignment.

```
architecture encoder_1hot_4to2_arch of encoder_1hot_4to2 is

 begin

  with (ABCD) select
    YZ <= "00" when "0001",
          "01" when "0010",
          "10" when "0100",
          "11" when "1000",
          "00" when others;

end architecture;
```

Example 3.19
4-to-2 binary encoder – VHDL modeling using selected signal assignments

3.4.5 Selected Signal Assignment Example: Multiplexer

Example 3.20 shows the process of designing a 4-to-1 multiplexer using selected signal assignments.

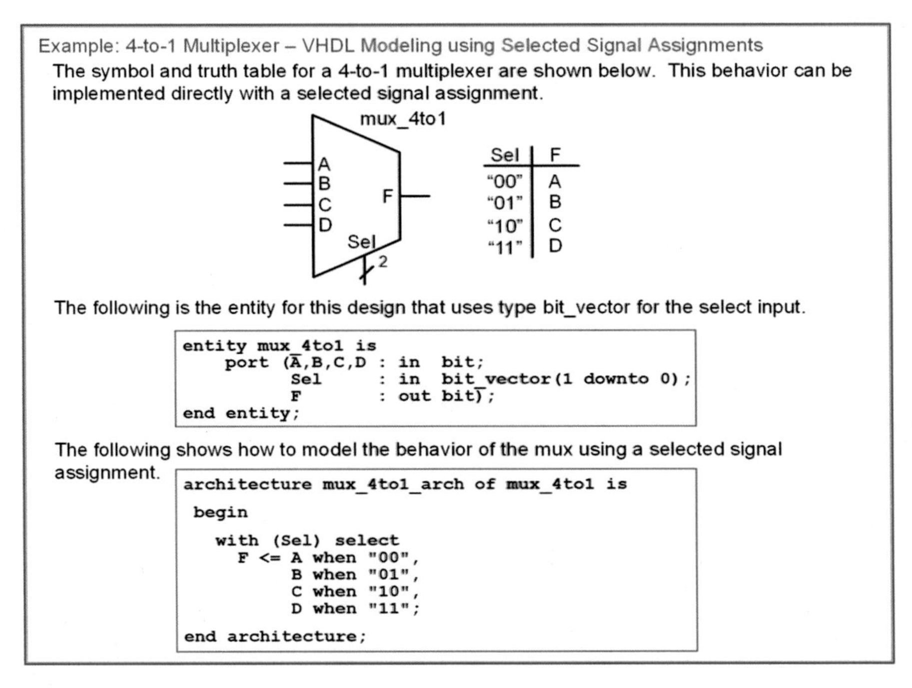

Example: 4-to-1 Multiplexer – VHDL Modeling using Selected Signal Assignments

The symbol and truth table for a 4-to-1 multiplexer are shown below. This behavior can be implemented directly with a selected signal assignment.

Sel	F
"00"	A
"01"	B
"10"	C
"11"	D

The following is the entity for this design that uses type bit_vector for the select input.

```
entity mux_4to1 is
    port (A,B,C,D : in  bit;
          Sel     : in  bit_vector(1 downto 0);
          F       : out bit);
end entity;
```

The following shows how to model the behavior of the mux using a selected signal assignment.

```
architecture mux_4to1_arch of mux_4to1 is

 begin

   with (Sel) select
     F <= A when "00",
          B when "01",
          C when "10",
          D when "11";

end architecture;
```

Example 3.20
4-to-1 multiplexer – VHDL modeling using selected signal assignments

3.4.6 Selected Signal Assignment Example: Demultiplexer

Example 3.21 shows the process of designing a 1-to-4 demultiplexer using selected signal assignments.

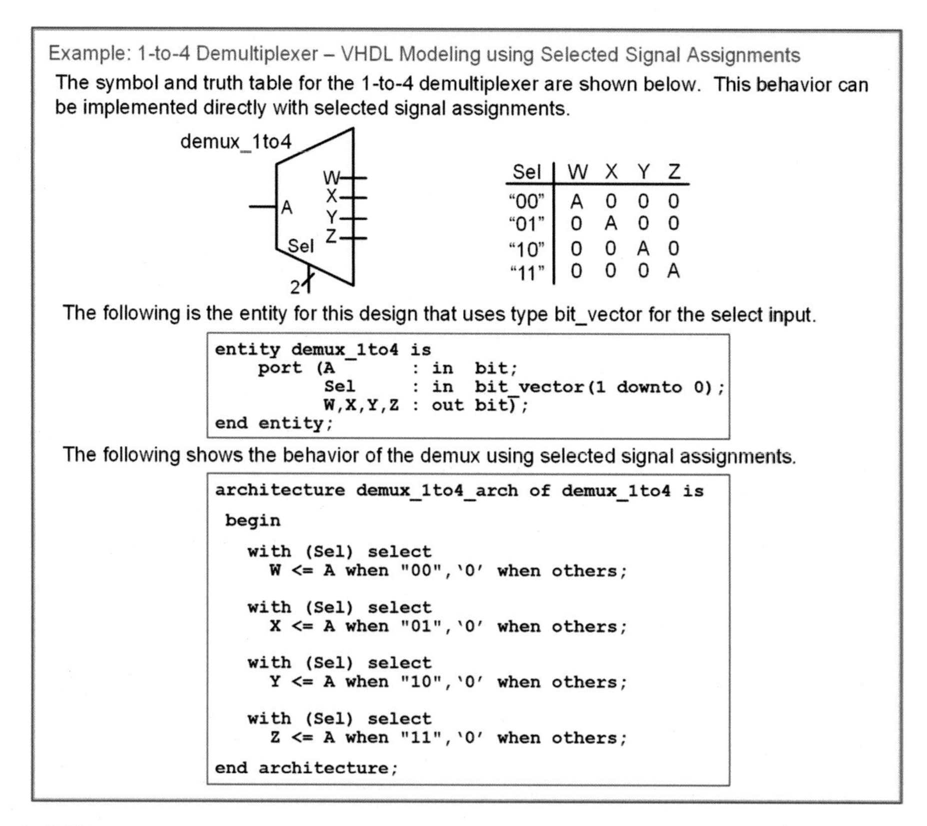

Example: 1-to-4 Demultiplexer – VHDL Modeling using Selected Signal Assignments

The symbol and truth table for the 1-to-4 demultiplexer are shown below. This behavior can be implemented directly with selected signal assignments.

Sel	W	X	Y	Z
"00"	A	0	0	0
"01"	0	A	0	0
"10"	0	0	A	0
"11"	0	0	0	A

The following is the entity for this design that uses type bit_vector for the select input.

```
entity demux_1to4 is
    port (A       : in  bit;
          Sel     : in  bit_vector(1 downto 0);
          W,X,Y,Z : out bit);
end entity;
```

The following shows the behavior of the demux using selected signal assignments.

```
architecture demux_1to4_arch of demux_1to4 is
 begin
   with (Sel) select
     W <= A when "00",'0' when others;

   with (Sel) select
     X <= A when "01",'0' when others;

   with (Sel) select
     Y <= A when "10",'0' when others;

   with (Sel) select
     Z <= A when "11",'0' when others;

end architecture;
```

Example 3.21
1-to-4 demultiplexer – VHDL modeling using selected signal assignments

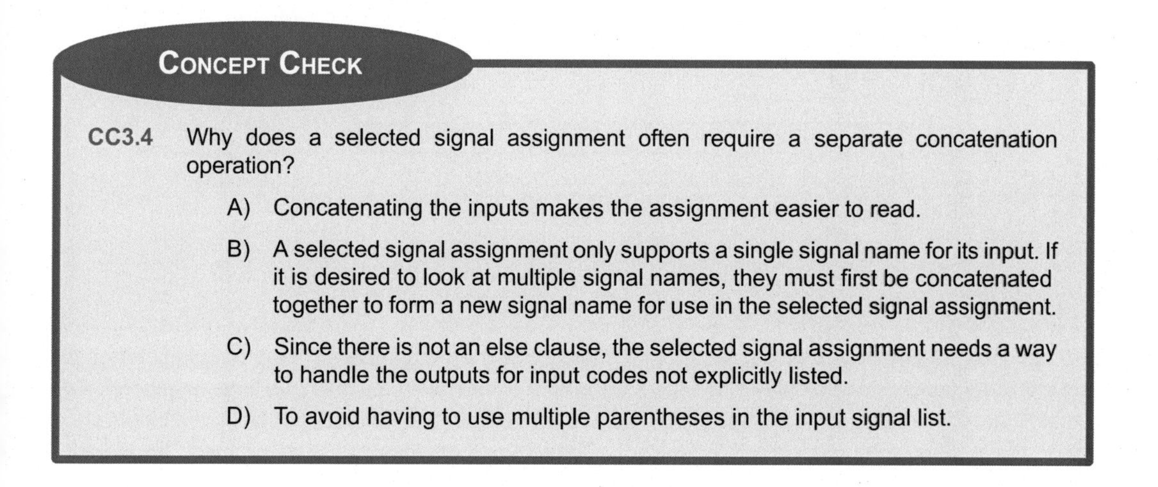

Concept Check

CC3.4 Why does a selected signal assignment often require a separate concatenation operation?

A) Concatenating the inputs makes the assignment easier to read.

B) A selected signal assignment only supports a single signal name for its input. If it is desired to look at multiple signal names, they must first be concatenated together to form a new signal name for use in the selected signal assignment.

C) Since there is not an else clause, the selected signal assignment needs a way to handle the outputs for input codes not explicitly listed.

D) To avoid having to use multiple parentheses in the input signal list.

3.5 Delayed Signal Assignments

3.5.1 Inertial Delay

VHDL provides the ability to delay a concurrent signal assignment in order to more accurately model the behavior of real gates. The keyword **after** is used to delay an assignment by a certain amount of time. The magnitude of the delay is provided as type time. The syntax for delaying an assignment is as follows:

```
signal_name <= <expression> after <time>;
```

Example:

```
A <= B after 3us;
C <= D and E after 10ns;
```

If an input pulse is shorter in duration than the amount of the delay, the input pulse is ignored. This is called the *inertial delay model*. Example 3.22 shows how to design a VHDL model with a delayed signal assignment using the inertial delay model.

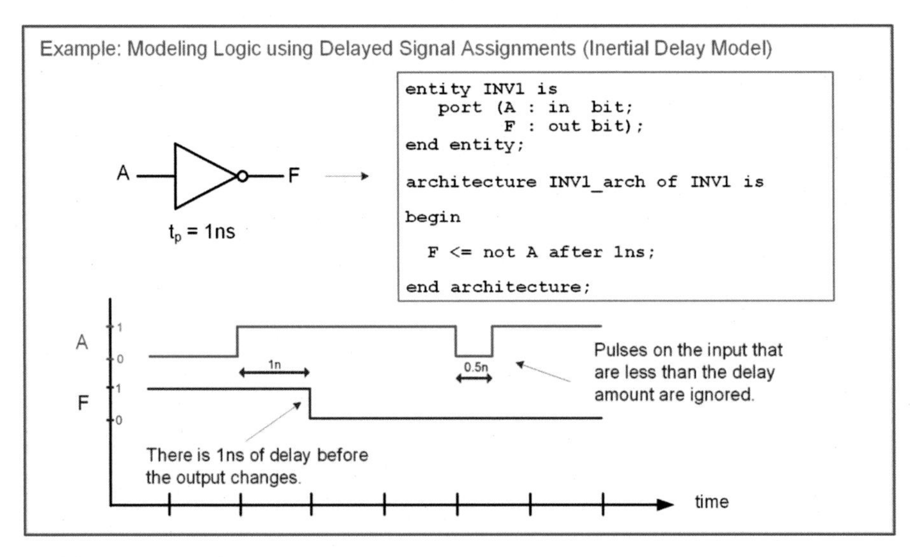

Example 3.22
Modeling logic using delayed signal assignments (inertial delay model)

3.5.2 Transport Delay

Ignoring brief input pulses on the input accurately models the behavior of on-chip gates. When the input pulse is faster than the delay of the gate, the output of the gate does not have time to respond. As a result, there will not be a logic change on the output. If it is desired to have all pulses on the inputs show up on the outputs when modeling the behavior of other types of digital logic, the keyword **transport** is used in conjunction with the after statement. This is called the *transport delay model*.

```
signal_name <= transport <expression> after <time>;
```

Example 3.23 shows how to perform a delayed signal assignment using the transport delay model.

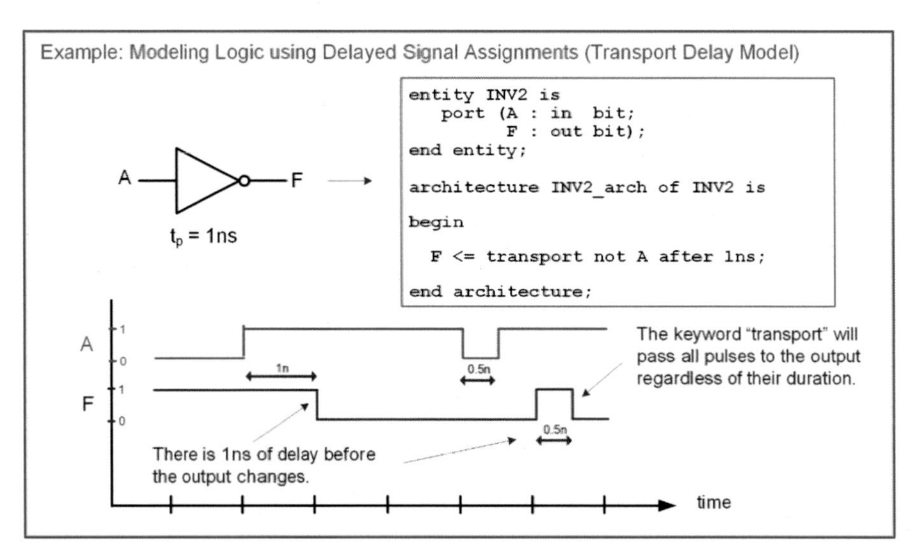

Example 3.23
Modeling logic using delayed signal assignments (transport delay model)

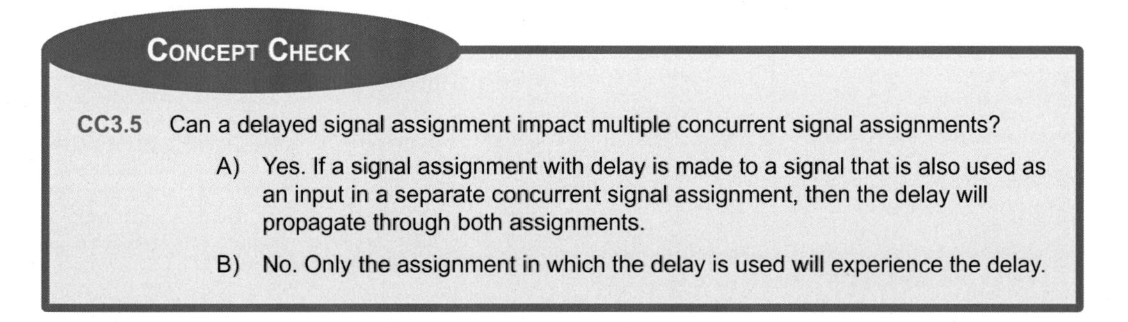
CONCEPT CHECK

CC3.5 Can a delayed signal assignment impact multiple concurrent signal assignments?

A) Yes. If a signal assignment with delay is made to a signal that is also used as an input in a separate concurrent signal assignment, then the delay will propagate through both assignments.

B) No. Only the assignment in which the delay is used will experience the delay.

Summary

- VHDL operators are defined to work on specific data types. Not all operators work on all types within a package.
- *Concurrency* is the term that describes operations being performed in parallel. This allows real-world system behavior to be modeled.
- VHDL contains three direct techniques to model concurrent logic behavior. These are *concurrent signal assignments with logical operators*, *conditional signal assignments*, and *selected signal assignments*.
- Delay can be modeled in VHDL using either the *inertial* or *transport* model. Inertial delay will ignore pulses that are shorter than the delay amount, while transport delay will pass all transitions.

Exercise Problems

Section 3.1: VHDL Operators

3.1.1 What data types do the logical operators in the standard package work on?

3.1.2 Which logical operator has the highest priority when evaluating the order of precedence of operations?

3.1.3 If parentheses are not used in a signal assignment with logical operators, how is the order of precedence determined?

3.1.4 What data types do the numerical operators in the standard package work on?

3.1.5 What is the return type of a relational operator?

Section 3.2: Concurrent Signal Assignments with Logical Operators

3.2.1 Design a VHDL model to implement the behavior described by the 3-input minterm list shown in Fig. 3.1. Use concurrent signal assignments and logical operators. Declare your entity to match the block diagram provided. Use the type bit for your ports.

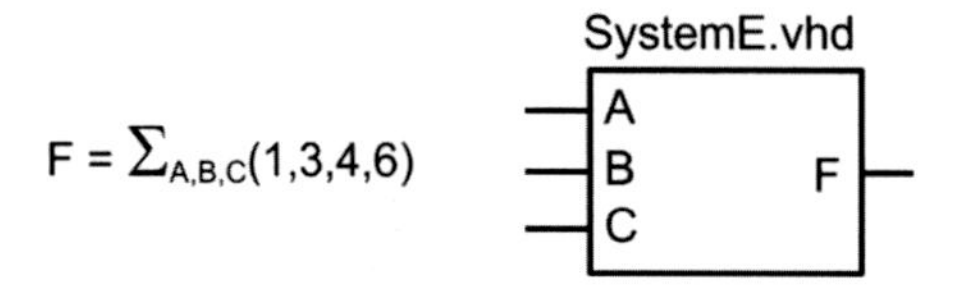

Fig. 3.1
System E functionality

3.2.2 Design a VHDL model to implement the behavior described by the 3-input maxterm list shown in Fig. 3.2. Use concurrent signal assignments and logical operators. Declare your entity to match the block diagram provided. Use the type bit for your ports.

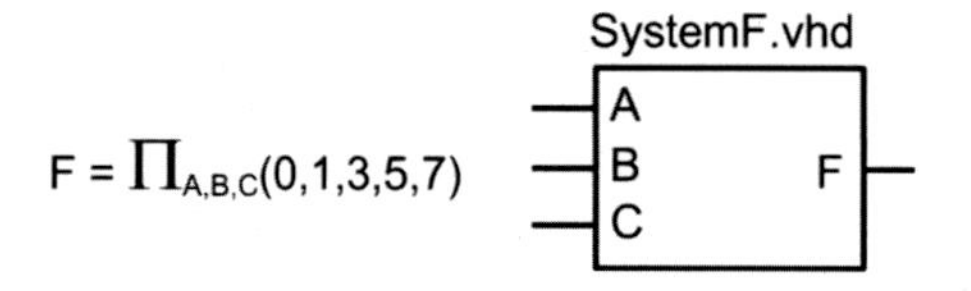

Fig. 3.2
System F functionality

3.2.3 Design a VHDL model to implement the behavior described by the 3-input truth table shown in Fig. 3.3. Use concurrent signal assignments and logical operators. Declare your entity to match the block diagram provided. Use the type bit for your ports.

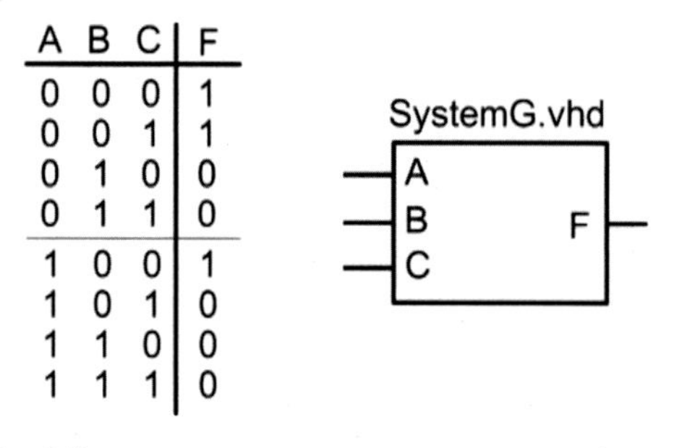

A	B	C	F
0	0	0	1
0	0	1	1
0	1	0	0
0	1	1	0
1	0	0	1
1	0	1	0
1	1	0	0
1	1	1	0

Fig. 3.3
System G functionality

3.2.4 Design a VHDL model to implement the behavior described by the 4-input minterm list shown in Fig. 3.4. Use concurrent signal assignments and logical operators. Declare your entity to match the block diagram provided. Use the type bit for your ports.

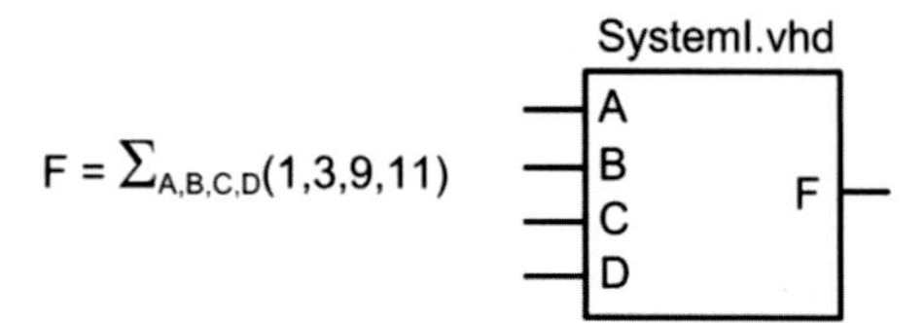

Fig. 3.4
System I functionality

3.2.5 Design a VHDL model to implement the behavior described by the 4-input maxterm list shown in Fig. 3.5. Use concurrent signal assignments and logical operators. Declare your entity to match the block diagram provided. Use the type bit for your ports.

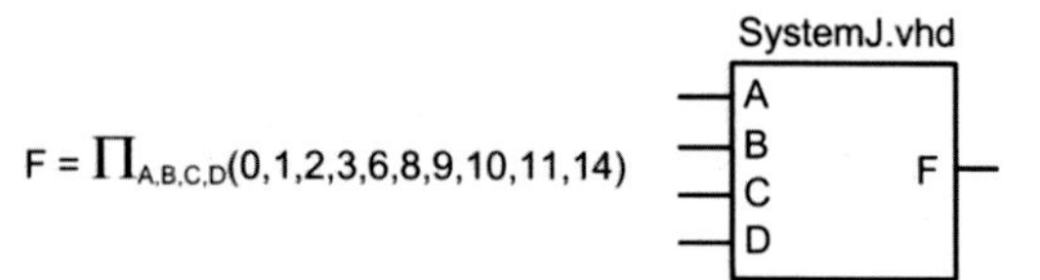

Fig. 3.5
System J functionality

3.2.6 Design a VHDL model to implement the behavior described by the 4-input truth table shown in Fig. 3.6. Use concurrent signal assignments and logical operators. Declare your entity to match the block diagram provided. Use the type bit for your ports.

A	B	C	D	F
0	0	0	0	1
0	0	0	1	1
0	0	1	0	1
0	0	1	1	0
0	1	0	0	1
0	1	0	1	1
0	1	1	0	1
0	1	1	1	0
1	0	0	0	1
1	0	0	1	1
1	0	1	0	1
1	0	1	1	0
1	1	0	0	1
1	1	0	1	1
1	1	1	0	1
1	1	1	1	0

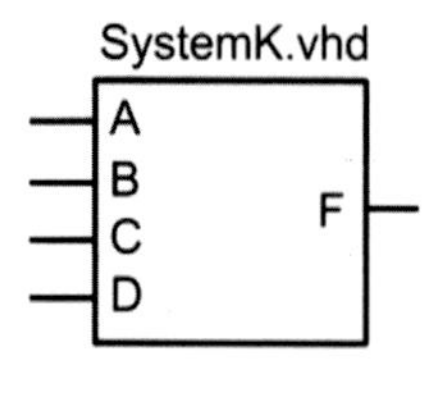

Fig. 3.6
System K functionality

Section 3.3: Conditional Signal Assignments

3.3.1 Design a VHDL model to implement the behavior described by the 3-input minterm list shown in Fig. 3.1. Use conditional signal assignments. Declare your entity to match the block diagram provided. Use the type bit for your ports.

3.3.2 Design a VHDL model to implement the behavior described by the 3-input maxterm list shown in Fig. 3.2. Use conditional signal assignments. Declare your entity to match the block diagram provided. Use the type bit for your ports.

3.3.3 Design a VHDL model to implement the behavior described by the 3-input truth table shown in Fig. 3.3. Use conditional signal assignments. Declare your entity to match the block diagram provided. Use the type bit for your ports.

3.3.4 Design a VHDL model to implement the behavior described by the 4-input minterm list shown in Fig. 3.4. Use conditional signal assignments. Declare your entity to match the block diagram provided. Use the type bit for your ports.

3.3.5 Design a VHDL model to implement the behavior described by the 4-input maxterm list shown in Fig. 3.5. Use conditional signal assignments. Declare your entity to match the block diagram provided. Use the type bit for your ports.

3.3.6 Design a VHDL model to implement the behavior described by the 4-input truth table shown in Fig. 3.6. Use conditional signal assignments. Declare your entity to match the block diagram provided. Use the type bit for your ports.

Section 3.4: Selected Signal Assignments

3.4.1 Design a VHDL model to implement the behavior described by the 3-input minterm list shown in Fig. 3.1. Use selected signal assignments. Declare your entity to match the block diagram provided. Use the type bit for your ports.

3.4.2 Design a VHDL model to implement the behavior described by the 3-input maxterm list shown in Fig. 3.2. Use selected signal assignments. Declare your entity to match the block diagram provided. Use the type bit for your ports.

3.4.3 Design a VHDL model to implement the behavior described by the 3-input truth table shown in Fig. 3.3. Use selected signal assignments. Declare your entity to match the block diagram provided. Use the type bit for your ports.

3.4.4 Design a VHDL model to implement the behavior described by the 4-input minterm list shown in Fig. 3.4. Use selected signal assignments. Declare your entity to match the block diagram provided. Use the type bit for your ports.

3.4.5 Design a VHDL model to implement the behavior described by the 4-input maxterm list shown in Fig. 3.5. Use selected signal assignments. Declare your entity to match the block diagram provided. Use the type bit for your ports.

3.4.6 Design a VHDL model to implement the behavior described by the 4-input truth table shown in Fig. 3.6. Use selected signal assignments. Declare your entity to match the block diagram provided. Use the type bit for your ports.

Section 3.5: Delayed Signal Assignments

3.5.1 Design a VHDL model to implement the behavior described by the 3-input minterm list shown in Fig. 3.1. Use concurrent signal assignments and logical operators. Create the model so that every logic operation has 1ns of inertial delay. Declare your entity to match the block diagram provided. Use the type bit for your ports.

3.5.2 Design a VHDL model to implement the behavior described by the 3-input maxterm list shown in Fig. 3.2. Use concurrent signal assignments and logical operators. Create the model so that every logic operation has 1ns of inertial delay Declare your entity to match the block diagram provided. Use the type bit for your ports.

3.5.3 Design a VHDL model to implement the behavior described by the 3-input truth table shown in Fig. 3.3. Use concurrent signal assignments and logical operators. Create the model so that every logic operation has 1ns of inertial delay Declare your entity to match the block diagram provided. Use the type bit for your ports.

3.5.4 Design a VHDL model to implement the behavior described by the 4-input minterm list shown in Fig. 3.4. Use concurrent signal assignments and logical operators. Create the model so that every logic operation has 1ns of transport delay Declare your entity to match the block diagram provided. Use the type bit for your ports.

3.5.5 Design a VHDL model to implement the behavior described by the 4-input maxterm list shown in Fig. 3.5. Use concurrent signal assignments and logical operators. Create the model so that every logic operation has 1ns of transport delay Declare your entity to match the block diagram provided. Use the type bit for your ports.

3.5.6 Design a VHDL model to implement the behavior described by the 4-input truth table shown in Fig. 3.6. Use concurrent signal assignments and logical operators. Create the model so that every logic operation has 1ns of transport delay Declare your entity to match the block diagram provided. Use the type bit for your ports.

Chapter 4: Structural Design and Hierarchy

This chapter describes how to accomplish hierarchy within VHDL using lower-level sub-systems. Structural design in VHDL refers to including lower-level sub-systems within a higher-level system in order to produce the desired functionality. A purely structural VHDL design would not contain any behavioral modeling in the architecture such as signal assignments but instead just contain the instantiation and interconnections of other sub-systems.

Learning Outcomes—After completing this chapter, you will be able to:

4.1 Instantiate and map the ports of a lower-level component in VHDL.
4.2 Design a VHDL model for a system that uses hierarchy.

4.1 Components

4.1.1 Component Instantiation

A sub-system is called a **component** in VHDL. For any component that is going to be used in an architecture, it must be declared before the begin statement. Refer to Sect. 2.2.3.3 for the syntax of declaring a component. A specific component only needs to be declared once. After the begin statement, it can be used as many times as necessary. Each component is executed concurrently.

The term *instantiation* refers to the *use* or *inclusion* of the component in the VHDL system. When a component is instantiated, it needs to be given a unique identifying name. This is called the *instance name*. To instantiate a component, the instance name is given first, followed by a colon and then the component name. The last part of instantiating a component is connecting signals to its ports. The way in which signals are connected to the ports of the component is called the **port map**. The syntax for instantiating a component is as follows:

```
instance_name : <component name>
  port map (<port connections>);
```

4.1.2 Port Mapping

There are two techniques to connect signals to the ports of the component, *explicit port mapping* and *positional port mapping*.

4.1.2.1 Explicit Port Mapping

In explicit port mapping, the name of each port of the component is given, followed by the connection indicator **=>**, followed by the signal it is connected to. The port connections can be listed in any order since the details of the connection (i.e., port name to signal name) are explicit. Each connection name is separated by a comma. The syntax for explicit port mapping is as follows:

```
instance_name : <component name>
  port map (port1 => signal1, port2 => signal2, ...);
```


B. J. LaMeres, *Quick Start Guide to VHDL*, https://doi.org/10.1007/978-3-031-42543-1_4

Example 4.1 shows how to design a VHDL model of a combinational logic circuit using structural VHDL and explicit port mapping. Note that this example again uses the same truth table as in Examples 3.1, 3.10, and 3.16 to illustrate a comparison between approaches.

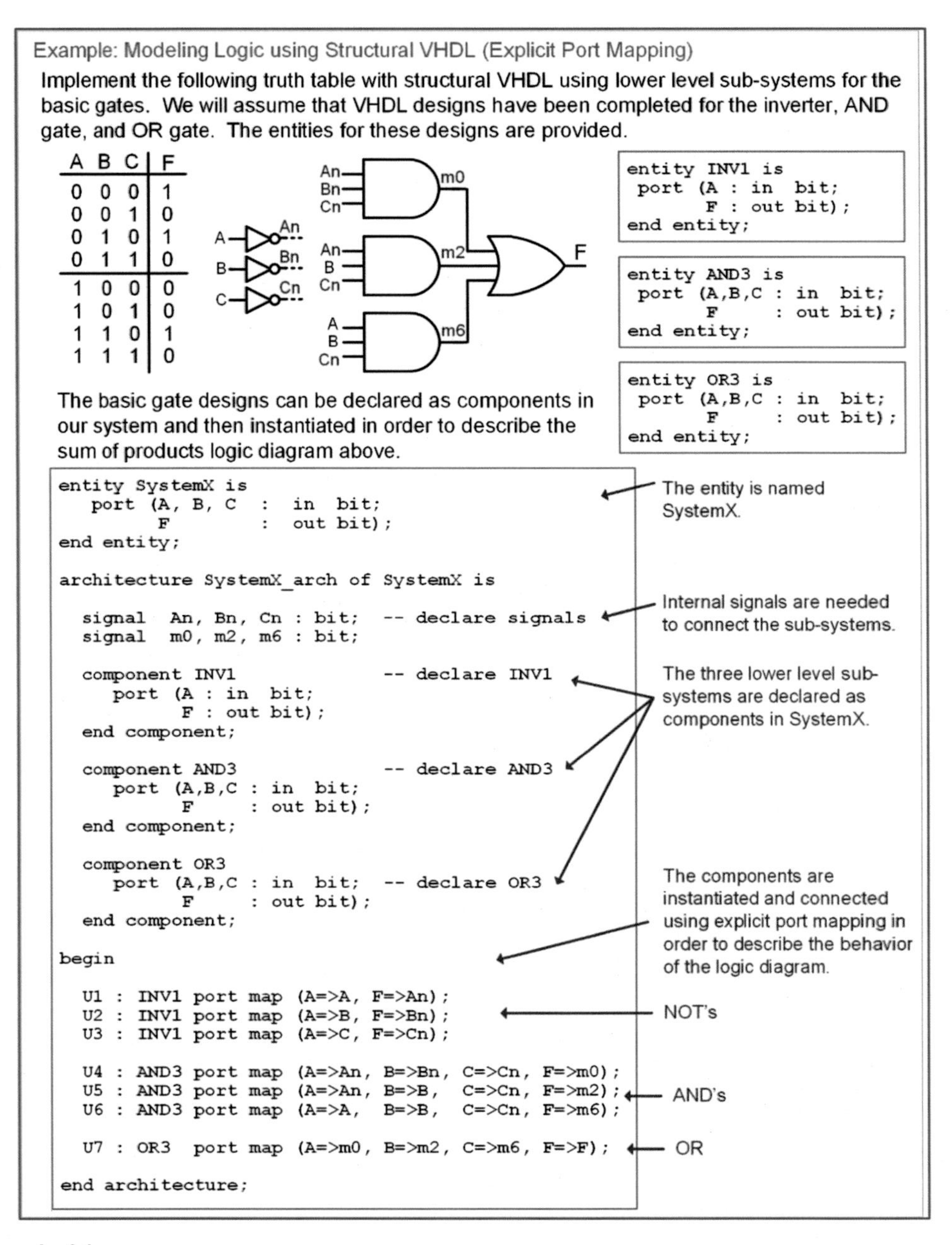

Example 4.1
Modeling logic using structural VHDL (explicit port mapping)

4.1.2.2 Positional Port Mapping

In positional port mapping, the names of the ports of the component are not explicitly listed. Instead, the signals are listed in the same order that the ports of the component were defined. Each signal name is separated by a comma. This approach requires less text to describe but can also lead to misconnections due to mismatches in the order of the signals being connected. The syntax for positional port mapping is as follows:

```
instance_name : <component name>
  port map (signal1, signal2, ...);
```

Example 4.2 shows how to create the same structural VHDL model as in Example 4.1 but using positional port mapping instead.

Example: Modeling Logic using Structural VHDL (Positional Port Mapping)

In positional port mapping the port names are not listed in the component instantiation. Instead, the signals are simply listed in the same order as the ports were defined. The signal listed first will be connected to the port defined first. The signal listed second will be connected to the port defined second, etc.

Explicit Port Mapping

```
begin

  U1 : INV1 port map (A=>A, F=>An);
  U2 : INV1 port map (A=>B, F=>Bn);
  U3 : INV1 port map (A=>C, F=>Cn);

  U4 : AND3 port map (A=>An, B=>Bn, C=>Cn, F=>m0);
  U5 : AND3 port map (A=>An, B=>B,  C=>Cn, F=>m2);
  U6 : AND3 port map (A=>A,  B=>B,  C=>Cn, F=>m6);

  U7 : OR3  port map (A=>m0, B=>m2, C=>m6, F=>F);
```

Positional Port Mapping of Same System

```
begin

  U1 : INV1 port map (A, An);
  U2 : INV1 port map (B, Bn);
  U3 : INV1 port map (C, Cn);

  U4 : AND3 port map (An, Bn, Cn, m0);
  U5 : AND3 port map (An, B,  Cn, m2);
  U6 : AND3 port map (A,  B,  Cn, m6);

  U7 : OR3  port map (m0, m2, m6, F);
```

Example 4.2
Modeling logic using structural VHDL (positional port mapping)

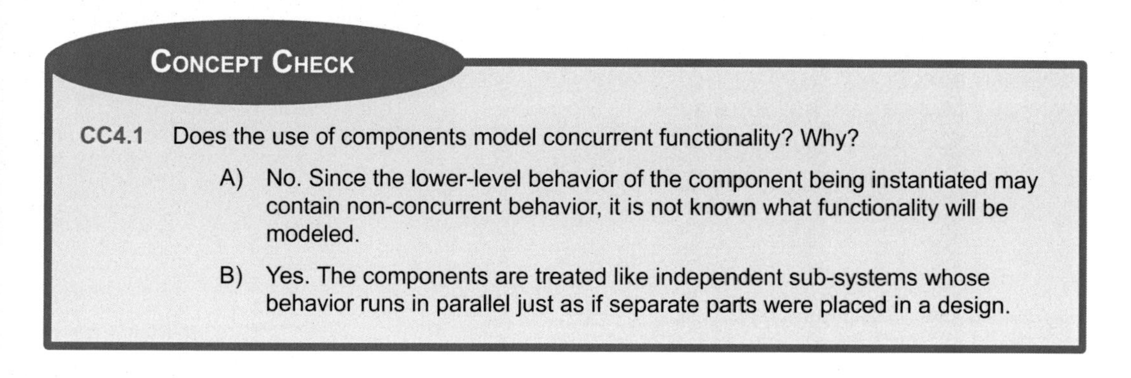
CONCEPT CHECK

CC4.1 Does the use of components model concurrent functionality? Why?

A) No. Since the lower-level behavior of the component being instantiated may contain non-concurrent behavior, it is not known what functionality will be modeled.

B) Yes. The components are treated like independent sub-systems whose behavior runs in parallel just as if separate parts were placed in a design.

4.2 Structural Design Examples: Ripple Carry Adder

This section gives an example of a structural design that implements a simple binary adder.

4.2.1 Half Adders

When creating a binary adder, it is desirable to design incremental sub-systems that can be re-used. This reduces design effort and minimizes troubleshooting complexity. The most basic component in the adder is called a *half adder*. This circuit computes the sum and carry out on two input arguments. The reason it is called a half adder instead of a full adder is because it does not accommodate a *carry in* during the computation; thus, it does not provide all the necessary functionality required for the positional adder. Example 4.3 shows the design of a half adder. Notice that two combinational logic circuits are required in order to produce the sum (the XOR gate) and the carry out (the AND gate). These two gates are in parallel to each other; thus, the delay through the half adder is due to only one level of logic.

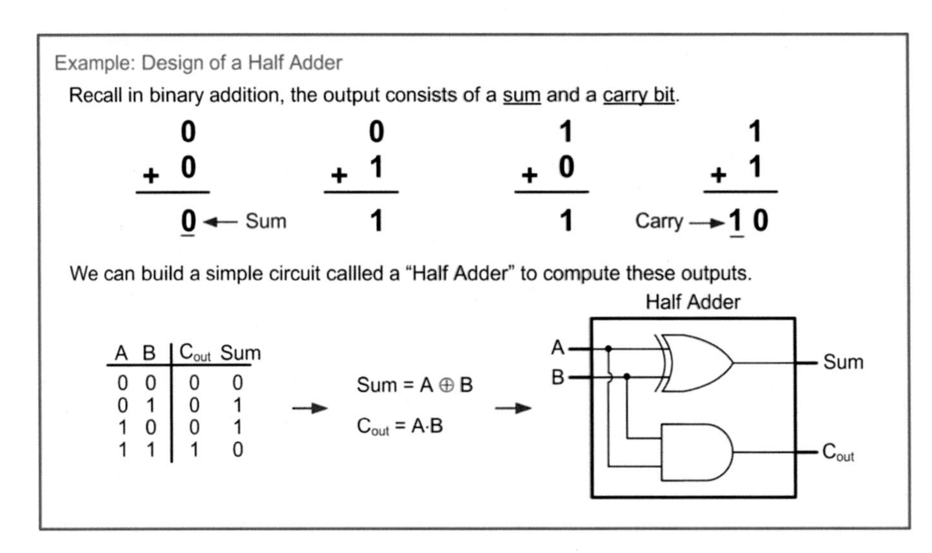

Example 4.3
Design of a half adder

4.2.2 Full Adders

A full adder is a circuit that still produces a sum and carry out but considers three inputs in the computations (A, B, and C_{in}). Example 4.4 shows the design of a full adder using the classical design approach. This step is shown to illustrate why it is possible to re-use half adders to create the full adder. In order to do this, it is necessary to have the minimal sum-of-products logic expression.

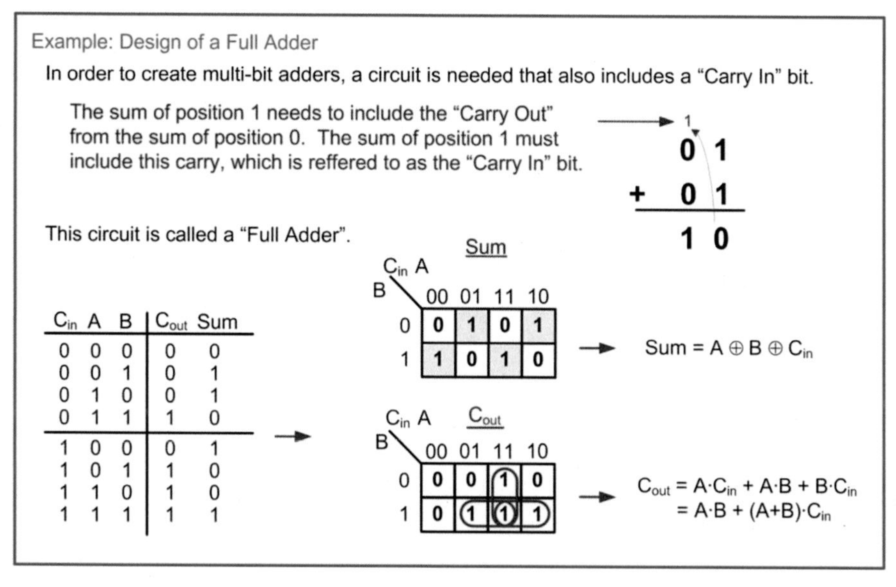

Example 4.4
Design of a full adder

As mentioned before, it is desirable to re-use design components as we construct more complex systems. One such design re-use approach is to create a full adder using two half adders. This is straightforward for the sum output since the logic is simply two cascaded XOR gates ($Sum = A \oplus B \oplus Cin$). The carry out is not as straightforward. Notice that the expression for Cout derived in Example 4.4 contains the term (A + B). If this term could be manipulated to use an XOR gate instead, it would allow the full adder to take advantage of existing circuitry in the system. Figure 4.1 shows a derivation of an equivalency that allows (A + B) to be replaced with ($A \oplus B$) in the Cout logic expression.

A Useful Logic Equivalency that can be Exploited in Arithmetic Circuits

The logic expression for the carry out of a full adder was given as: $C_{out} = A \cdot B + (A + B) \cdot C_{in}$. It turns out that the exact same output is produced by the expression $A \cdot B + (A \oplus B) \cdot C_{in}$. Let's examine how this is possible by breaking down the expressions into their individual parts and solving at each step.

FA Inputs	Desired Output	$C_{out} = A \cdot B + (A + B) \cdot C_{in}$			$C_{out} = A \cdot B + (A \oplus B) \cdot C_{in}$		
C_{in} A B	C_{out}	$A \cdot B$	$(A+B) \cdot C_{in}$	$A \cdot B + (A+B) \cdot C_{in}$	$A \cdot B$	$(A \oplus B) \cdot C_{in}$	$A \cdot B + (A \oplus B) \cdot C_{in}$
0 0 0	0	0	0	0	0	0	0
0 0 1	0	0	0	0	0	0	0
0 1 0	0	0	0	0	0	0	0
0 1 1	1	1	0	1	1	0	1
1 0 0	0	0	0	0	0	0	0
1 0 1	1	0	1	1	0	1	1
1 1 0	1	0	1	1	0	1	1
1 1 1	1	1	1	1	1	0	1

$$C_{out} = A \cdot B + (A + B) \cdot C_{in} = A \cdot B + (A \oplus B) \cdot C_{in}$$

Equivalent !

Fig. 4.1
A useful logic equivalency that can be exploited in arithmetic circuits

The ability to implement the carry out logic using the expression $C_{out} = A{\cdot}B + (A{\oplus}B){\cdot}C_{in}$ allows us to implement a full adder with two half adders and the addition of a single OR gate. Example 4.5 shows this approach. In this new configuration, the sum is produced in two levels of logic, while the carry out is produced in three levels of logic.

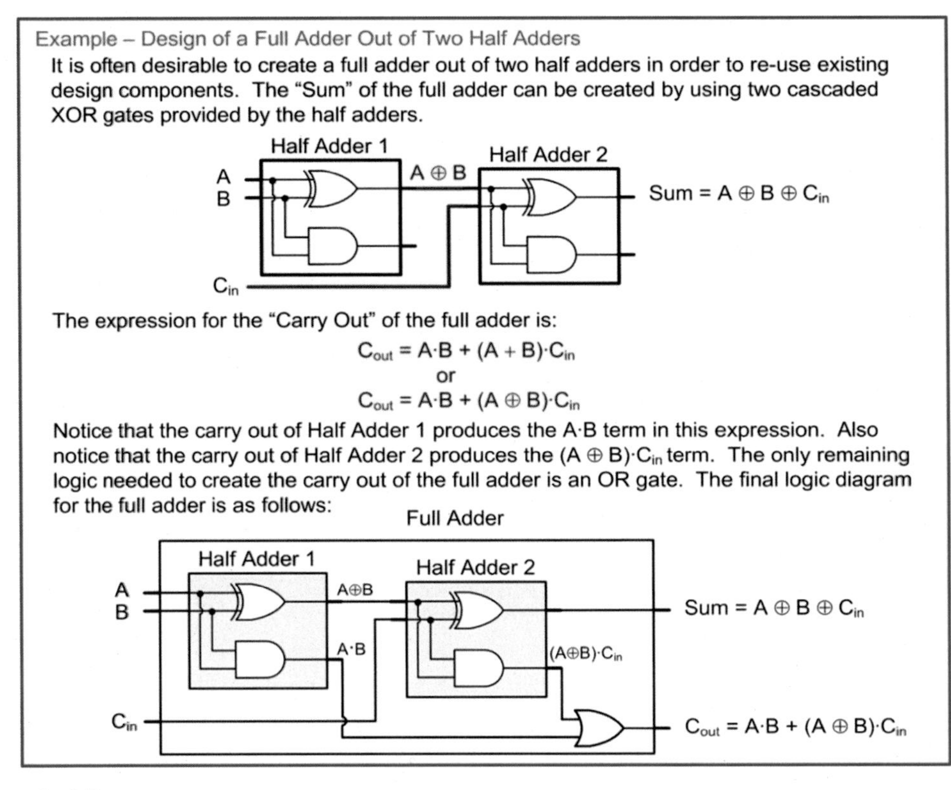

Example 4.5
Design of a full adder out of half adders

4.2.3 Ripple Carry Adder (RCA)

The full adder can now be used in the creation of multi-bit adders. The simplest architecture exploiting the full adder is called a *ripple carry adder* (RCA). In this approach, full adders are used to create the sum and carry out of each bit position. The carry out of each full adder is used as the carry in for the next higher position. Since each subsequent full adder needs to wait for the carry to be produced by the preceding stage, the carry is said to *ripple* through the circuit, thus giving this approach its name. Example 4.6 shows how to design a 4-bit ripple carry adder using a chain of full adders. Notice that the carry in for the full adder in position 0 is tied to a logic 0. The 0 input has no impact on the result of the sum but enables a full adder to be used in the 0th position.

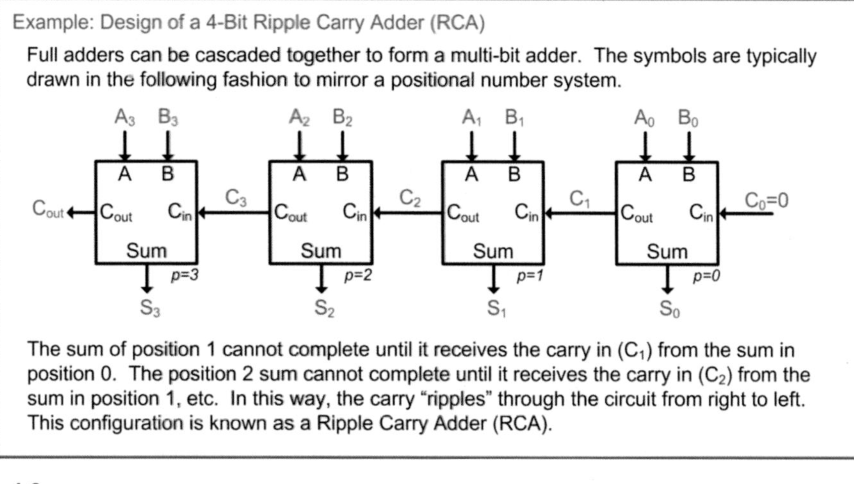

Example 4.6
Design of a 4-bit ripple carry adder (RCA)

4.2.4 Structural Model of a Ripple Carry Adder in VHDL

Now that the hierarchical design of the RCA is complete, we can now model it in VHDL as a system of lower-level components. Example 4.7 shows the structural model for a full adder in VHDL consisting of two half adders. The full adder is created by instantiating two versions of the half adder as components. In this example, all gates are modeled with a delay of 1ns.

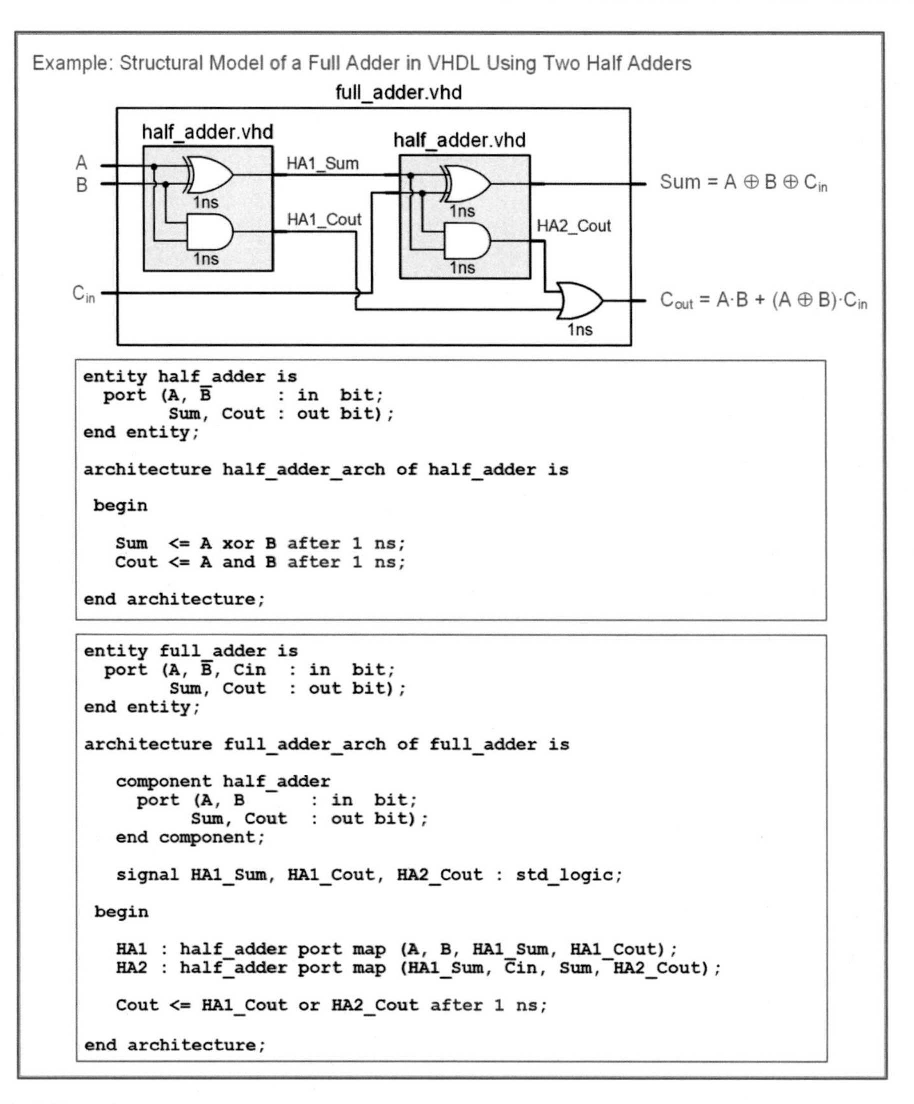

Example 4.7
Structural model of a full adder in VHDL using two half adders

Example 4.8 shows the structural model of a 4-bit ripple carry adder in VHDL. The RCA is created by instantiating four full adders. Notice that a logic 0 can be directly inserted into the port map of the first full adder to model the behavior of $C_0 = 0$.

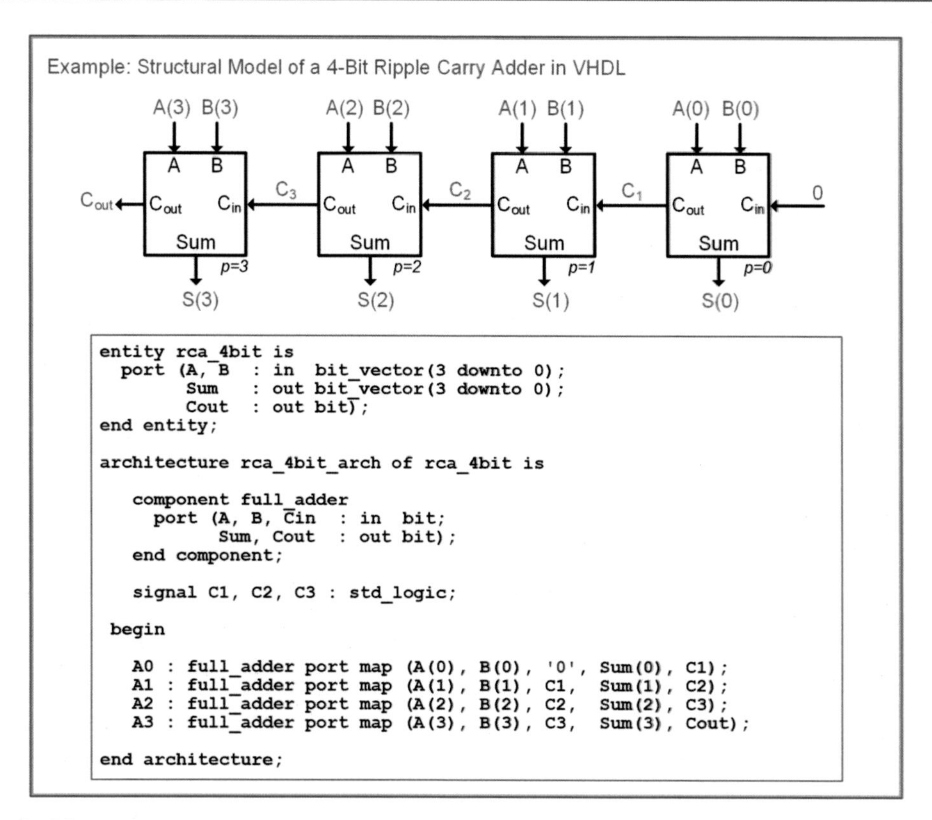


```
entity rca_4bit is
  port (A, B  : in  bit_vector(3 downto 0);
        Sum   : out bit_vector(3 downto 0);
        Cout  : out bit);
end entity;

architecture rca_4bit_arch of rca_4bit is

   component full_adder
     port (A, B, Cin  : in  bit;
           Sum, Cout  : out bit);
   end component;

   signal C1, C2, C3 : std_logic;

 begin

   A0 : full_adder port map (A(0), B(0), '0', Sum(0), C1);
   A1 : full_adder port map (A(1), B(1), C1,  Sum(1), C2);
   A2 : full_adder port map (A(2), B(2), C2,  Sum(2), C3);
   A3 : full_adder port map (A(3), B(3), C3,  Sum(3), Cout);

end architecture;
```

Example 4.8
Structural model of a 4-bit ripple carry adder in VHDL

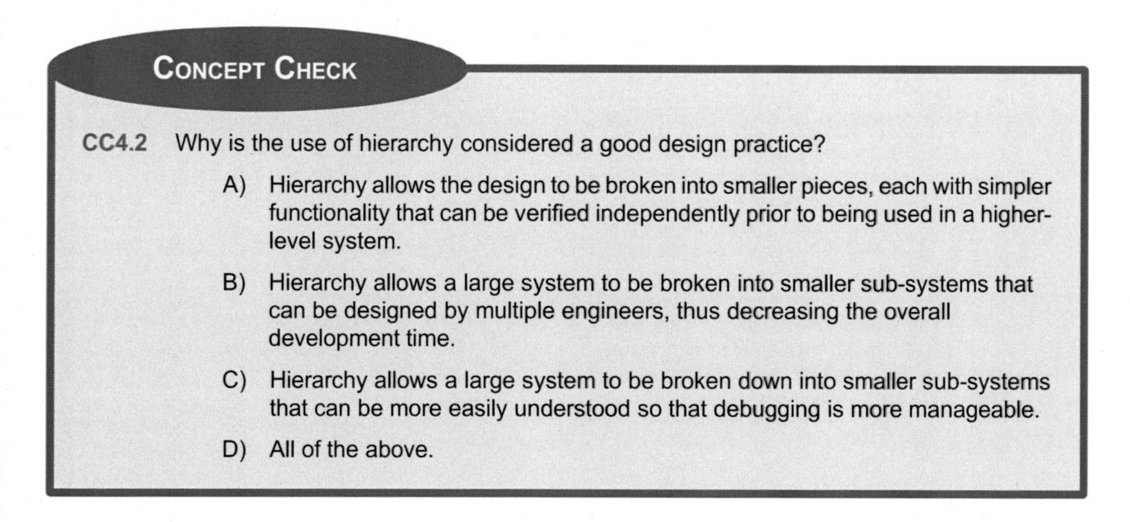

CONCEPT CHECK

CC4.2 Why is the use of hierarchy considered a good design practice?

A) Hierarchy allows the design to be broken into smaller pieces, each with simpler functionality that can be verified independently prior to being used in a higher-level system.

B) Hierarchy allows a large system to be broken into smaller sub-systems that can be designed by multiple engineers, thus decreasing the overall development time.

C) Hierarchy allows a large system to be broken down into smaller sub-systems that can be more easily understood so that debugging is more manageable.

D) All of the above.

Summary

- A *component* is how a VHDL system uses another VHDL file as a sub-system.
- VHDL components are treated as concurrent sub-systems.
- To use a component, it must first be *declared*, which defines the name and entity of the sub-system to be used. This occurs before the *begin* statement in the architecture.

- A component can be *instantiated* one or more times, which includes one or more copies of the sub-system in the higher-levels system. This occurs after the *begin* statement in the architecture.
- The ports of the component can be connected using either *explicit* or *positional port mapping*.
- Explicit port mapping involves listing both the names of the lower-level component's ports along with the higher-level signals that form the connection. The connections in explicit port mapping can be listed in any order. Explicit port mapping is less prone to mistaken connections.
- Positional port mapping involves listing only the names of the higher-level signals during instantiation. The order in which the signals are listed will be connected to the ports of the lower-level sub-system in the order that the ports were declared. Positional port mapping provides a more compact approach to port mapping. Positional port mapping is more prone to mistaken connections due to potentially listing the signals in the wrong order during mapping.

Exercise Problems

Section 4.1: Components

4.1.1 How many times does a component need to be declared within an architecture?

4.1.2 How many times can a component be instantiated?

4.1.3 Does declaring a component occur before or after the *begin* statement in the architecture?

4.1.4 Does instantiating a component occur before or after the *begin* statement in the architecture?

4.1.5 Which port mapping technique is more compact, explicit or positional?

4.1.6 Which port mapping technique is less prone to connection errors because the names of the lower-level ports are listed within the mapping?

Section 4.2: Structural Design Examples

4.2.1 Design a VHDL model to implement the behavior described by the 3-input minterm list shown in Fig. 3.1. Use a structural design approach and basic gates. You will need to create whatever basic gates are needed for your design (e.g., INV1, AND2, OR4, etc.) and then instantiate them in your upper-level architecture to create the desired functionality. The lower-level gates can be implemented with concurrent signal assignments and logical operators (e.g., F <= not A). Declare your entity to match the block diagram provided. Use the type bit for your ports.

4.2.2 Design a VHDL model to implement the behavior described by the 3-input maxterm list shown in Fig. 3.2. Use a structural design approach and basic gates. You will need to create whatever basic gates are needed for your design (e.g., INV1, AND2, OR4, etc.) and then instantiate them in your upper-level architecture to create the desired functionality. The lower-level gates can be implemented with concurrent signal assignments and logical operators (e.g., F <= not A). Declare your entity to match the block diagram provided. Use the type bit for your ports.

4.2.3 Design a VHDL model to implement the behavior described by the 3-input truth table shown in Fig. 3.3. Use a structural design approach and basic gates. You will need to create whatever basic gates are needed for your design (e.g., INV1, AND2, OR4, etc.) and then instantiate them in your upper-level architecture to create the desired functionality. The lower-level gates can be implemented with concurrent signal assignments and logical operators (e.g., F <= not A). Declare your entity to match the block diagram provided. Use the type bit for your ports.

4.2.4 Design a VHDL model to implement the behavior described by the 4-input minterm list shown in Fig. 3.4. Use a structural design approach and basic gates. You will need to create whatever basic gates are needed for your design (e.g., INV1, AND2, OR4, etc.) and then instantiate them in your upper-level architecture to create the desired functionality. The lower-level gates can be implemented with concurrent signal assignments and logical operators (e.g., F <= not A). Declare your entity to match the block diagram provided. Use the type bit for your ports.

4.2.5 Design a VHDL model to implement the behavior described by the 4-input maxterm list shown in Fig. 3.5. Use a structural design approach and basic gates. You will need to create whatever basic gates are needed for your design (e.g., INV1, AND2, OR4, etc.) and then instantiate them in your upper-level architecture to create the desired functionality. The lower-level gates can be implemented with concurrent signal assignments and logical operators (e.g., F <= not A). Declare your entity to match the block diagram provided. Use the type bit for your ports.

4.2.6 Design a VHDL model to implement the behavior described by the 4-input truth table shown in Fig. 3.6. Use a structural design approach and basic gates. You will need to create whatever basic gates are needed for your design (e.g., INV1, AND2, OR4, etc.) and then instantiate them in your upper-level architecture to create the desired functionality. The lower-level gates can be implemented with concurrent signal assignments and logical operators (e.g., F <= not A). Declare your entity to match the block diagram provided. Use the type bit for your ports.

Chapter 5: Modeling Sequential Functionality

In Chap. 3 techniques were presented to describe the behavior of concurrent systems. The modeling techniques presented were appropriate for combinational logic because these types of circuits have outputs dependent only on the current values of their inputs. This means a model that continuously performs signal assignments provides an accurate model of this circuit behavior. When we start looking at sequential circuits (i.e., D-flip-flops, registers, finite-state machines, and counters), these devices only update their outputs based upon an event, most often the edge of a clock signal. The modeling techniques presented in Chap. 3 are unable to accurately describe this type of behavior. In this chapter we describe the VHDL constructs to model signal assignments that are triggered by an event to accurately model sequential logic. We can then use these techniques to describe more complex sequential logic circuits such as finite-state machines and register transfer-level systems.

Learning Outcomes—After completing this chapter, you will be able to:

5.1 Describe the behavior of a VHDL process and how it is used to model sequential logic circuits.
5.2 Model combinational logic circuits using a process and conditional programming constructs.
5.3 Describe how and why signal attributes are used in VHDL models.

5.1 The Process

VHDL uses a *process* to model signal assignments that are based on an event. A process is a technique to model behavior of a system; thus, a process is placed in the VHDL architecture after the begin statement. The signal assignments within a process have unique characteristics that allow them to accurately model sequential logic. First, the signal assignments do not take place until the process ends or is suspended. Second, the signal assignments will be made only once each time the process is triggered. Finally, the signal assignments will be executed in the order that they appear within the process. This assignment behavior is called a *sequential signal assignment*. Sequential signal assignments allow a process to model register transfer-level behavior where a signal can be used as both the operand of an assignment and the destination of a different assignment within the same process. VHDL provides two techniques to trigger a process, the *sensitivity list* and the *wait statement*.

5.1.1 Sensitivity Lists

A *sensitivity list* is a mechanism to control when a process is triggered (or started). A sensitivity list contains a list of signals that the process is sensitive to. If there is a transition on any of the signals in the list, the process will be triggered, and the signal assignments in the process will be made. The following is the syntax for a process that uses a sensitivity list:

B. J. LaMeres, *Quick Start Guide to VHDL*, https://doi.org/10.1007/978-3-031-42543-1_5

```
process_name : process (<signal_name1>, <signal_name2>, ...)

    -- variable declarations

    begin

        sequential_signal_assignment_1
        sequential_signal_assignment_2
                      :
  end process;
```

Let's look at a simple model for a flip-flop.

Example:

```
FlipFlop : process (Clock)
    begin
        Q <= D;
end process;
```

In this example, a transition on the signal Clock (LOW to HIGH or HIGH to LOW) will trigger the process. The signal assignment of D to Q will be executed once the process ends. When the signal Clock is not transitioning, the process will not trigger, and no assignments will be made to Q, thus modeling the behavior of Q holding its last value. This behavior is close to modeling the behavior of a real D-flip-flop, but more constructs are needed to model behavior that is sensitive to only a particular type of transition (i.e., rising or falling edge). These constructs will be covered later.

5.1.2 Wait Statements

A *wait statement* is a mechanism to suspend (or stop) a process and allow signal assignments to be executed without the need for the process to end. When using a wait statement, a sensitivity list is not used. Without a sensitivity list, the process will immediately trigger. Within the process, the wait statement is used to stop and start the process. There are three ways in which wait statements can be used. The first is an indefinite wait. In the following example, the process does not contain a sensitivity list, so it will trigger immediately. The keyword **wait** is used to suspend the process. Once this statement is reached, the signal assignments to Y1 and Y2 will be executed, and the process will suspend indefinitely.

Example:

```
Proc_Ex1 : process
    begin
        Y1 <= '0';
        Y2 <= '1';
        wait;
end process;
```

The second technique to use a wait statement to suspend a process is to use it in conjunction with the keyword **for** and a time expression. In the following example, the process will trigger immediately since it does not contain a sensitivity list. Once the process reaches the wait statement, it will suspend and execute the first signal assignment to CLK (CLK <= '0'). After 5 ns, the process will start again. Once it reaches the second wait statement, it will suspend and execute the second signal assignment to CLK (CLK <= '1'). After another 5 ns, the process will start again and immediately end due to the *end process* statement. After the process ends, it will immediately trigger again due to the lack of a sensitivity list and repeat the behavior just described. This behavior will continue indefinitely. This example creates a square wave called CLK with a period of 10ns.

Example:

```
Proc_Ex2 : process
    begin
        CLK <= '0'; wait for 5 ns;
        CLK <= '1'; wait for 5 ns;
end process;
```

The third technique to use a wait statement to suspend a process is to use it in conjunction with the keyword **until** and a Boolean condition. In the following example, the process will again trigger immediately because there is not a sensitivity list present. The process will then immediately suspend and only resume once a Boolean condition becomes true (i.e., Counter > 15). Once this condition is True, the process will start again. Once it reaches the second wait statement, it will execute the first signal assignment to RollOver (RollOver <= '1'). After 1ns, the process will resume. Once the process ends, it will execute the second signal assignment to RollOver (RollOver <= '0').

Example:

```
Proc_Ex3 : process
    begin
        wait until (Counter > 15);           -- first wait statement
        RollOver <= '1'; wait for 1 ns;      -- second wait statement
        RollOver <= '0';
end process;
```

Wait statements are typically not synthesizable and are most often used for creating stimulus patterns in test benches.

5.1.3 Sequential Signal Assignments

One of the more confusing concepts of a process is how sequential signal assignments behave. The rules of signal assignments within a process are as follows:

- Signals cannot be declared within a process.
- Signal assignments do not take place until the process ends or suspends.
- Signal assignments are executed in the sequence they appear in the process (once the process ends or process suspends).

Let's look at an example of how signals behave in a process. Example 5.1 shows the behavior of sequential signal assignments when executed within a process. Intuitively, we would assume that F will be the complement of A; however, due to the way that sequential signal assignments are performed within a process, this is not the case. In order to understand this behavior, let's look at the situation where A transitions from a 0 to a 1 with B = 0 and F = 0 initially. This transition triggers the process since A is listed in the sensitivity list. When the process triggers, A = 1 since this is where the input resides after the triggering transition. The first signal assignment (B <= A) will cause B = 1, but this assignment occurs only after the process ends. This means that when the second signal assignment is evaluated (F <= not B), it uses the initial value of B from when the process triggered (B = 0) since B is not updated to a 1 until the process ends. The second assignment yields F = 1. When the process ends, A = 1, B = 1, and F = 1. The behavior of this process will always result in A = B = F. This is counterintuitive because the statement F <= not B leads us to believe that F will always be the complement of A and B; however, this is not the case due to the way that signal assignments are only updated in a process upon suspension or when the process ends.

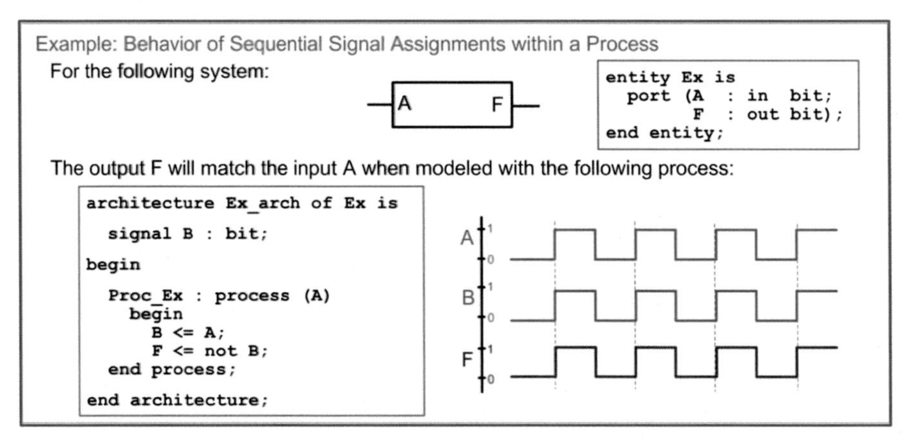

Example 5.1
Behavior of sequential signal assignments within a process

Now let's consider how these assignments behave when executed as concurrent signal assignments. Example 5.2 shows the behavior of the same signal assignments as in Example 5.1, but this time outside of a process. In this model, the statements are executed concurrently and produce the expected behavior of F being the complement of A.

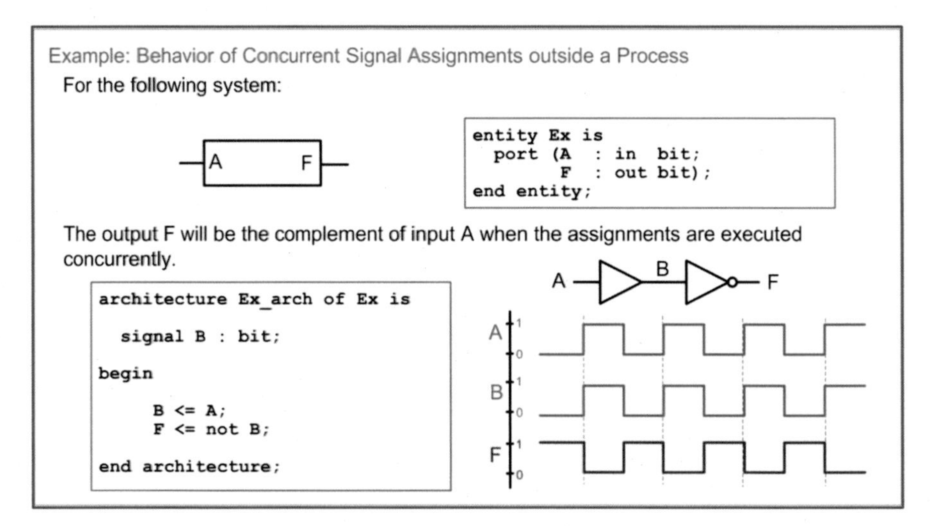

Example 5.2
Behavior of concurrent signal assignments outside a process

While the behavior of the sequential signal assignments initially seems counterintuitive, it is necessary to model the behavior of sequential storage devices and will become clear once more VHDL constructs have been introduced.

5.1.4 Variables

There are situations inside of processes in which it is desired for assignments to be made instantaneously instead of when the process suspends. For these situations, VHDL provides the concept of a *variable*. A variable has the following characteristics:

- Variables only exist within a process.
- Variables are defined in a process before the begin statement.
- Once the process ends, variables are removed from the system. This means that assignments to variables cannot be made by systems outside of the process.
- Assignments to variables are made using the ":=" operator.
- Assignments to variables are made instantaneously.

A variable is declared before the begin statement in a process. The syntax for declaring a variable is as follows:

```
variable variable_name : <type> := <initial_value>;
```

Let's reconsider the example in Example 5.1, but this time we'll use a variable in order to accomplish instantaneous signal assignments within the process. Example 5.3 shows this approach to model the behavior where F is the complement of A.

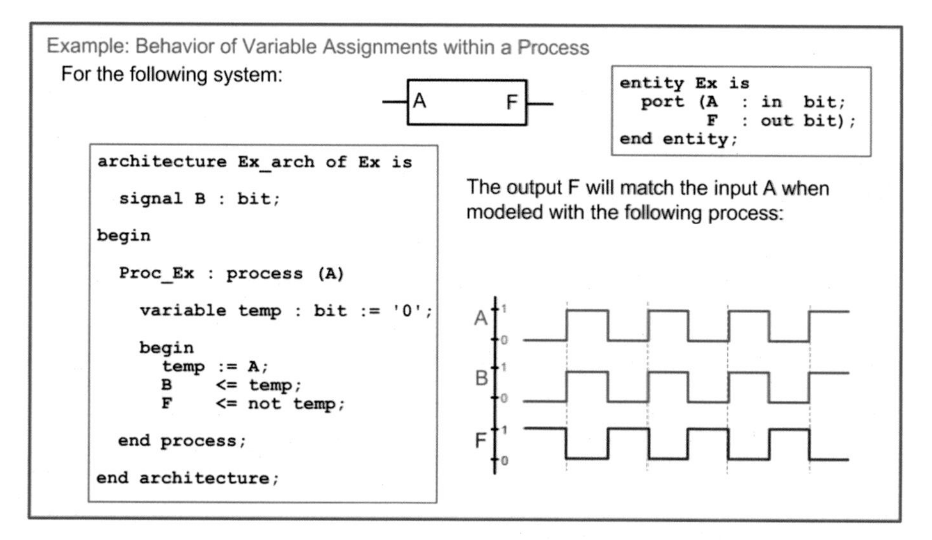

Example 5.3
Variable assignment behavior

Concept Check

CC5.1 If a model of a combinational logic circuit excludes one of its inputs from the sensitivity list, what is the implied behavior?

(A) A storage element because the output will be held at its last value when the unlisted input transitions.

(B) An infinite loop.

(C) A don't care will be used to form the minimal logic expression.

(D) Not applicable because this syntax will not compile.

5.2 Conditional Programming Constructs

One of the more powerful features that processes provide in VHDL is the ability to use conditional programming constructs such as if/then clauses, case statements, and loops. These constructs are only available within a process, but their use is not limited to modeling sequential logic. As we'll see, the characteristics of a process also support modeling of combinational logic circuits, so these conditional constructs are a very useful tool in VHDL. This provides the ability to model both combinational and sequential logic using the more familiar programming language constructs.

5.2.1 If/Then Statements

An *if/then* statement provides a way to make conditional signal assignments based on Boolean conditions. The **if** portion of statement is followed by a Boolean condition that if evaluated TRUE will cause the signal assignment after the **then** statement to be performed. If the Boolean condition is evaluated FALSE, no assignment is made. VHDL provides multiple variants of the if/then statement. An *if/then/else* statement provides a final signal assignment that will be made if the Boolean condition is evaluated false. An *if/then/elsif* statement allows multiple Boolean conditions to be used. The syntax for the various forms of the VHDL if/then statement are as follows:

```
if boolean_condition then sequential_statement
end if;

if boolean_condition then sequential_statement_1
else sequential_statement_2
end if;

if boolean_condition_1 then sequential_statement_1
elsif boolean_condition_2 then sequential_statement_2
  :
  :
elsif boolean_condition_n then sequential_statement_n
end if;

if boolean_condition_1 then sequential_statement_1
elsif boolean_condition_2 then sequential_statement_2
  :
  :
elsif boolean_condition_n then sequential_statement_n
else sequential_statement_n+1
end if;
```

Let's take a look at using an if/then statement to describe the behavior of a combinational logic circuit. Recall that a combinational logic circuit is one in which the output depends on the instantaneous values of the inputs. This behavior can be modeled by placing all of the inputs to the circuit in the sensitivity list of a process. A change on any of the inputs in the sensitivity list will trigger the process and cause the output to be updated. Example 5.4 shows how to model a 3-input combinational logic circuit using if/then statements within a process.

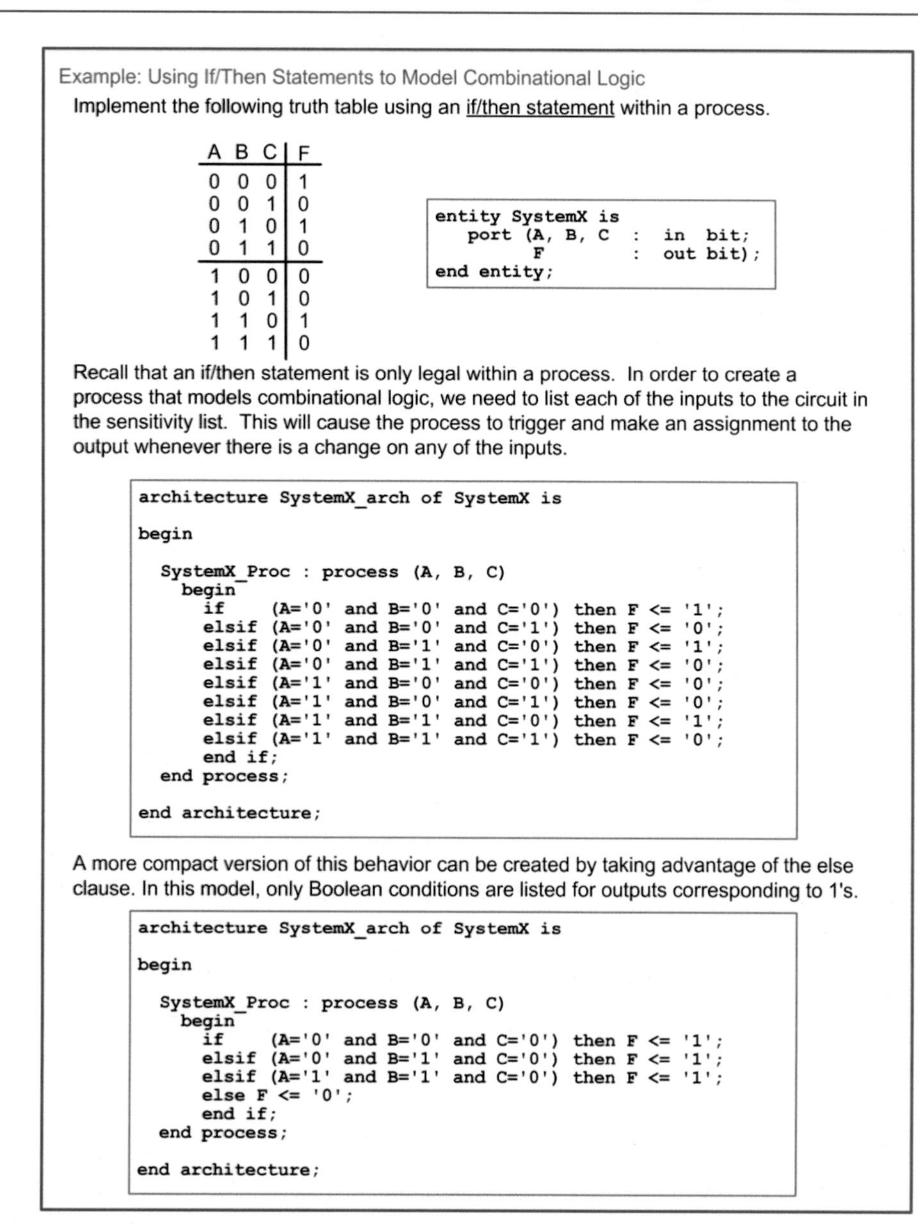

Example: Using If/Then Statements to Model Combinational Logic

Implement the following truth table using an if/then statement within a process.

A	B	C	F
0	0	0	1
0	0	1	0
0	1	0	1
0	1	1	0
1	0	0	0
1	0	1	0
1	1	0	1
1	1	1	0

```
entity SystemX is
   port (A, B, C   :  in  bit;
            F      :  out bit);
end entity;
```

Recall that an if/then statement is only legal within a process. In order to create a process that models combinational logic, we need to list each of the inputs to the circuit in the sensitivity list. This will cause the process to trigger and make an assignment to the output whenever there is a change on any of the inputs.

```
architecture SystemX_arch of SystemX is

begin

  SystemX_Proc : process (A, B, C)
    begin
      if    (A='0' and B='0' and C='0') then F <= '1';
      elsif (A='0' and B='0' and C='1') then F <= '0';
      elsif (A='0' and B='1' and C='0') then F <= '1';
      elsif (A='0' and B='1' and C='1') then F <= '0';
      elsif (A='1' and B='0' and C='0') then F <= '0';
      elsif (A='1' and B='0' and C='1') then F <= '0';
      elsif (A='1' and B='1' and C='0') then F <= '1';
      elsif (A='1' and B='1' and C='1') then F <= '0';
      end if;
  end process;

end architecture;
```

A more compact version of this behavior can be created by taking advantage of the else clause. In this model, only Boolean conditions are listed for outputs corresponding to 1's.

```
architecture SystemX_arch of SystemX is

begin

  SystemX_Proc : process (A, B, C)
    begin
      if    (A='0' and B='0' and C='0') then F <= '1';
      elsif (A='0' and B='1' and C='0') then F <= '1';
      elsif (A='1' and B='1' and C='0') then F <= '1';
      else F <= '0';
      end if;
  end process;

end architecture;
```

Example 5.4
Using if/then statements to model combinational logic

5.2.2 Case Statements

A *case* statement is another technique to model signal assignments based on Boolean conditions. As with the if/then statement, a case statement can only be used inside of a process. The statement begins with the keyword **case** followed by the input signal name that assignments will be based off of. The input signal name can be optionally enclosed in parentheses for readability. The keyword **when** is used to specify a particular value (or choice) of the input signal that will result in associated sequential signal assignments. The assignments are listed after the **=>** symbol. The following is the syntax for a case statement:

```
case (input_name) is
   when choice_1 => sequential_statement(s);
   when choice_2 => sequential_statement(s);
                         :
                         :
   when choice_n => sequential_statement(s);
end case;
```

When not all the possible input conditions (or choices) are specified, a **when others** clause is used to provide signal assignments for all other input conditions. The following is the syntax for a case statement that uses a *when others* clause:

```
case (input_name) is
   when choice_1 => sequential_statement(s);
   when choice_2 => sequential_statement(s);
                         :
                         :
   when others  => sequential_statement(s);
end case;
```

Multiple choices that correspond to the same signal assignments can be pipe-delimited in the case statement. The following is the syntax for a case statement with pipe-delimited choices:

```
case (input_name) is
   when choice_1 | choice_2 => sequential_statement(s);
   when others              => sequential_statement(s);
end case;
```

The input signal for a case statement must be a single signal name. If multiple scalars are to be used as the input expression for a case statement, they should be concatenated either outside of the process resulting in a new signal vector or within the process resulting in a new variable vector. Example 5.5 shows how to model a 3-input combinational logic circuit using case statements within a process.

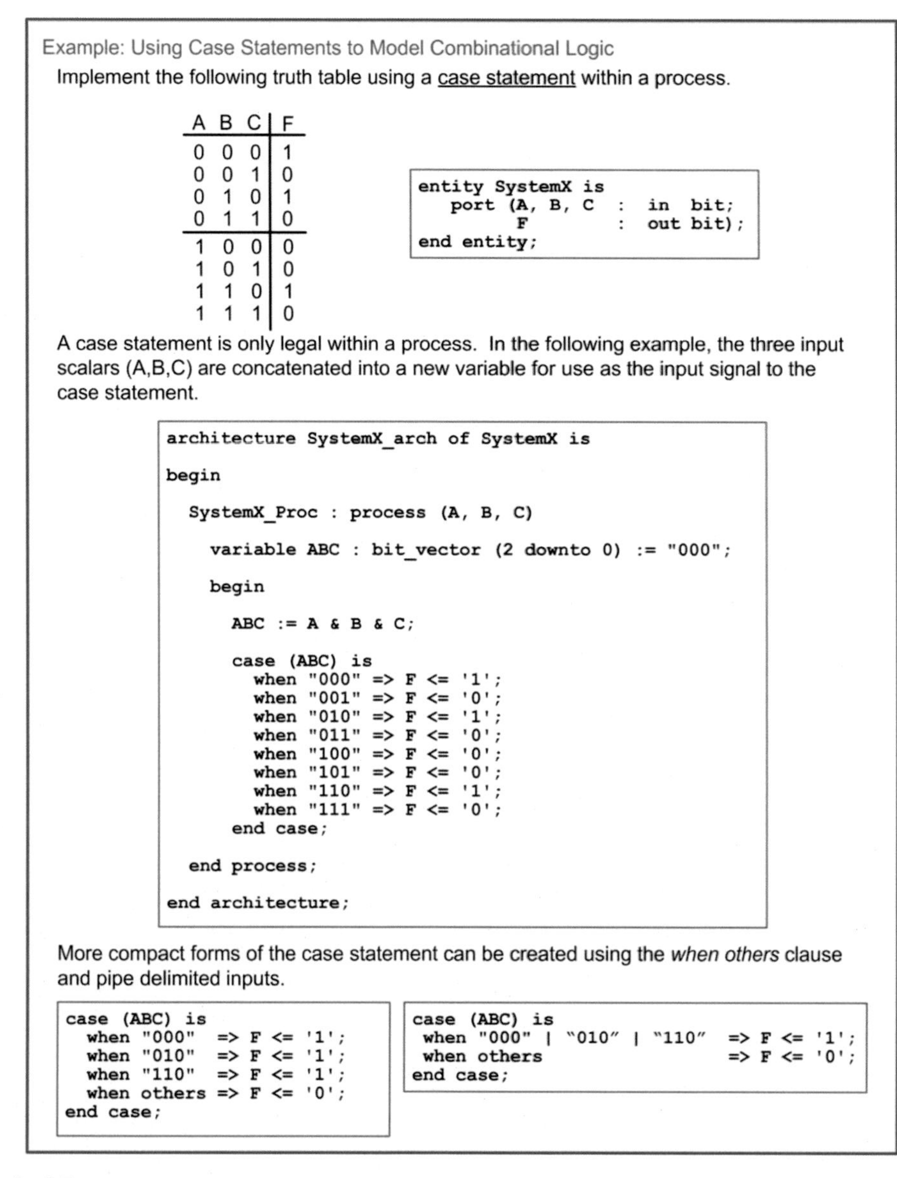

Example: Using Case Statements to Model Combinational Logic

Implement the following truth table using a case statement within a process.

A	B	C	F
0	0	0	1
0	0	1	0
0	1	0	1
0	1	1	0
1	0	0	0
1	0	1	0
1	1	0	1
1	1	1	0

```
entity SystemX is
   port (A, B, C  :  in  bit;
         F        :  out bit);
end entity;
```

A case statement is only legal within a process. In the following example, the three input scalars (A,B,C) are concatenated into a new variable for use as the input signal to the case statement.

```
architecture SystemX_arch of SystemX is

begin

  SystemX_Proc : process (A, B, C)

    variable ABC : bit_vector (2 downto 0) := "000";

    begin

      ABC := A & B & C;

      case (ABC) is
        when "000" => F <= '1';
        when "001" => F <= '0';
        when "010" => F <= '1';
        when "011" => F <= '0';
        when "100" => F <= '0';
        when "101" => F <= '0';
        when "110" => F <= '1';
        when "111" => F <= '0';
      end case;

  end process;

end architecture;
```

More compact forms of the case statement can be created using the *when others* clause and pipe delimited inputs.

```
case (ABC) is
  when "000"   => F <= '1';
  when "010"   => F <= '1';
  when "110"   => F <= '1';
  when others => F <= '0';
end case;
```

```
case (ABC) is
 when "000" | "010" | "110"   => F <= '1';
 when others                  => F <= '0';
end case;
```

Example 5.5
Using case statements to model combinational logic

If/then statements can be embedded within a case statement, and, conversely, case statements can be embedded within an if/then statement.

5.2.3 Infinite Loops

A *loop* within VHDL provides a mechanism to perform repetitive assignments infinitely. This is useful in test benches for creating stimulus such as clocks or other periodic waveforms. A loop can only be used within a process. The keyword **loop** is used to signify the beginning of the loop. Sequential signal assignments are then inserted. The end of the loop is signified with the keywords **end loop**. Within the loop, the *wait for*, *wait until*, and *after* statements are all legal. Signal assignments within a loop will be executed repeatedly forever unless an **exit** or **next** statement is encountered. The *exit* clause provides a Boolean condition that will force the loop to end if the condition is evaluated true. When using the exit statement, an additional signal assignment is typically placed after the loop to provide the desired behavior when the loop is not active. Using flow control statements such as *wait for* and *wait after* provides a means to avoid having the loop immediately executed again after exiting. The *next* clause provides a way to skip the remaining signal assignments and begin the next iteration of the loop. The following is the syntax for an infinite loop in VHDL:

```
loop
   exit when boolean_condition;        -- optional exit statement
   next when boolean_condition;        -- optional next statement
   sequential_statement(s);
end loop;
```

Consider the following example of an infinite loop that generates a clock signal (CLK) with a period of 100ns. In this example, the process does not contain a sensitivity list, so a wait statement must be used to control the signal assignments. This process in this example will trigger immediately and then enter the infinite loop and never exit.

Example:

```
Clock_Proc1 : process
  begin
    loop
       CLK <= not CLK;
       wait for 50 ns;
    end loop;
 end process;
```

Now consider the following loop example that will generate a clock signal with a period of 100 ns with an enable (EN) line. This loop will produce a periodic clock signal as long as EN = 1. When EN = 0, the clock output will remain at CLK = 0. An exit condition is placed at the beginning of the loop to check if EN = 0. If this condition is true, the loop will exit, and the clock signal will be assigned a 0. The process will then wait until EN = 1. Once EN = 1, the process will end and then immediately trigger again and reenter the loop. When EN = 1, the clock signal will be toggled (CLK <= not CLK) and then wait for 50 ns. This toggling behavior will repeat as long as EN = 1.

Example:

```
Clock_Proc2 : process
  begin
    loop
       exit when EN='0';
       CLK <= not CLK;
       wait for 50 ns;
    end loop;

    CLK <= '0';
    wait until EN='1';

 end process;
```

It is important to keep in mind that infinite loops that continuously make signal assignments without the use of sensitivity lists or wait statements will cause logic simulators to hang.

5.2.4 While Loops

A *while loop* provides a looping structure with a Boolean condition that controls its execution. The loop will only execute as long as its condition is evaluated true. The following is the syntax for a VHDL while loop:

```
while boolean_condition loop
   sequential_statement(s);
end loop;
```

Let's implement the previous example of a loop that generates a clock signal (CLK) with a period of 100 ns as long as EN = 1. The Boolean condition for the while loop is EN = 1. When EN = 1, the loop will be executed indefinitely. When EN = 0, the while loop will be skipped. In this case, an additional signal assignment is necessary to model the desired behavior when the loop is not used (i.e., CLK = 0).

Example:

```
Clock_Proc3 : process
   begin
     while (EN='1') loop
        CLK <= not CLK;
        wait for 50 ns;
   end loop;

   CLK <= '0';
   wait until EN='1';

 end process;
```

5.2.5 For Loops

A *for loop* provides the ability to create a loop that will execute a predefined number of times. The range of the loop is specified with integers (*min, max*) at the beginning of the for loop. A *loop variable* is implicitly declared in the loop that will increment (or decrement) from *min* to *max* of the range. The loop variable is of type integer. If it is desired to have the loop variable increment from min to max, the keyword **to** is used when specifying the range of the loop. If it is desired to have the loop variable decrement max to min, the keyword **downto** is used when specifying the range of the loop. The loop variable can be used within the loop as an index for vectors; thus, the for loop is useful for automatically accessing and assigning multiple signals within a single loop structure. The following is the syntax for a VHDL for loop in which the loop variable will increment from min to max of the range:

```
for loop_variable in min to max loop
   sequential_statement(s);
end loop;
```

The following is the syntax of a for loop in which the loop variable will decrement from max to min of the range:

```
for loop_variable in max downto min loop
   sequential_statement(s);
end loop;
```

For loops are useful for test benches in which a range of patterns are to be created. For loops are also synthesizable as long as the complete behavior of the desired system is described by the loop. The following is an example of creating a simple counter using the loop variable. The signal Count_Out in this example is of type integer. This allows the loop variable *i* to be assigned to Count_Out each time through the loop since the loop variable is also of type integer. This counter will count from 0 to 15 and then repeat. The count will increment every 50ns.

Example:

```
Counter_Proc : process
  begin
    for i in 0 to 15 loop
      Count_Out <= i;
      wait for 50 ns;
    end loop;
 end process;
```

CONCEPT CHECK

CC5.2 When using an if/then statement to model a combinational logic circuit, is using the *else* clause the same as using *don't cares* when minimizing a logic expression with a K-map?

A) Yes. The else clause allows the synthesizer to assign whatever output values are necessary in order to create the most minimal circuit.

B) No. The else clause explicitly states the output values for all input codes not listed in the if/elsif portion of the if/then construct. This is the same as filling in the truth table with specific values for all input codes covered by the else clause and the synthesizer will create the logic expression accordingly.

5.3 Signal Attributes

There are situations where we want to describe behavior that is based on more than just the current value of a signal. For example, a real D-flip-flop will only update its outputs on a particular type of transition (i.e., rising or falling). In order to model this behavior, we need to specify more information about the signal. This is accomplished by using *attributes*. Attributes provide additional information about a signal other than just its present value. An attribute can provide information such as past values, whether an assignment was made to a signal, or when the last time an assignment resulted in a value change. A signal attribute is implemented by placing an apostrophe (') after the signal name and then listing the VHDL attribute keyword. Different attributes will result in different output types. Attributes that yield Boolean output types can be used as inputs to Boolean decision conditions for other VHDL constructs. Other attributes can be used to define the range of new vectors by referencing the size of existing vectors or automatically defining the number of iterations in a loop. Finally, some attributes can be used to create self-checking test benches that monitor the impact of circuit delays on the functionality of a system. The following are a list of the commonly used, pre-defined VHDL signal attributes. The example signal name *A* is used to illustrate how scalar attributes operate. The example signal **B** is used to illustrate how vector attributes operate with type bit_vector (7 downto 0).

Attribute	Information returned	Type returned
A'**event**	true when signal A changes, false otherwise	boolean
A'**active**	true when an assignment is made to A, false otherwise	boolean
A'**last_event**	time when signal A last changed	time
A'**last_active**	time when signal A was last assigned to	time
A'**last_value**	the previous value of A	same type as A

Attribute	Information returned	Type returned
B'**length**	size of the vector (e.g., 8)	integer
B'**left**	left bound of the vector (e.g., 7)	integer
B'**right**	right bound of the vector (e.g., 0)	integer
B'**range**	range of the vector "(7 downto 0)"	string

Signal attributes can be used to model edge-sensitive behavior. Let's look at the model for a simple D-flip-flop. A process is used to model the synchronous behavior of the D-flip-flop. The sensitivity list contains only the *Clock* input. The *D* input is not included in the sensitivity list because a change on D should not trigger the process. Attributes and logical operators are not allowed in the sensitivity list of a process. As a result, the process will trigger on every edge of the clock signal. Within the process, an if/then statement is used with the Boolean condition **(Clock'event and Clock = '1')** in order to make signal assignments only on a rising edge of the clock. The syntax for this Boolean condition is understood and is synthesizable by all CAD tools. An else clause is not included in the if/then statement. This implies that when there is not a rising edge, no assignments will be made to the outputs and they will simply hold their last value. Example 5.6 shows how to model a simple D-flip-flop using attributes. Note that this example does not model the reset behavior of a real D-flip-flop.

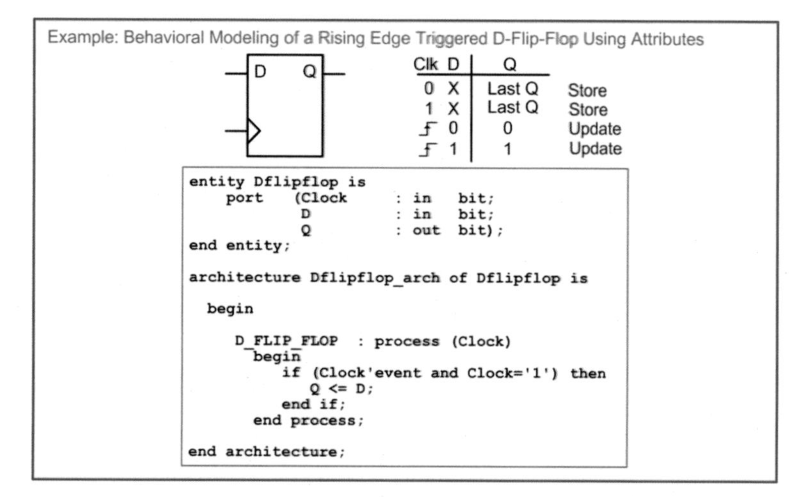

Example 5.6
Behavioral modeling of a rising edge-triggered D-flip-flop using attributes

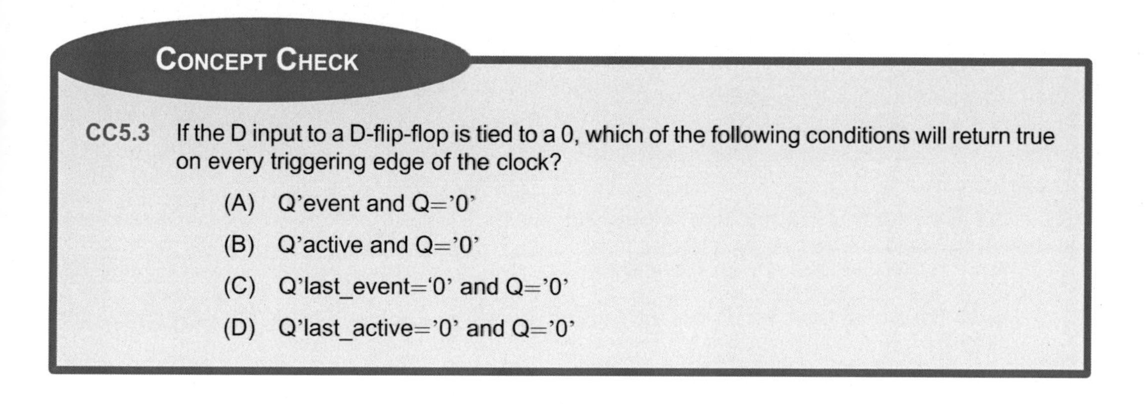
Concept Check

CC5.3 If the D input to a D-flip-flop is tied to a 0, which of the following conditions will return true on every triggering edge of the clock?

(A) Q'event and Q='0'

(B) Q'active and Q='0'

(C) Q'last_event='0' and Q='0'

(D) Q'last_active='0' and Q='0'

Summary

- To model sequential logic, an HDL needs to be able to trigger signal assignments based on a triggering event. This is accomplished in VHDL using a *process*.
- A *sensitivity* list is a way to control when a VHDL process is triggered. A sensitivity list contains a list of signals. If any of the signals in the sensitivity list transition, it will cause the process to trigger. If a sensitivity list is omitted, the process will trigger immediately.
- Signal assignments are made when a process suspends. There are two techniques to suspend a process. The first is using the *wait* statement. The second is simply ending the process.
- Sensitivity lists and wait statements are never used at the same time. Sensitivity lists are used to model synthesizable logic, while wait statements are used for test benches.
- When signal assignments are made in a process, they are made in the order they are listed in the process. If assignments are made to the same signal within a process, only the last assignment will take place when the process suspends.
- If assignments are needed to occur prior to the process suspending, a *variable* is used. In VHDL, variables only exist within a process. Variables are defined when a process triggers and deleted when the process ends.
- Processes also allow more advanced modeling constructs in VHDL. These include *if/then statements*, *case statements*, *infinite loops*, *while loops*, and *for loops*.
- *Signal attributes* allow additional information to be observed about a signal other than its value.

Exercise Problems

Section 5.1: The Process

5.1.1 When using a sensitivity list in a process, what will cause the process to *trigger*?

5.1.2 When using a sensitivity list in a process, what will cause the process to *suspend*?

5.1.3 When a sensitivity list is not used in a process, when will the process trigger?

5.1.4 Can a sensitivity list and a wait statement be used in the same process at the same time?

5.1.5 Does a wait statement *trigger* or *suspend* a process?

5.1.6 When are signal assignments officially made in a process?

5.1.7 Why are assignments in a process called *sequential signal assignments?*

5.1.8 Can signals be declared in a process?

5.1.9 Are variables declared within a process visible to the rest of the VHDL model (e.g., are they visible outside of the process)?

5.1.10 What happens to a variable when a process ends?

5.1.11 What is the assignment operator for variables?

Section 5.2: Conditional Programming Constructs

5.2.1 Design a VHDL model to implement the behavior described by the 4-input truth table in Fig. 5.1. Use a process and an if/then statement. Use std_logic and std_logic_vector types for your signals. Declare the entity to match the block diagram provided. Hint: Notice that there are far more input codes producing F = 0 than producing F = 1. Can you use this to your advantage to make your VHDL model simpler?

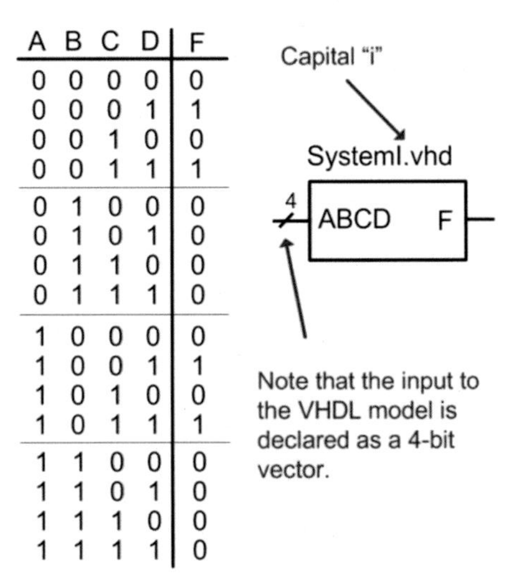

A	B	C	D	F
0	0	0	0	0
0	0	0	1	1
0	0	1	0	0
0	0	1	1	1
0	1	0	0	0
0	1	0	1	0
0	1	1	0	0
0	1	1	1	0
1	0	0	0	0
1	0	0	1	1
1	0	1	0	0
1	0	1	1	1
1	1	0	0	0
1	1	0	1	0
1	1	1	0	0
1	1	1	1	0

Fig. 5.1
System I functionality

5.2.2 Design a VHDL model to implement the behavior described by the 4-input truth table in Fig. 5.1. Use a process and a case statement. Use std_logic and std_logic_vector types for your signals. Declare the entity to match the block diagram provided.

5.2.3 Design a VHDL model to implement the behavior described by the 4-input minterm list in Fig. 5.2. Use a process and an if/then statement. Use std_logic and std_logic_vector types for your signals. Declare the entity to match the block diagram provided.

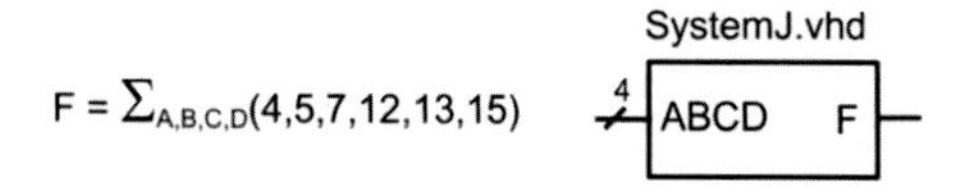

Fig. 5.2
System J functionality

5.2.4 Design a VHDL model to implement the behavior described by the 4-input minterm list in Fig. 5.2. Use a process and a case statement. Use std_logic and std_logic_vector types for your signals. Declare the entity to match the block diagram provided.

5.2.5 Design a VHDL model to implement the behavior described by the 4-input maxterm list in Fig. 5.3. Use a process and an if/then statement. Use std_logic and std_logic_vector types for your signals. Declare the entity to match the block diagram provided.

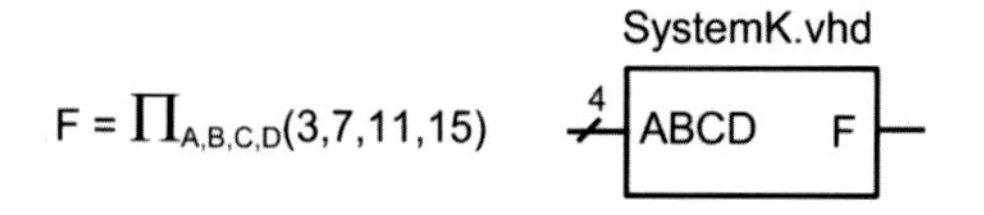

Fig. 5.3
System K functionality

5.2.6 Design a VHDL model to implement the behavior described by the 4-input maxterm list in Fig. 5.3. Use a process and a case statement. Use std_logic and std_logic_vector types for your signals. Declare the entity to match the block diagram provided.

5.2.7 Design a VHDL model to implement the behavior described by the 4-input truth table in Fig. 5.4. Use a process and an if/then statement. Use std_logic and std_logic_vector types for your signals. Declare the entity to match the block diagram provided. Hint: Notice that there are far more input codes producing F = 1 than producing F = 0. Can you use this to your advantage to make your VHDL model simpler?

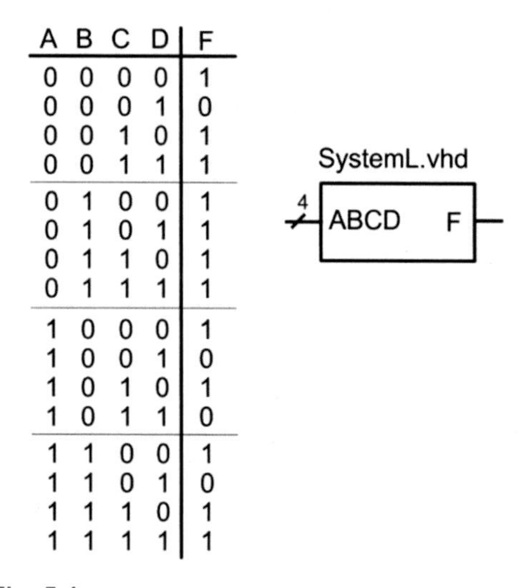

A	B	C	D	F
0	0	0	0	1
0	0	0	1	0
0	0	1	0	1
0	0	1	1	1
0	1	0	0	1
0	1	0	1	1
0	1	1	0	1
0	1	1	1	1
1	0	0	0	1
1	0	0	1	0
1	0	1	0	1
1	0	1	1	0
1	1	0	0	1
1	1	0	1	0
1	1	1	0	1
1	1	1	1	1

Fig. 5.4
System L functionality

5.2.8 Design a VHDL model to implement the behavior described by the 4-input truth table in Fig. 5.4. Use a process and a case statement. Use std_logic and std_logic_vector types for your signals. Declare the entity to match the block diagram provided.

Section 5.3 – Signal Attributes

5.3.1 What is the purpose of a signal attribute?

5.3.2 What is the data type returned when using the signal attribute *'event*?

5.3.3 What is the data type returned when using the signal attribute *'last_event*?

5.3.4 What is the data type returned when using the signal attribute *'length*?

Chapter 6: Packages

One of the drawbacks of the VHDL standard package is that it provides limited functionality in its synthesizable data types. The bit and bit_vector, while synthesizable, lack the ability to accurately model many of the topologies implemented in modern digital systems. Of primary interest are topologies that involve multiple drivers connected to a single wire. The standard package will not permit this type of connection; however, this type of topology is a common way to interface multiple nodes on a shared interconnection. Furthermore, the standard package does not provide many useful features for these types, such as don't cares, arithmetic using the + and − operators, type conversion functions, or the ability to read/write external files. To increase the functionality of VHDL, packages are included in the design. This chapter introduces the most common packages used in modern VHDL models.

Learning Outcomes—After completing this chapter, you will be able to:

6.1 Describe the capabilities of the STD_LOGIC_1164 package that allow more accurate models of modern digital systems to be described.
6.2 Describe the capabilities of the NUMERIC_STD package that allow behavioral models of arithmetic circuits to be described including operations using data types from the STD_LOGIC_1164 package.
6.3 Describe how text reporting using external I/O can is handled by the TEXTIO and STD_LOGIC_TEXTIO packages.
6.4 Describe the capabilities of some of the other common packages provided in the IEEE library.

6.1 STD_LOGIC_1164

In the late 1980s, the IEEE 1164 standard was released that added functionality to VHDL to allow a multi-valued logic system (i.e., a signal can take on more values than just 0 and 1). This standard also provided a mechanism for multiple drivers to be connected to the same signal. An updated release in 1993 called IEEE 1164-1993 was the most significant update to this standard and contains the majority of functionality used in VHDL today. Nearly all systems described in VHDL include the 1164 standard as a package. This package is included by adding the following syntax at the beginning of the VHDL file:

```
library IEEE;
use IEEE.std_logic_1164.all;
```

This package defines four new data types: **std_ulogic**, **std_ulogic_vector**, **std_logic**, and **std_logic_vector**. The std_ulogic and std_logic are enumerated, scalar types that can provide a multi-valued logic system. The types std_ulogic_vector and std_logic_vector are vector types containing a linear array of scalar types std_ulogic and std_logic respectively. The scalar types can take on nine different values as described below.

Value	Description	Notes
U	Uninitialized	Default initial value
X	Forcing unknown	
0	Forcing 0	
1	Forcing 1	
Z	High impedance	
W	Weak unknown	
L	Weak 0	Pull-down
H	Weak 1	Pull-up
-	Don't care	Used for synthesis only

B. J. LaMeres, *Quick Start Guide to VHDL*, https://doi.org/10.1007/978-3-031-42543-1_6

These values can be assigned to signals by enclosing them in single quotes (scalars) or double quotes (vectors).

Example:

```
A <= 'X';        -- assignment to a scalar (std_ulogic or std_logic)
V <= "01ZH";     -- assignment to a 4-bit vector (std_ulogic_vector
                    or std_logic_vector)
```

The type std_ulogic is *unresolved* (note: the "u" standard for "unresolved"). This means that if a signal is being driven by two circuits with type std_ulogic, the VHDL simulator will not be able to *resolve* the conflict, and it will result in a compiler error. The std_logic type is *resolved*. This means that if a signal is being driven by two circuits with type std_logic, the VHDL simulator *will* be able to resolve the conflict and will allow the simulation to continue. Figure 6.1 shows an example of a shared signal topology and how conflicts are handled when using various data types.

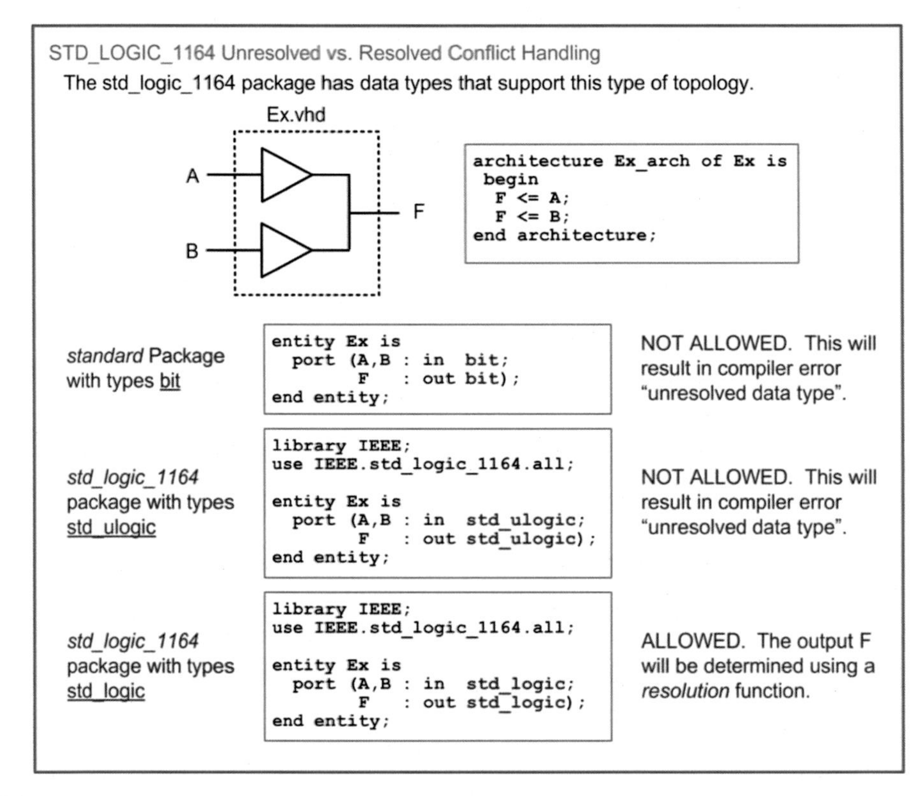

Fig. 6.1
STD_LOGIC_1164 unresolved vs. resolved conflict handling

6.1.1 STD_LOGIC_1164 Resolution Function

The std_logic_1164 will resolve signal conflict of type std_logic using a **resolution function**. The nine allowed values each have a relative drive strength that allows a resolution to be made in the event of conflict. Whenever there is a conflict, the simulator will consult the resolution function to determine the value of the signal. Figure 6.2 shows the relative drive strengths of the nine possible signal values provided by the std_logic_1164 package and the resolution function table.

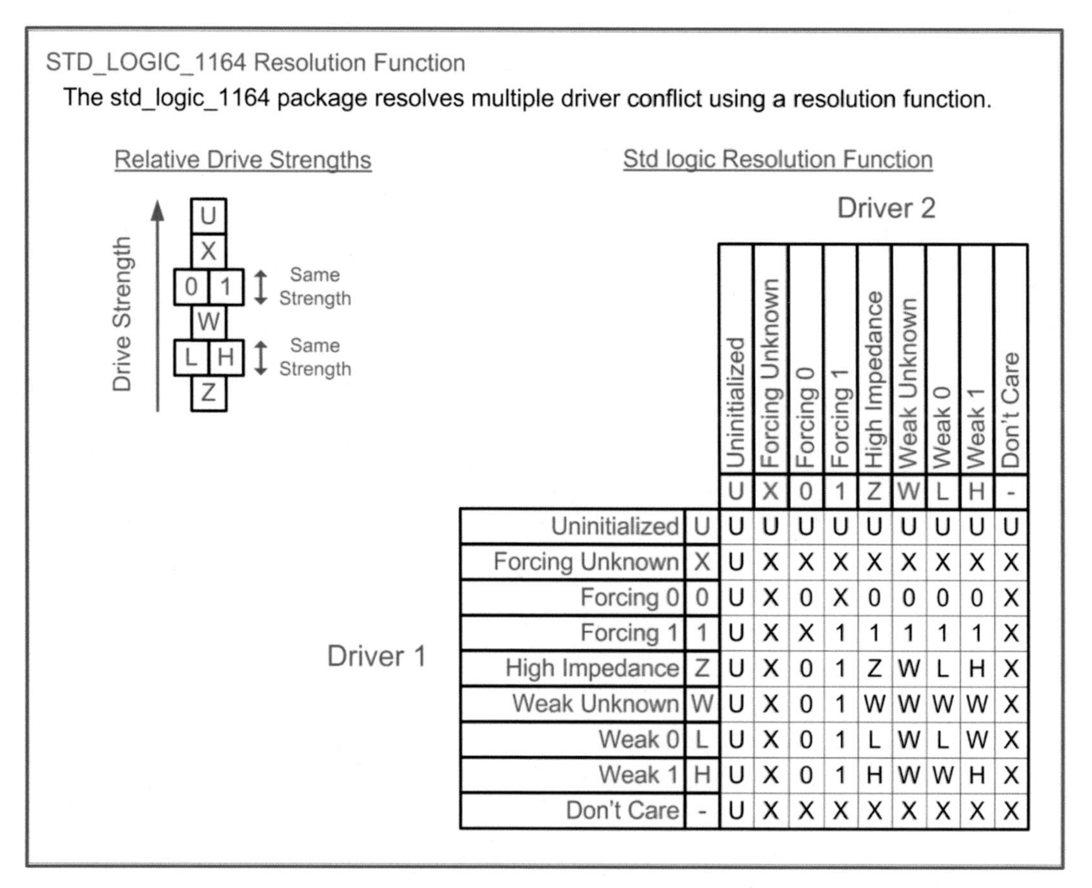

Driver 1 \ Driver 2		Uninitialized U	Forcing Unknown X	Forcing 0 0	Forcing 1 1	High Impedance Z	Weak Unknown W	Weak 0 L	Weak 1 H	Don't Care -
Uninitialized	U	U	U	U	U	U	U	U	U	U
Forcing Unknown	X	U	X	X	X	X	X	X	X	X
Forcing 0	0	U	X	0	X	0	0	0	0	X
Forcing 1	1	U	X	X	1	1	1	1	1	X
High Impedance	Z	U	X	0	1	Z	W	L	H	X
Weak Unknown	W	U	X	0	1	W	W	W	W	X
Weak 0	L	U	X	0	1	L	W	L	W	X
Weak 1	H	U	X	0	1	H	W	W	H	X
Don't Care	-	U	X	X	X	X	X	X	X	X

Fig. 6.2
STD_LOGIC_1164 resolution function

6.1.2 STD_LOGIC_1164 Logical Operators

The std_logic_1164 also contains new definitions for all of the logical operators (**and, nand, or, nor, xor, xnor, not)** for types std_ulogic and std_logic. These are required since these data types can take on more logic values than just a 0 or 1; thus, the logical operator definitions from the standard package are not sufficient.

6.1.3 STD_LOGIC_1164 Edge Detection Functions

The std_logic_1164 also provides functions for the detection of rising or falling transitions on a signal. The functions **rising_edge()** and **falling_edge()** provide a more readable form of this functionality compared to the (Clock'event and Clock='1') approach. Example 6.1 shows the use of the rising_edge() function to model the behavior of a rising edge-triggered D-flip-flop.

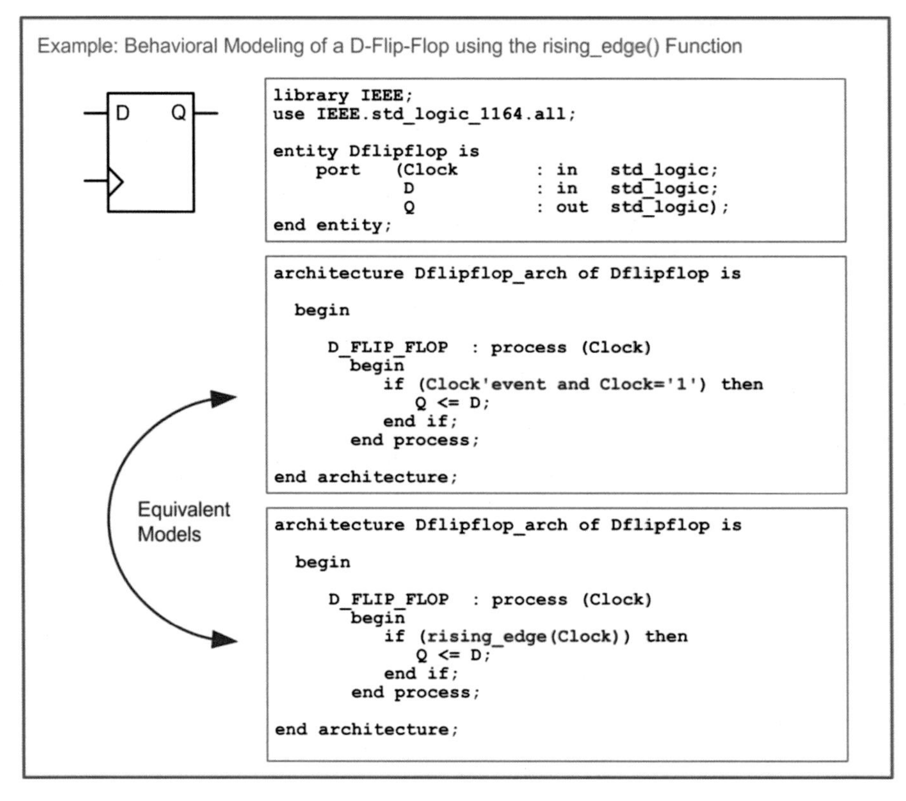

Example 6.1
Behavioral modeling of a D-flip-flop using the rising_edge() function

6.1.4 STD_LOGIC_1164 Type Conversion Functions

The std_logic_1164 package also provides functions to convert between data types. Functions exist to convert between bit, std_ulogic, and std_logic. Functions also exist to convert between these types' vector forms (bit_vector, std_ulogic_vector, and std_logic_vector). The functions are listed below.

Name	Input type	Return type
To_bit()	std_ulogic	bit
To_bitvector()	std_ulogic_vector	bit_vector
To_bitvector()	std_logic_vector	bit_vector
To_StdULogic()	bit	std_ulogic
To_StdULogicVector()	bit_vector	std_ulogic_vector
To_StdULogicVector()	std_logic_vector	std_ulogic_vector
To_StdLogicVector()	bit_vector	std_logic_vector
To_StdLogicVector()	std_ulogic_vector	std_logic_vector

When using these functions, the function name and input signal are placed to the right of the assignment operator, and the target signal is placed on the left.

Example:

```
A <= To_bit(B);                -- B is type std_ulogic, A is type bit
V <= To_StdLogicVector(C);     -- C is type bit_vector, V is std_logic_vector
```

When identical function names exist that can have different input data types, the VHDL compiler will automatically decide which function to use based on the input argument type. For example, the function "To_bitvector" exists for an input of std_ulogic_vector *and* std_logic_vector. When using this function, the compiler will automatically detect which input type is being used and select the corresponding function variant. No additional syntax is required by the designer in this situation.

CONCEPT CHECK

CC6.1 What is the primary contribution of the STD_LOGIC_1164 package?

A) Arithmetic operators for the types bit and bit_vector.

B) Functions that allow all operators in the standard package to be used on all data types from the same package.

C) The ability to read and write from external files.

D) New data types that can take on more values beyond 0's and 1's in order to more accurately model modern digital systems.

6.2 NUMERIC_STD

The *numeric_std* package provides numerical computation for types std_logic and std_logic_vector. When performing binary arithmetic, the results of arithmetic operations and comparisons vary greatly depending on whether the binary number is unsigned or signed. As a result, the numeric_std package defines two new data types, **unsigned** and **signed**. An unsigned type is defined to have its MSB in the leftmost position of the vector and the LSB in the rightmost position of the vector. A signed number uses two's complement representation with the leftmost bit of the vector being the sign bit. When declaring a signal to be one of these types, it is implied that these represent the encoding of an underlying native type of std_logic/std_logic_vector. The use of unsigned/signed types provides the interpretation of how arithmetic, logical, and comparison operators will perform. This also implies that the numeric_std package requires the std_logic_1164 to always be included. While the numeric_std package includes an inclusion call of the std_logic_1164 package, it is common to explicitly include both the std_logic_1164 and the numeric_std packages in the main VHDL file. The VHDL compiler will ignore redundant package statements. The syntax for including these packages is as follows:

```
library IEEE;
use IEEE.std_logic_1164.all;   -- defines types std_ulogic and std_logic
use IEEE.numeric_std.all;      -- defines types unsigned and signed
```

6.2.1 NUMERIC_STD Arithmetic Functions

The numeric_std package provides support for a variety of arithmetic functions for the types unsigned and signed. These include **+**, **–**, *****, **/**, **mod**, **rem**, and **abs** functions. These arithmetic operations behave differently for the unsigned versus signed types, but the VHDL compiler will automatically use the correct operation based on the types of the input arguments.

Most synthesis tools support the addition, subtraction, and multiplication operators in this package. This provides a higher level of abstraction when modeling arithmetic circuitry. Recall that the VHDL standard package does not support addition, subtraction, and multiplication of types bit/bit_vector using the +, – and, * operators. Using the numeric_std package gives the ability to model these arithmetic operations with a synthesizable data type using the more familiar mathematical operators. The division, modulo, remainder, and absolute value functions are not synthesizable directly from this package.

Example:

```
F <= A + B;         -- A, B, F are type unsigned(3 downto 0)
F <= A - B;
```

The numeric_std package gives the ability to model arithmetic at a higher level of abstraction. Let's look at an example of implementing an adder circuit using the "+" operator. While this operator is supported for the type integer in the std_logic_1164 package, modeling adders using integers can be onerous due to the multiple levels of casting, range checking, and manual handling of carry out. A simpler approach to modeling adder behavior is to use the types unsigned/signed and the "+" operator provided in the numeric_std package. Temporary signals or variables of these types are required to model the adder behavior with the "+" sign. Also, type casting is still required when assigning the values back to the output ports. One advantage of this approach is that range checking is eliminated because rollover is automatically handled with these types.

Example 6.2 shows a behavioral model for a 4-bit adder in VHDL. In this model, a 5-bit unsigned vector is created (Sum_uns). The two inputs, A and B, are concatenated with a leading zero in order to facilitate assigning the sum to this 5-bit vector. The advantage of this approach is that the carry out of the adder is automatically included in the sum as the highest position bit. Since A and B are of type std_logic_vector, they must be converted to unsigned before the addition with the "+" operator can take place. The concatenation, type conversion, and addition can all take place in a single assignment.

Example:

```
Sum_uns <= unsigned(('0' & A)) + unsigned(('0' & B));
```

The 5-bit vector Sum_uns now contains the 4-bit sum and carry out. The final step is to assign the separate components of this vector to the output ports of the system. The 4-bit sum portion requires a type conversion back to std_logic_vector before it can be assigned to the output port *Sum*. Since the *Cout* port is a scalar, an unsigned signal can be assigned to it directly without the need for a conversion.

Example:

```
Sum <= std_logic_vector(Sum_uns(3 downto 0));
Cout <= Sum_uns(4);
```

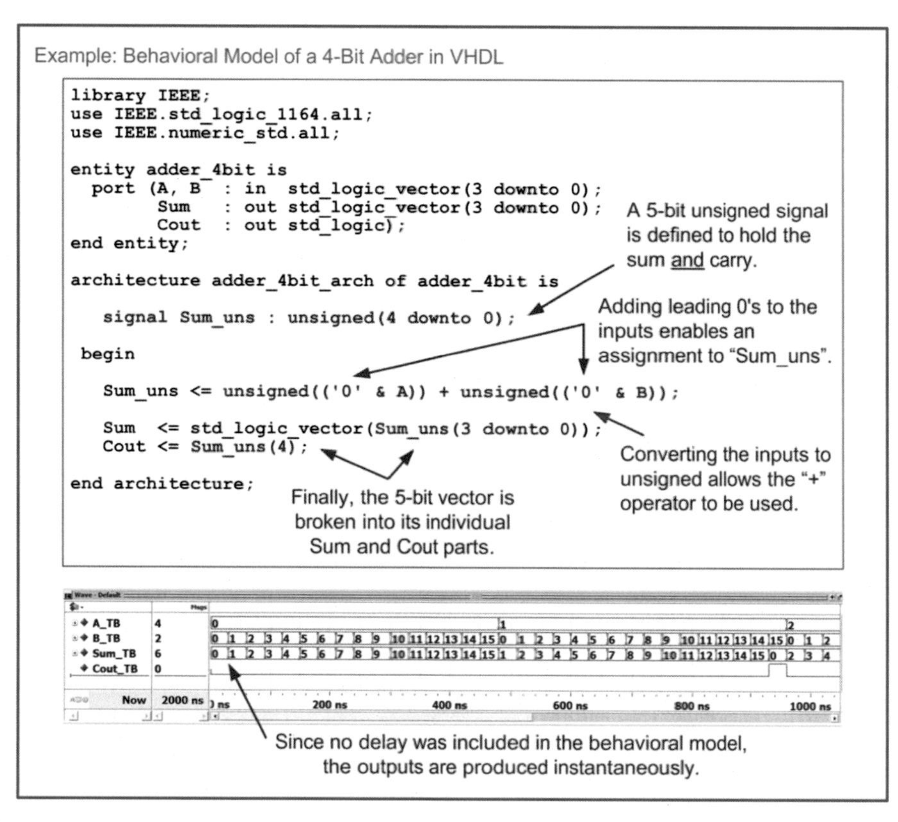
Example: Behavioral Model of a 4-Bit Adder in VHDL

```
library IEEE;
use IEEE.std_logic_1164.all;
use IEEE.numeric_std.all;

entity adder_4bit is
  port (A, B  : in  std_logic_vector(3 downto 0);
        Sum   : out std_logic_vector(3 downto 0);
        Cout  : out std_logic);
end entity;

architecture adder_4bit_arch of adder_4bit is

   signal Sum_uns : unsigned(4 downto 0);

 begin

   Sum_uns <= unsigned(('0' & A)) + unsigned(('0' & B));

   Sum  <= std_logic_vector(Sum_uns(3 downto 0));
   Cout <= Sum_uns(4);

end architecture;
```


Example 6.2
Behavioral model of a 4-bit adder in VHDL

6.2.2 NUMERIC_STD Logical Functions

The numeric_std package provides support for all of the logical operators (**and, nand, or, nor, xor, xnor, not)** for types unsigned and signed. It also provides two new shift functions **shift_left()** and **shift_right()**. These shift functions will fill the vacant position in the vector after the shift with a 0; thus, these are *logical shifts*. This package also provides two new rotate functions **rotate_left()** and **rotate_right()**.

6.2.3 NUMERIC_STD Comparison Functions

The numeric_std package provides support for all of the comparison functions for types unsigned and signed. These include **>**, **<**, **<=**, **>=**, =, and **/=**. These comparisons return type Boolean.

Example: (A = "0000", B = "1111")

```
if (A < B) then   -- This condition is TRUE if A and B are UNSIGNED
      :

if (A < B) then   -- This condition is FALSE if A and B are SIGNED
```

6.2.4 NUMERIC_STD Edge Detection Functions

The numeric_std also provides the functions **rising_edge()** and **falling_edge()** for the detection of rising or falling edge transition detection for types unsigned and signed.

6.2.5 NUMERIC_STD Conversion Functions

The numeric_std package contains a variety of useful conversion functions. Of particular usefulness are functions between the type *integer* and to/from *unsigned/signed*. This allows behavioral models for counters, adders, and subtractors to be implemented using the more readable type integer. After the functionally has been described, a conversion can be used to turn the result into types unsigned or signed to provide a synthesizable output. When converting an integer to a vector, a *size* argument is included. The size argument is of type integer and provides the number of bits in the vector that the integer will be converted to.

Name	Input type	Return type
To_integer()	unsigned	integer
To_integer()	signed	integer
To_unsigned()	integer, <size>	unsigned (size-1 downto 0)
To_signed()	integer, <size>	signed (size-1 downto 0)

6.2.6 NUMERIC_STD Type Casting

VHDL contains a set of built-in *type casting* operations that are commonly used with the numeric_std package to convert between *std_logic_vector* and *unsigned/signed*. Since the types unsigned/signed are based on the underlying type std_logic_vector, the conversion is simply known as casting. The following are the built-in type casting capabilities in VHDL:

Name	Input type	Return type
std_logic_vector()	unsigned	std_logic_vector
std_logic_vector()	signed	std_logic_vector
unsigned()	std_logic_vector	unsigned
signed()	std_logic_vector	signed

When using these type casts, they are placed on the right-hand side of the assignment exactly as a conversion function.

Example:

```
A <= std_logic_vector(B);  -- B is unsigned, A is std_logic_vector
C <= unsigned(D);          -- D is std_logic_vector, C is unsigned
```

Type casts and conversion functions can be compounded in order to perform multiple conversions in one assignment. This is useful when converting between types that do not have a direct cast or conversion function. Let's look at the example of converting an integer to an 8-bit std_logic_vector where the number being represented is unsigned. The first step is to convert the integer to an unsigned type. This can be accomplished with the *to_unsigned* function defined in the numeric_std package. This can be embedded in a second cast from unsigned to std_logic_vector. In the following example, E is the target of the operation and is of type std_logic vector. F is the argument of assignment and is of type integer. Recall that the to_unsigned conversions require both the input integer name and the size of the unsigned vector being converted to.

Example:

```
E <= std_logic_vector(to_unsigned(F, 8));
```

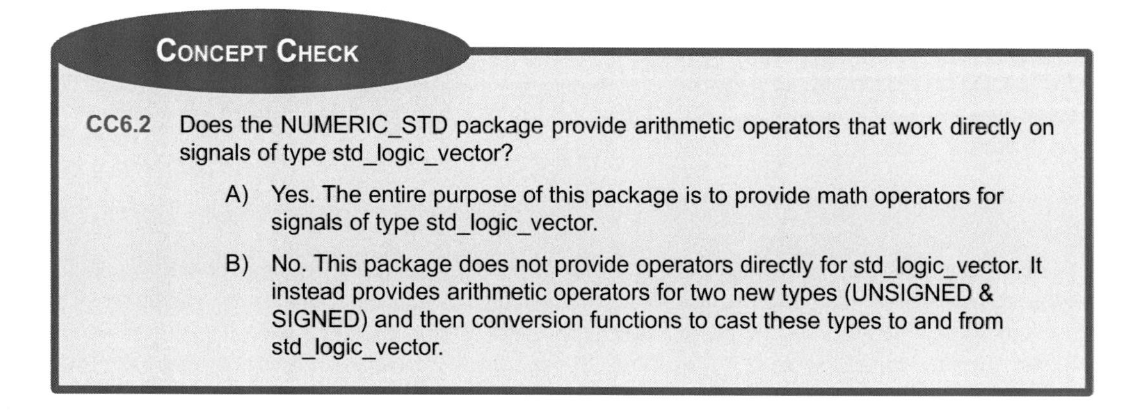

CONCEPT CHECK

CC6.2 Does the NUMERIC_STD package provide arithmetic operators that work directly on signals of type std_logic_vector?

A) Yes. The entire purpose of this package is to provide math operators for signals of type std_logic_vector.

B) No. This package does not provide operators directly for std_logic_vector. It instead provides arithmetic operators for two new types (UNSIGNED & SIGNED) and then conversion functions to cast these types to and from std_logic_vector.

6.3 TEXTIO and STD_LOGIC_TEXTIO

The *textio* package provides the ability to read and write to/from external input/output (I/O). External I/O refers to items such as files or the standard input/output of a computer. This package contains functions that allow the values of signals and variables to be read and written in addition to strings. This allows more sophisticated output messages to be created compared to the *report* statement alone, which can only output strings. The ability to read in values from a file allows sophisticated test patterns to be created outside of VHDL and then read in during simulation for testing a system. It is important to keep in mind that the term "I/O" refers to external files or the transcript window, not the inputs and outputs of a system model. The textio package is not synthesizable and is only used in test benches. The textio package is within the STD library and is included in a VHDL design using the following syntax:

```
library STD;
use STD.textio.all;
```

This package by itself only supports reading and writing types bit, bit_vector, integer, character, and string. Since the majority of synthesizable designs use types std_logic and std_logic_vector, an additional package was created that added support for these types. The package is called *std_logic_textio* and is located within the IEEE library. The syntax for including this package is as below.

```
library IEEE;
use IEEE.std_logic_textio.all;
```

The textio package defines two new types for interfacing with external I/O. These types are **file** and **line**. The type *file* is used to identify or create a file for reading/writing within the VHDL design. The syntax for declaring a file is as follows:

```
file file_handle : <file_type> open <file_mode> is <"filename">;
```

Declaring a file will automatically open the file and keep it open until the end of the process that is using it. The *file_handle* is a unique identifier for the file that is used in subsequent procedures. The file handle name is user-defined. A file handle eliminates the need to specify the entire file name each time a file access procedure is called. The *file_type* describes the information within the file. There are two supported file types, **TEXT** and **INTF**. A *TEXT* file is one that contains strings of characters. This is the most common type of file used as there are functions that can convert between types string, bit/bit_vector, and std_logic/std_logic_vector. This allows all of the information in the file to be stored as characters, which makes the file readable by other programs. An *INTF* file type contains only integer values, and the information is stored as a 32-bit, signed binary number. The *file_mode* describes whether the file will be read from or written to. There are two supported modes, **WRITE_MODE** and **READ_MODE**. The *filename* is given within double quotes and is user-defined. It is common to enter an extension on the file so that it can be opened by other programs (e.g., output.txt). Declaring a file always takes place within a process before the process begin statement. The following are examples of how to declare files:

```
file Fout: TEXT open WRITE_MODE is "output_file.txt";
file Fin:  TEXT open READ_MODE  is "input_file.txt";
```

The information within a file is accessed (either read or written) using the concept of a *line*. In the textio package, a file is interpreted as a sequence of lines, each containing either a string of characters or an integer value. The type *line* is used as a temporary buffer when accessing a line within the file. When accessing a file, a variable is created of type *line*. This variable is then used either to hold information that is *read* from a line in the file or to hold the information that is to be *written* to a line in the file. A variable is necessary for this behavior since assignments to/from the file must be made immediately. As such, a line variable is always declared within a process before the process begin statement. The syntax for declaring a variable of type line is as follows:

```
variable <line_variable_name> : line;
```

There are two procedures that allow information to be transferred between a line variable in VHDL and a line in a file. These procedures are **readline()** and **writeline()**. Their syntax is as follows:

```
readline(<file_handle>, <line_variable_name>);
writeline(<file_handle>, <line_variable_name>);
```

The transfer of information between a line variable and a line in a file using these procedures is accomplished on the entire line. There is no mechanism to read or write only a portion of the line in a file. Once a file is opened/created using a file declaration, the lines are accessed in the order they appear in the file. The first procedure called (either readline() or writeline()) will access the first line of the file. The next time a procedure is called, it will access the second line of the file. This will continue until all of the lines have been accessed. The textio package provides a function to indicate when the end of the file has been reached when performing a readline(). This function is called **endfile()** and returns type Boolean. This function will return True once the end of the file has been reached. Figure 6.3 shows a graphical representation of how the textio package handles external file access.

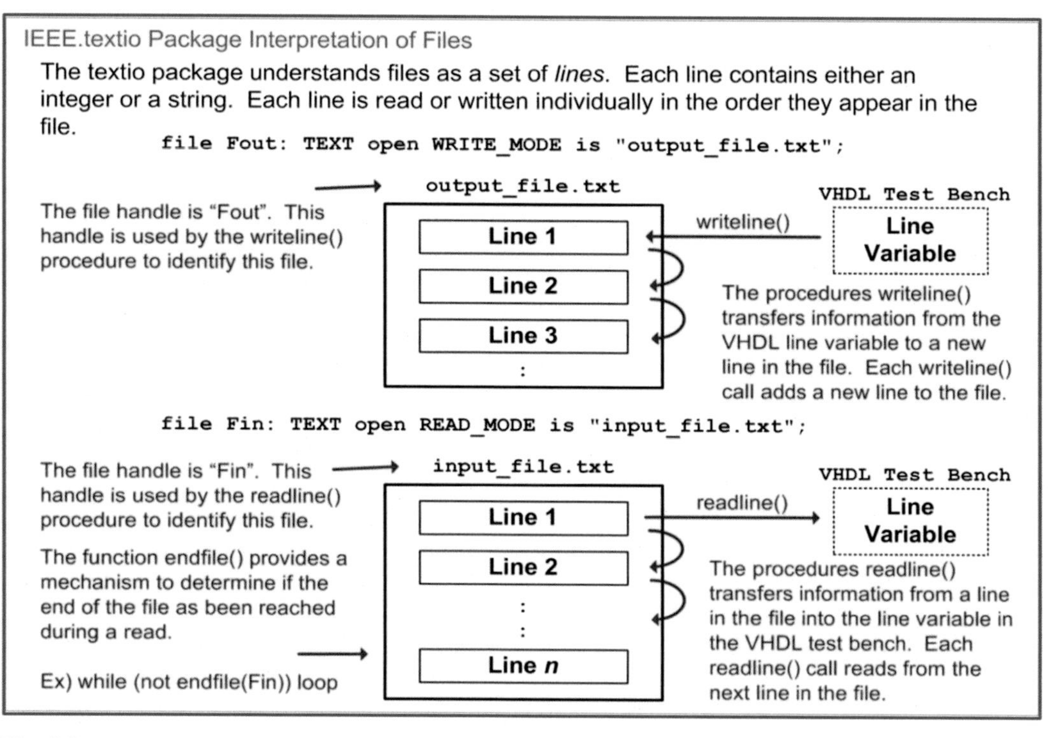

Fig. 6.3
IEEE.textio package interpretation of files

Two additional procedures are provided to add or retrieve information to/from the line variable within the VHDL test bench. These procedures are **read()** and **write()**. The syntax for these procedures is as follows:

```
read(<line_variable_name>, <destination_variable>);
write(<line_variable_name>, <source_variable>);
```

When using the read() procedure, the information in the line variable is treated as *space-delimited*. This means that each read() procedure will retrieve the information from the line variable until it reaches a white space. This allows multiple read() procedures to be used in order to parse the information into separate *destination_variable* names. The destination_variable must be of the appropriate type and size of the information being read from the file. For example, if the field in the line being read is a 4-character string ("wxyz"), then a destination variable must be defined of type *string(1 to 4)*. If the field being read is a 2-bit std_logic_vector, then a destination variable must be defined of type *std_logic_vector(1 downto 0)*. The read() procedure will ignore the delimiting white space character.

When using the write() procedure, the *source_destination* is assumed to be of type bit, bit_vector, integer, std_logic, or std_logic_vector. If it is desired to enter a text string directly, then the function **string** is used with the format *string'<"characters...">*. Multiple write() procedures can be used to insert information into the line variable. Each subsequent write procedure appends the information to the end of the string. This allows different types of information to be interleaved (e.g., text, signal value, text, etc.).

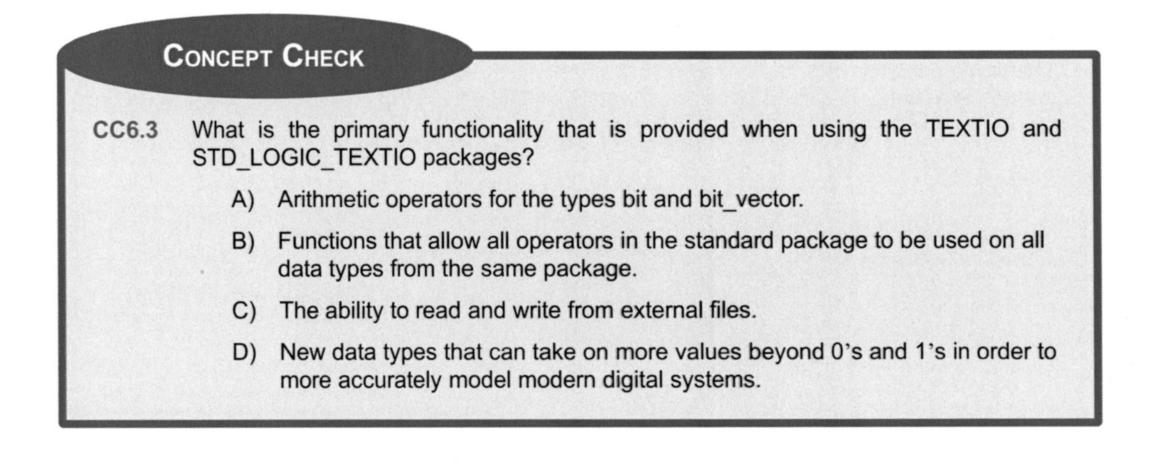

CONCEPT CHECK

CC6.3 What is the primary functionality that is provided when using the TEXTIO and STD_LOGIC_TEXTIO packages?

A) Arithmetic operators for the types bit and bit_vector.

B) Functions that allow all operators in the standard package to be used on all data types from the same package.

C) The ability to read and write from external files.

D) New data types that can take on more values beyond 0's and 1's in order to more accurately model modern digital systems.

6.4 Other Common Packages

6.4.1 NUMERIC_STD_UNSIGNED

When using the numeric_std package, the data types unsigned and signed must be used in order to get access to the numeric operators. While this provides ultimate control over the behavior of the signal operations and comparisons, many designs may only use unsigned types. In order to provide a mechanism to treat all vectors as unsigned while leaving their type as std_logic_vector, the numeric_std_unsigned package was created. When this package is used, it will treat all std_logic_vectors in the design as unsigned. This package requires the std_logic_1164 and numeric_std packages to be previously included. When used, all signals and ports can be declared as std_logic/std_logic_vector, and they will be treated as unsigned when performing arithmetic operations and comparisons. The following is an example of how to include this package:

```
library IEEE;
use IEEE.std_logic_1164.all;
use IEEE.numeric_std.all;
use IEEE.numeric_std_unsigned.all;
```

The numeric_std_unsigned package contains a few more type conversions beyond the numeric_std package. These additional conversions are as follows:

Name	Input type	Return type
To_Integer	std_logic_vector	integer
To_StdLogicVector	unsigned	std_logic_vector

6.4.2 NUMERIC_BIT

The *numeric_bit* package provides numerical computation for types bit and bit_vector. Since the vast majority of VHDL designs today use types std_logic/std_logic_vector instead of bit/bit_vector, this package is rarely used. This package is included by adding the following syntax at the beginning of the VHDL file in the design:

```
library IEEE;
use IEEE.numeric_bit.all;    -- defines types unsigned and signed
```

The numeric_bit package is nearly identical to numeric_std. It defines data types **unsigned** and **signed**, which provide information on the encoding style of the underlying data types bit and bit_vector. All of the arithmetic, logical, and comparison functions defined in numeric_std are supported in numeric_bit (**+**, **–**, *****, **/**, **mod**, **rem**, **abs**, **and**, **nand**, **or**, **nor**, **xor**, **xnor**, **not**, **>**, **<**, **<=**, **>=**, **=**, **/=**) for types unsigned and signed. This package also provides the same edge detection (**rising_edge()**, **falling_edge()**), shift (**shift_left()**, **shift_right()**), and rotate (**rotate_left()**, **rotate_right()**) functions for types unsigned and signed.

The primary difference between numeric_bit and numeric_std is that numeric_bit also provides support for the shift/rotate operators from the standard package (**sll**, **srl**, **rol**, **ror**). Also, the conversion functions are defined only for conversions between integer, unsigned, and signed.

Name	Input type	Return type
To_integer	unsigned	integer
To_integer	signed	integer
To_unsigned	integer, <size>	unsigned (size-1 downto 0)
To_signed	integer, <size>	signed (size-1 downto 0)

6.4.3 NUMERIC_BIT_UNSIGNED

The numeric_bit_unsigned package provides a way to treat all bit/bit_vectors in a design as unsigned numbers. The syntax for including the numeric_bit_unsigned package is shown below. In this example, all bit/bit_vectors will be treated as unsigned numbers for all arithmetic operations and comparisons.

```
library IEEE;
use IEEE.numeric_bit.all;
use IEEE.numeric_bit_unsigned.all;
```

The numeric_bit_unsigned package contains a few more type conversions beyond the numeric_bit package. These additional conversions are as follows:

Name	Input type	Return type
To_integer	std_logic_vector	integer
To_BitVector	unsigned	bit_vector

6.4.4 MATH_REAL

The *math_real* package provides numerical computation for the type *real*. The type *real* is the VHDL type used to describe a 32-bit floating-point number. None of the operators provided in the math_real package are synthesizable. This package is primarily used for test benches. This package is included by adding the following syntax at the beginning of the VHDL file in the design:

```
library IEEE;
use IEEE.math_real.all;
```

The math_real package defines a set of commonly used constants, which are shown below.

Constant name	Type	Value	Description
MATH_E	real	2.718	Value of e
MATH_1_E	real	0.367	Value of 1/e
MATH_PI	real	3.141	Value of pi
MATH_1_PI	real	0.318	Value of 1/pi
MATH_LOG_OF_2	real	0.693	Natural log of 2
MATH_LOG_OF_10	real	2.302	Natural log of10
MATH_LOG2_OF_E	real	1.442	Log base 2 of e
MATH_LOG10_OF_E	real	0.434	Log base 10 of e
MATH_SQRT2	real	1.414	Sqrt of 2
MATH_SQRT1_2	real	0.707	Sqrt of 1/2
MATH_SQRT_PI	real	1.772	Sqrt of pi
MATH_DEG_TO_RAD	real	0.017	Conversion factor from degree to radian
MATH_RAD_TO_DEG	real	57.295	Conversion factor from radian to degree

Only three digits of accuracy are shown in this table; however, the constants defined in the math_real package have full 32-bit accuracy. The math_real package provides a set of commonly used floating-point operators for the type real.

Function name	Return type	Description
SIGN	real	Returns sign of input
CEIL	real	Returns smallest integer value
FLOOR	real	Returns largest integer value
ROUND	real	Rounds input up/down to whole number
FMAX	real	Returns largest of two inputs
FMIN	real	Returns smallest of two inputs
SQRT	real	Returns square root of input
CBRT	real	Returns cube root of input
**	real	Raise to power of (X**Y)
EXP	real	e^X
LOG	real	log(X)
SIN	real	sin(X)
COS	real	cos(X)
TAN	real	tan(X)
ASIN	real	asin(X)
ACOS	real	acos(X)
ATAN	real	atan(X)
ATAN2	real	atan(X/Y)
SINH	real	sinh(X)
COSH	real	cosh(X)
TANH	real	tanh(X)
ASINH	real	asinh(X)
ACOSH	real	acosh(X)
ATANH	real	atanh(X)

6.4.5 MATH_COMPLEX

The *math_complex* package provides numerical computation for complex numbers. Again, nothing in this package is synthesizable and is typically used only for test benches. This package is included by adding the following syntax at the beginning of the VHDL file in the design:

```
library IEEE;
use IEEE.math_complex.all;
```

This package defines three new data types, **complex**, **complex_vector**, and **complex_polar**. The type *complex* is defined with two fields, *real* and *imaginary*. The type *complex_vector* is a linear array of type complex. The type *complex_polar* is defined with two fields, *magnitude* and *angle*. This package provides a set of common operations for use with complex numbers. This package also supports the arithmetic operators **+**, −, *****, and **/** for the type complex.

Function name	Return type	Description
CABS	real	Absolute value of complex number
CARG	real (radians)	Returns angle of complex number
CMPLX	complex	Returns complex number form of input
CONJ	complex or complex_polar	Returns complex conjugate
CSQRT	real	Returns square root
CEXP	real	Returns e^Z of complex input
COMPLEX_TO_POLAR	complex_polar	Convert complex to complex_polar
POLAR_TO_COMPLEX	complex	Convert complex_polar to complex

6.4.6 Legacy Packages (STD_LOGIC_ARITH / UNSIGNED / SIGNED)

Prior to the release of the *numeric_std* package by the IEEE, Synopsys Inc. created a set of packages to provide computational operations for types std_logic and std_logic_vector. Since these arithmetic packages were defined very early in the life of VHDL, they were widely adopted. Unfortunately, due to these packages not being standardized through a governing body such as the IEEE, vendors began modifying the packages to meet proprietary needs. This led to a variety of incompatibility issues that have plagued these packages. As a result, all new designs requiring computational operations should be based on the IEEE *numeric_std* package. While the IEEE standard is the recommended numerical package for VHDL, the original Synopsys packages are still commonly found in designs and in design examples, so providing an overview of their functionality is necessary.

Synopsis Inc. created the **std_logic_arith** package to provide computational operations for types std_logic and std_logic_vector. Just as with the *numeric_std* package, this package defines two new types, **unsigned** and **signed**. Arithmetic, comparison, and shift operators are provided for these types that include **+**, −, *****, **abs**, **>**, **<**, **<=**, **>=**, =, **/=**, **shl**, and **shr**. This package also provides a set of conversion functions between types unsigned, signed, std_logic_vector, and integer. The syntax for these conversions are as follows:

Name	Input type	Return type
CONV_INTEGER	unsigned	integer
CONV_INTEGER	signed	integer
CONV_UNSIGNED	integer, <size>	unsigned
CONV_UNSIGNED	signed	unsigned
CONV_SIGNED	integer, <size>	signed
CONV_SIGNED	unsigned	signed
CONV_STD_LOGIC_VECTOR	integer, <size>	std_logic_vector(size-1 downto 0)

Name	Input type	Return type
CONV_STD_LOGIC_VECTOR	unsigned, <size>	std_logic_vector(size-1 downto 0)
CONV_STD_LOGIC_VECTOR	signed, <size>	std_logic_vector(size-1 downto 0)

The Synopsys packages have the ability to treat all std_logic_vectors in a design as either unsigned or signed by including an additional package. The **std_logic_unsigned** package, when included in conjunction with the std_logic_arith package, will treat all std_logic_vectors in the design as unsigned numbers. The syntax for using the Synopsys arithmetic packages on unsigned numbers is as follows. The *std_logic_1164* package is required to define types std_logic/std_logic_vector. The *std_logic_arith* package provides the computational operators for types std_logic/std_logic_vector. Finally, the *std_logic_unsigned* package treats all std_logic/std_logic_vector types as unsigned numbers when performing arithmetic operations.

```
library IEEE;
use IEEE.std_logic_1164.all;
use IEEE.std_logic_arith.all;
use IEEE.std_logic_unsigned.all;
```

The **std_logic_signed** package works in a similar manner with the exception that it treats all std_logic/std_logic_vector types as signed numbers when performing arithmetic operations. The *std_logic_unsigned* and *std_logic_signed* packages are never used together since they will conflict with each other.

The syntax for using the std_logic_signed package is as follows:

```
library IEEE;
use IEEE.std_logic_1164.all;
use IEEE.std_logic_arith.all;
use IEEE.std_logic_signed.all;
```

One of the more confusing aspects of the Synopsys packages is that they are included in the IEEE library. This means that both the *numeric_std* package (IEEE standard) and the *std_logic_arith* package (Synopsys, nonstandard) are part of the same library, but one is recommended, while the other is not. This is due to the fact that the Synopsys packages were developed first, and putting them into the IEEE library was the most natural location since this library was provided as part of the VHDL standard. When the numeric_std package was standardized by the IEEE, it also was naturally inserted into the IEEE library. As a result, today's IEEE library contains both styles of packages.

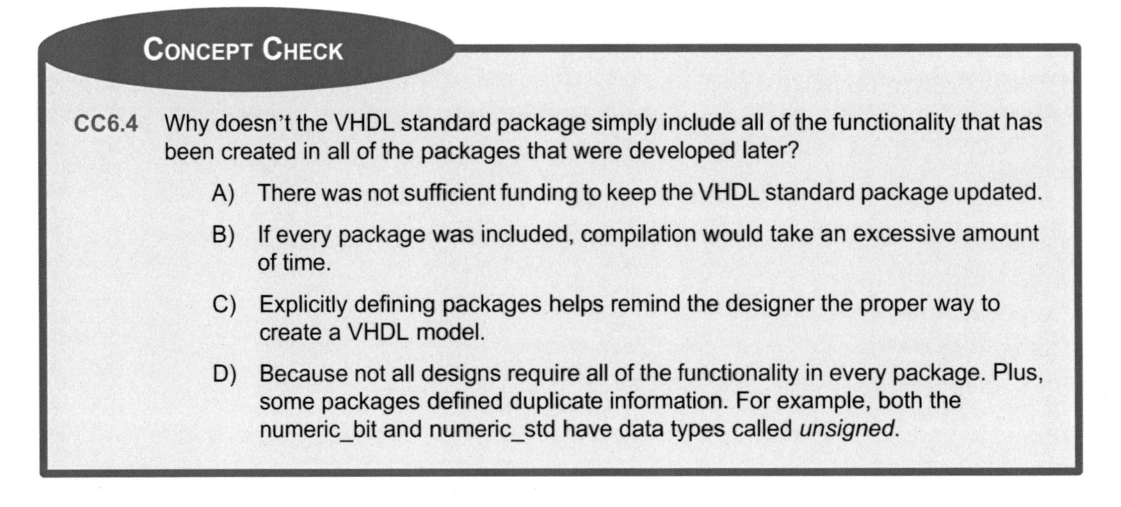

CONCEPT CHECK

CC6.4 Why doesn't the VHDL standard package simply include all of the functionality that has been created in all of the packages that were developed later?

A) There was not sufficient funding to keep the VHDL standard package updated.

B) If every package was included, compilation would take an excessive amount of time.

C) Explicitly defining packages helps remind the designer the proper way to create a VHDL model.

D) Because not all designs require all of the functionality in every package. Plus, some packages defined duplicate information. For example, both the numeric_bit and numeric_std have data types called *unsigned*.

Summary

- The IEEE STD_LOGIC_1164 package provides more realistic data types for modeling modern digital systems. This package provides the std_ulogic and std_logic data types. These data types can take on nine different values (U, X, 0, 1, Z, W, L, H, and –).
- The std_logic data type provides a resolution function that allows multiple outputs to be connected to the same signal. The resolution function will determine the value of the signal based on a predefined priority given in the function.
- The IEEE STD_LOGIC_1164 package provides logical operators and edge detection functions for the types std_ulogic and std_logic. It also provides conversion functions to and from the type *bit*.
- The IEEE NUMERIC_STD package provides the data types *unsigned* and *signed*. These types use the underlying data type std_logic. These types provide the ability to treat vectors as either unsigned or two's complement codes.
- The IEEE NUMERIC_STD package provides arithmetic operations for the types unsigned and signed. This package also provides conversions functions and type casts to and from the types integer and std_logic_vector.
- The TEXTIO and STD_LOGIC_TEXTIO packages provide the functionality to read and write to external files.

Exercise Problems

Section 6.1: STD_LOGIC_1164

6.1.1 What are all the possible values that a signal of type *std_logic* can take on?

6.1.2 What is the difference between the types *std_ulogic* and *std_logic*?

6.1.3 If a signal of type *std_logic* is assigned both a 0 and Z at the same time, what will the final signal value be?

6.1.4 If a signal of type *std_logic* is assigned both a 1 and X at the same time, what will the final signal value be?

6.1.5 If a signal of type *std_logic* is assigned both a 0 and L at the same time, what will the final signal value be?

6.1.6 Are any arithmetic operations provided for the type *std_logic_vector* in the STD_LOGIC_1164 package?

Section 6.2: NUMERIC_STD

6.2.1 If you declare a signal of type *unsigned* from the NUMERIC_STD package, what are all the possible values that the signal can take on?

6.2.2 If you declare a signal of type *signed* from the NUMERIC_STD package, what are all the possible values that the signal can take on?

6.2.3 If two signals (A and B) are declared of type *signed* from the NUMERIC_STD package and hold the values A <= "1111" and B <= "0000," which signal has a greater value?

6.2.4 If two signals (A and B) are declared of type *unsigned* from the NUMERIC_STD package and hold the values A <= "1111" and B <= "0000," which signal has a greater value?

6.2.5 If you are using the NUMERIC_STD package, what is the syntax to convert a signal of type *unsigned* into *std_logic_vector*?

6.2.6 If you are using the NUMERIC_STD package, what is the syntax to convert a signal of type *integer* into *std_logic_vector*?

Section 6.3: TEXTIO and STD_LOGIC_TEXTIO

6.3.1 What does the keyword *file* accomplish?

6.3.2 What is the difference between the commands *write* and *writeline*?

6.3.3 Can two different types of information be written to a line variable in one command?

6.3.4 What is the name of the special file handle reserved for the standard output of a computer?

Section 6.4: Other Common Packages

6.4.1 What is the impact of including the NUMERIC_STD_UNSIGNED package?

6.4.2 Does the NUMERIC-BIT package support resolved data types?

6.4.3 Are the functions in the MATH_REAL and MATH_COMPLEX package synthesizable?

6.4.4 Can the NUMERIC_STD and STD_LOGIC_ARITH packages be used at the same time? why or why not?

Chapter 7: Test Benches

The functional verification of VHDL designs is accomplished through simulation using a *test bench*. A test bench is a VHDL system that instantiates the system to be tested as a component and then generates the input patterns and observes the outputs. VHDL provides a variety of capability to design test benches that can automate stimulus generation and provide automated output checking. These capabilities can be expanded by including packages that take advantage of reading/writing to external I/O. This chapter provides the details of VHDL's built-in capabilities that allow test benches to be created and some examples of automated stimulus generation and using external files.

Learning Outcomes—After completing this chapter, you will be able to:

7.1 Design a VHDL test bench that manually creates each stimulus pattern using a series of signal assignments and wait statements within a process.

7.2 Design a VHDL test bench that uses for loops to automatically generate an exhaustive set of stimulus patterns.

7.3 Design a VHDL test bench that automatically checks the outputs of the system being tested using report and assert statements.

7.4 Design a VHDL test bench that uses external I/O as part of the testing procedures including reading stimulus patterns from, and writing the results to, external files.

7.1 Test Bench Overview

Creating the testing strategy for a design is a critical piece of the digital design process. In HDL-based testing, the system being tested is often called a *device under test (DUT)* or *unit under test (UUT)*. Test benches are only used for simulation so we can use abstract modeling techniques that are unsynthesizable to generate the stimulus patterns. VHDL also contains specific functionality to report on the status of a test and also automatically check that the outputs are correct. Example 7.1 shows how to create a simple test bench to verify the operation of SystemX. The test bench does not have any inputs or outputs; thus, there are no ports declared in the entity. SystemX is declared as a component in the test bench and then instantiated (DUT1). Internal signals are declared to connect to the component under test (A_TB, B_TB, C_TB, F_TB). A process is then used to drive the inputs of SystemX. Within the process, wait statements are used to control the execution of the signal assignments; thus, the process does not have a sensitivity list. Each possible input code is generated within the process. The output (F_TB) is observed using a simulation tool in the form of either a waveform or a table listing.

B. J. LaMeres, *Quick Start Guide to VHDL*, https://doi.org/10.1007/978-3-031-42543-1_7

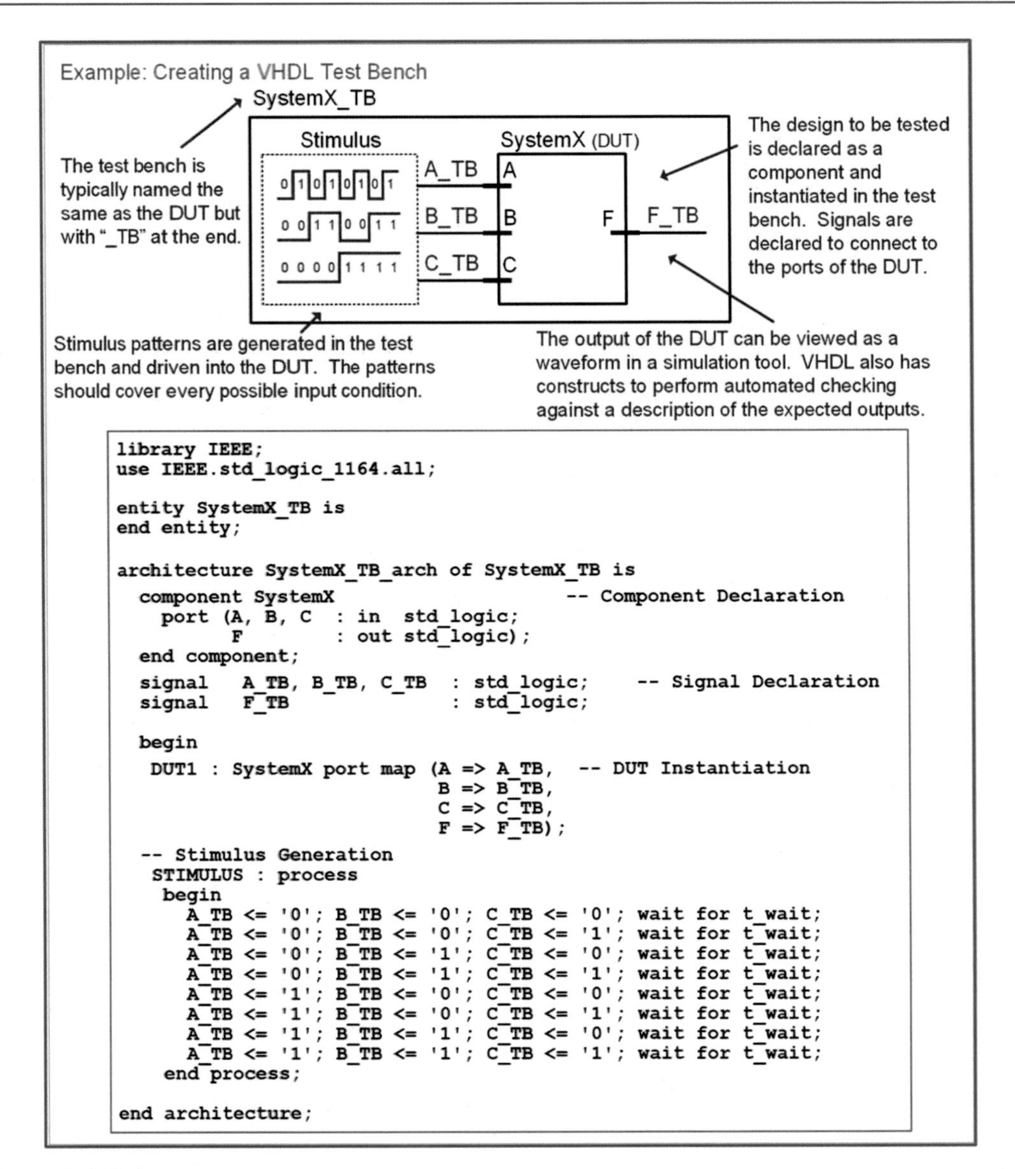


```
library IEEE;
use IEEE.std_logic_1164.all;

entity SystemX_TB is
end entity;

architecture SystemX_TB_arch of SystemX_TB is
  component SystemX                               -- Component Declaration
    port (A, B, C  : in  std_logic;
          F        : out std_logic);
  end component;

  signal   A_TB, B_TB, C_TB   : std_logic;     -- Signal Declaration
  signal   F_TB               : std_logic;

  begin
   DUT1 : SystemX port map (A => A_TB,   -- DUT Instantiation
                            B => B_TB,
                            C => C_TB,
                            F => F_TB);

  -- Stimulus Generation
   STIMULUS : process
    begin
      A_TB <= '0'; B_TB <= '0'; C_TB <= '0'; wait for t_wait;
      A_TB <= '0'; B_TB <= '0'; C_TB <= '1'; wait for t_wait;
      A_TB <= '0'; B_TB <= '1'; C_TB <= '0'; wait for t_wait;
      A_TB <= '0'; B_TB <= '1'; C_TB <= '1'; wait for t_wait;
      A_TB <= '1'; B_TB <= '0'; C_TB <= '0'; wait for t_wait;
      A_TB <= '1'; B_TB <= '0'; C_TB <= '1'; wait for t_wait;
      A_TB <= '1'; B_TB <= '1'; C_TB <= '0'; wait for t_wait;
      A_TB <= '1'; B_TB <= '1'; C_TB <= '1'; wait for t_wait;
    end process;

end architecture;
```

Example 7.1
Creating a VHDL test bench

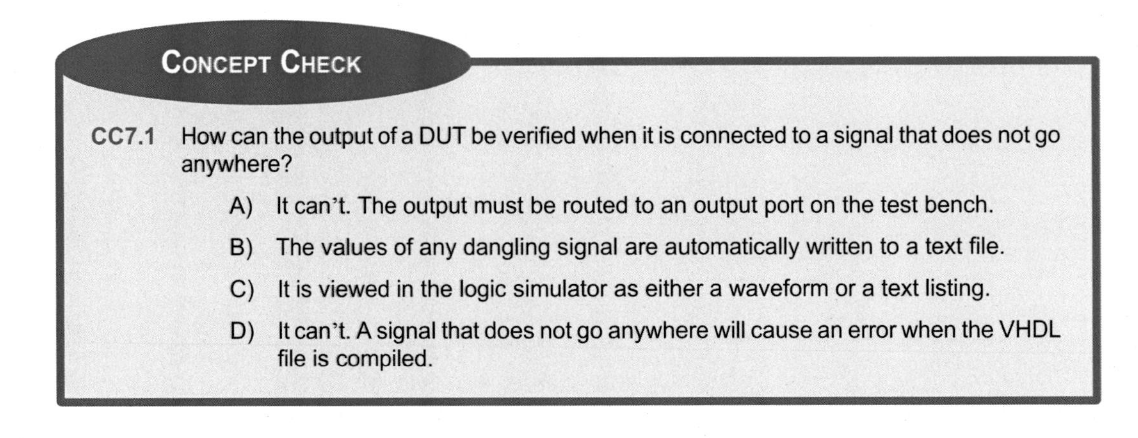

CONCEPT CHECK

CC7.1 How can the output of a DUT be verified when it is connected to a signal that does not go anywhere?

A) It can't. The output must be routed to an output port on the test bench.

B) The values of any dangling signal are automatically written to a text file.

C) It is viewed in the logic simulator as either a waveform or a text listing.

D) It can't. A signal that does not go anywhere will cause an error when the VHDL file is compiled.

7.2 Generating Stimulus Vectors Using For Loops

Typically, testing a DUT under all possible input conditions is necessary to verify functionality. Testing under each and every input condition can require a large number of input conditions. As a case study, consider an n-bit adder. To test an n-bit adder under each and every numeric input condition will take $(2^n)^2$ test vectors. For a simple 4-bit adder, this equates to 256 input patterns. Even for a small circuit such as this, the large number of input patterns precludes the use of manual signal assignments in the test bench to stimulate the circuit. One approach to automatically generate an exhaustive set of input test patterns is to use nested for loops. Example 7.2 shows a test bench that uses two nested for loops to generate the 256 unique input conditions for the 4-bit ripple carry adder designed back in Example 4.8. Note that the loop variables *i* and *j* are automatically created when the loops are declared. Since the loop variables are defined as integers, type conversions are required prior to driving the values into the RCA. The simulation waveform illustrates how the ripple carry adder has a noticeable delay before the output sum is produced. During the time the carry is rippling through the adder chain, glitches can appear on each of the sum bits in addition to the carry out signal. The values in this waveform are displayed as unsigned decimal symbols to make the results easier to interpret.

Example: VHDL Test Bench for a 4-Bit Ripple Carry Adder Using Nested For Loops

Nested for loops can be used in order to generate an exhaustive set of test vectors to stimulate the adder.

```
library IEEE;
use IEEE.std_logic_1164.all;
use IEEE.numeric_std.all;

entity rca_4bit_TB is
end entity;

architecture rca_4bit_TB_arch of rca_4bit_TB is

   component rca_4bit
     port (A, B  : in  std_logic_vector(3 downto 0);
           Sum   : out std_logic_vector(3 downto 0);
           Cout  : out std_logic);
   end component;

   signal A_TB, B_TB, Sum_TB  : std_logic_vector(3 downto 0);
   signal Cout_TB             : std_logic;

 begin

   DUT : rca_4bit port map (A_TB, B_TB, Sum_TB, Cout_TB);

   STIM : process
     begin

      for i in 0 to 15 loop
         for j in 0 to 15 loop
           A_TB <= std_logic_vector(to_unsigned(i,4));
           B_TB <= std_logic_vector(to_unsigned(j,4));
           wait for 30 ns;
         end loop;
      end loop;

   end process;

end architecture;
```

The simulation waveform for the ripple carry adder is as follows. The numbers are shown in unsigned decimal format for readability.

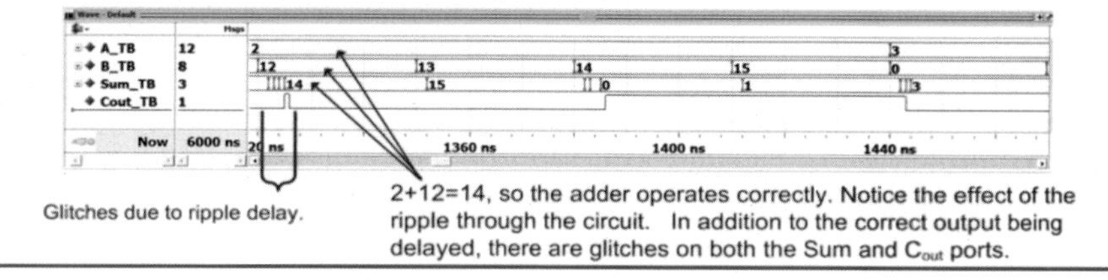

Example 7.2
VHDL test bench for a 4-bit ripple carry adder using nested for loops

CONCEPT CHECK

CC7.2 If you used two nested for loops to generate an exhaustive set of patterns for the inputs of an 8-bit adder, how many patterns would be generated? There is no carry-in bit.

A) 16 B) 256 C) 512 D) 65,536

7.3 Automated Checking Using Report and Assert Statements

7.3.1 Report Statement

The keyword **report** can be used within a test bench in order to provide the status of the current test. A report statement will print a string to the transcript window of the simulation tool. The report output also contains an optional severity level. There are four levels of severity (ERROR, WARNING, NOTE, and FAILURE). The severity level *FAILURE* will halt a simulation, while the levels *ERROR*, *WARNING*, and *NOTE* will allow the simulation to continue. If the severity level is omitted, the report is assumed to be a severity level of NOTE. The syntax for using a report statement is as follows:

```
report "string to be printed" severity <level>;
```

Let's look at how we could use the report function within the example test bench to print the current value of the input pattern to the transcript window of the simulator. Example 7.3 shows the new process and resulting transcript output of the simulator when using report statements.

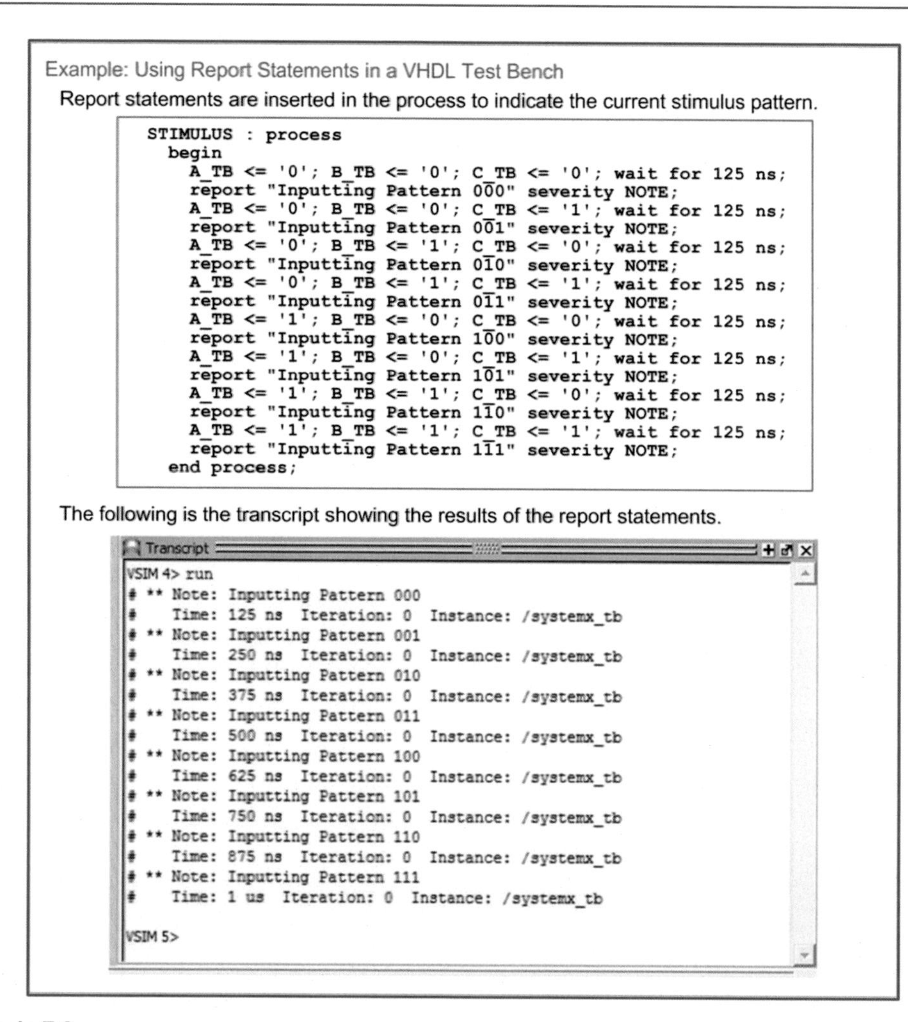

Example: Using Report Statements in a VHDL Test Bench

Report statements are inserted in the process to indicate the current stimulus pattern.

```
STIMULUS : process
  begin
    A_TB <= '0'; B_TB <= '0'; C_TB <= '0'; wait for 125 ns;
    report "Inputting Pattern 000" severity NOTE;
    A_TB <= '0'; B_TB <= '0'; C_TB <= '1'; wait for 125 ns;
    report "Inputting Pattern 001" severity NOTE;
    A_TB <= '0'; B_TB <= '1'; C_TB <= '0'; wait for 125 ns;
    report "Inputting Pattern 010" severity NOTE;
    A_TB <= '0'; B_TB <= '1'; C_TB <= '1'; wait for 125 ns;
    report "Inputting Pattern 011" severity NOTE;
    A_TB <= '1'; B_TB <= '0'; C_TB <= '0'; wait for 125 ns;
    report "Inputting Pattern 100" severity NOTE;
    A_TB <= '1'; B_TB <= '0'; C_TB <= '1'; wait for 125 ns;
    report "Inputting Pattern 101" severity NOTE;
    A_TB <= '1'; B_TB <= '1'; C_TB <= '0'; wait for 125 ns;
    report "Inputting Pattern 110" severity NOTE;
    A_TB <= '1'; B_TB <= '1'; C_TB <= '1'; wait for 125 ns;
    report "Inputting Pattern 111" severity NOTE;
  end process;
```

The following is the transcript showing the results of the report statements.

```
Transcript
VSIM 4> run
# ** Note: Inputting Pattern 000
#    Time: 125 ns  Iteration: 0  Instance: /systemx_tb
# ** Note: Inputting Pattern 001
#    Time: 250 ns  Iteration: 0  Instance: /systemx_tb
# ** Note: Inputting Pattern 010
#    Time: 375 ns  Iteration: 0  Instance: /systemx_tb
# ** Note: Inputting Pattern 011
#    Time: 500 ns  Iteration: 0  Instance: /systemx_tb
# ** Note: Inputting Pattern 100
#    Time: 625 ns  Iteration: 0  Instance: /systemx_tb
# ** Note: Inputting Pattern 101
#    Time: 750 ns  Iteration: 0  Instance: /systemx_tb
# ** Note: Inputting Pattern 110
#    Time: 875 ns  Iteration: 0  Instance: /systemx_tb
# ** Note: Inputting Pattern 111
#    Time: 1 us  Iteration: 0  Instance: /systemx_tb

VSIM 5>
```

Example 7.3
Using report statements in a VHDL test bench

7.3.2 Assert Statement

The **assert** statement provides a mechanism to check a Boolean condition before using the report statement. This allows report outputs to be selectively printed based on the values of signals in the system under test. This can be used to print either the successful operation or the failure of a system. If the Boolean condition associated with the assert statement is evaluated True, it *will not* execute the subsequent report statement. If the Boolean condition is evaluated False, it will execute the subsequent report statement. The assert statement is always used in conjunction with the report statement. The following is the syntax for the assert statement:

```
assert boolean_condition report "string" severity <level>;
```

Let's look at how we could use the assert function within the example test bench to check whether the output (F_TB) is correct. In the example in Example 7.4, the system passes the first pattern but fails the second.

Example: Using Assert Statements in a VHDL Test Bench

Assert statements are used to check the correctness of the system outputs.

```
STIMULUS : process
  begin
    A_TB <= '0'; B_TB <= '0'; C_TB <= '0'; wait for 125 ns;
    assert (F_TB='1') report "Failed test at 000" severity FAILURE;
    assert (F_TB='0') report "Passed test at 000" severity NOTE;

    A_TB <= '0'; B_TB <= '0'; C_TB <= '1'; wait for 125 ns;
    assert (F_TB='1') report "Failed test at 001" severity FAILURE;
    assert (F_TB='0') report "Passed test at 001" severity NOTE;
                                    :
  end process;
```

An intentional failure was introduced at the second input pattern to show how the simulation will end if a report statement is issued with a severity level of FAILURE. The following is the output of the transcript for this case.

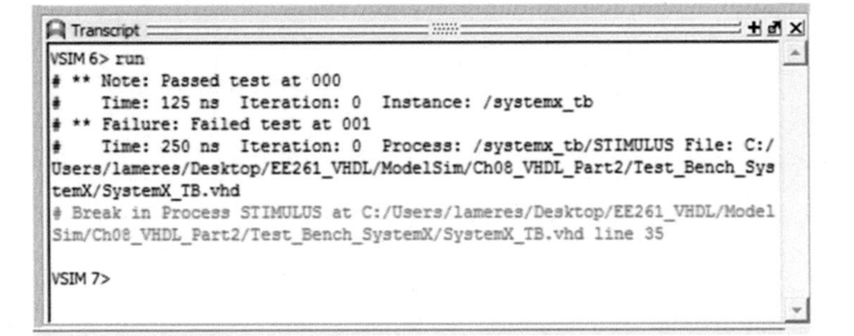

Example 7.4
Using assert statements in a VHDL test bench

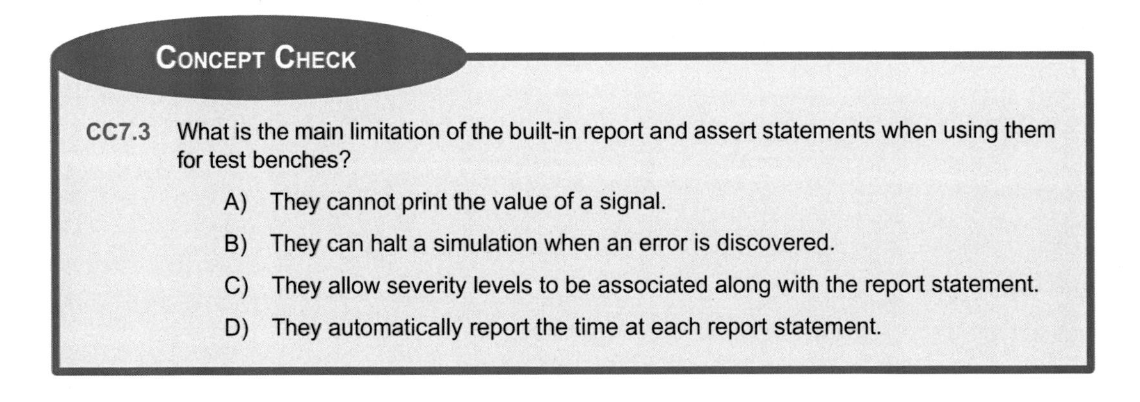
Concept Check

CC7.3 What is the main limitation of the built-in report and assert statements when using them for test benches?

A) They cannot print the value of a signal.

B) They can halt a simulation when an error is discovered.

C) They allow severity levels to be associated along with the report statement.

D) They automatically report the time at each report statement.

7.4 Using External I/O in Test Benches

7.4.1 Writing to an External File from a Test Bench

When it is desired to report larger amounts of data, writing to the transcript becomes impractical, and an external file is needed. In order to write to an external file from a test bench, the *textio* and *std_logic_textio* packages are needed. To illustrate how to do this, let's look at an example of a test bench that writes information about the tests being conducted to an external file. Example 7.5 shows the model for the system to be tested (SystemX) and an overview of the test bench approach (SystemX_TB). Note that the DUT does not need to include the *textio* and *std_logic_textio* packages as the file writing functionality exists within the test bench file.

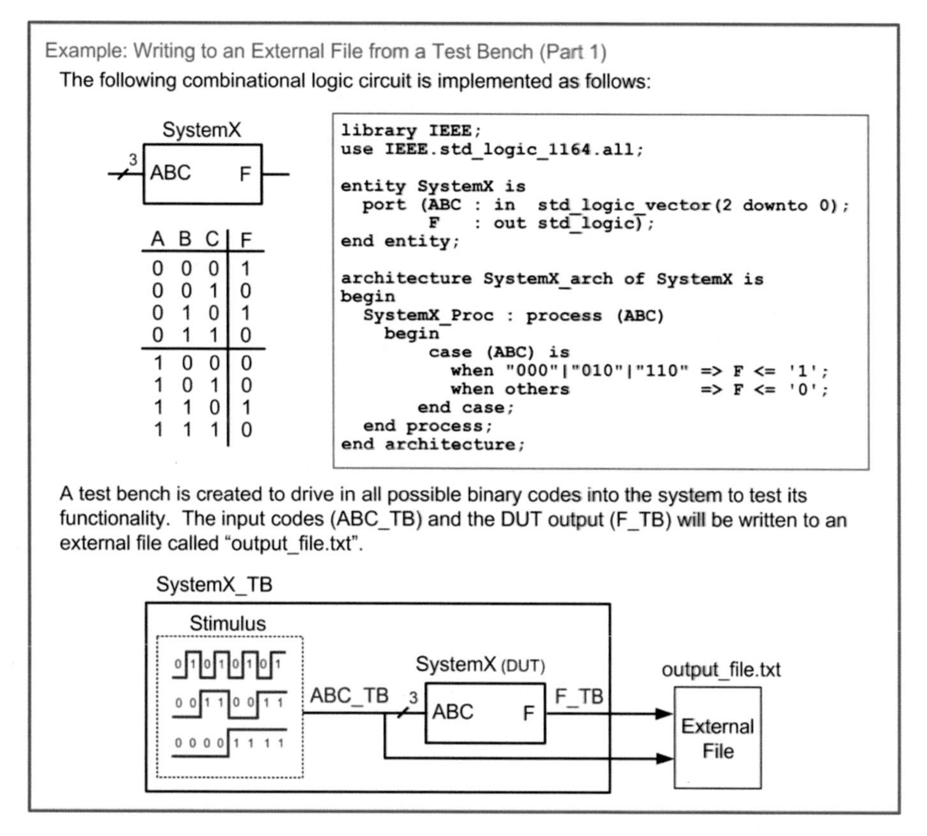

Example 7.5
Writing to an external file from a test bench (Part 1)

Example 7.6 shows the details of the test bench model. In this test bench, a file is declared in order to create "output_file.txt." This file is given the handle *Fout*. A line variable is also declared called *current_line* to act as a temporary buffer to hold information that will be written to the file. The procedure write() is used to add information to the line variable. The first write() procedure is used to create a text message ("Beginning Test. . ."). Notice that since the information to be written to the line variable is of type string, a conversion function must be used within the write() procedure (e.g., string'(Beginning Test. . .")). This message is written as the first line in the file using the writeline() procedure. After an input vector has been applied to the DUT, a new line is constructed containing the descriptive text, the input vector value, and the output value from the DUT. This message is repeated for each input code in the test bench.

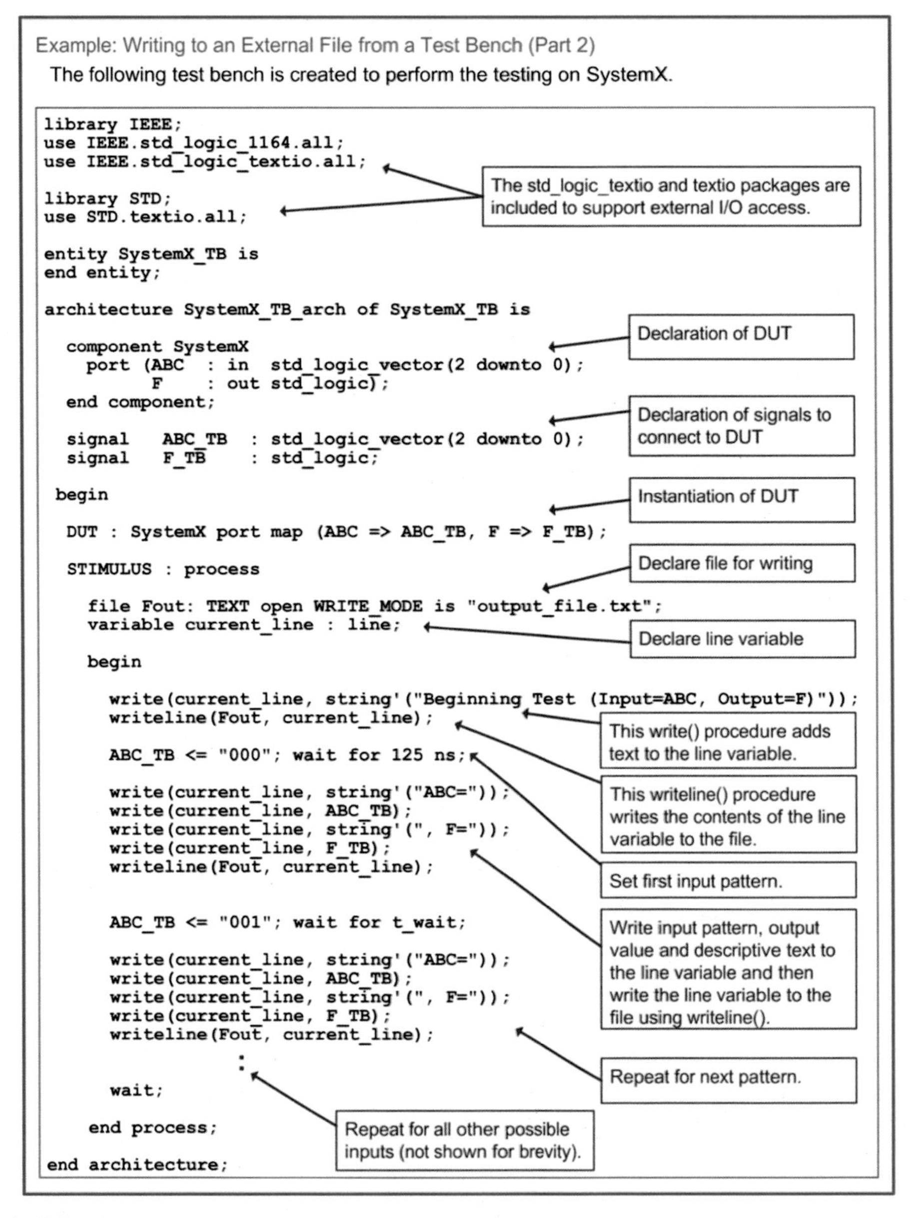

Example: Writing to an External File from a Test Bench (Part 2)

The following test bench is created to perform the testing on SystemX.

```
library IEEE;
use IEEE.std_logic_1164.all;
use IEEE.std_logic_textio.all;

library STD;
use STD.textio.all;

entity SystemX_TB is
end entity;

architecture SystemX_TB_arch of SystemX_TB is

  component SystemX
    port (ABC  : in  std_logic_vector(2 downto 0);
          F    : out std_logic);
  end component;

  signal   ABC_TB  : std_logic_vector(2 downto 0);
  signal   F_TB    : std_logic;

 begin

  DUT : SystemX port map (ABC => ABC_TB, F => F_TB);

  STIMULUS : process

    file Fout: TEXT open WRITE_MODE is "output_file.txt";
    variable current_line : line;

    begin

      write(current_line, string'("Beginning Test (Input=ABC, Output=F)"));
      writeline(Fout, current_line);

      ABC_TB <= "000"; wait for 125 ns;

      write(current_line, string'("ABC="));
      write(current_line, ABC_TB);
      write(current_line, string'(", F="));
      write(current_line, F_TB);
      writeline(Fout, current_line);


      ABC_TB <= "001"; wait for t_wait;

      write(current_line, string'("ABC="));
      write(current_line, ABC_TB);
      write(current_line, string'(", F="));
      write(current_line, F_TB);
      writeline(Fout, current_line);
                     :

      wait;

    end process;

end architecture;
```


Example 7.6
Writing to an external file from a test bench (Part 2)

Example 7.7 shows the resulting file that is created from this test bench.

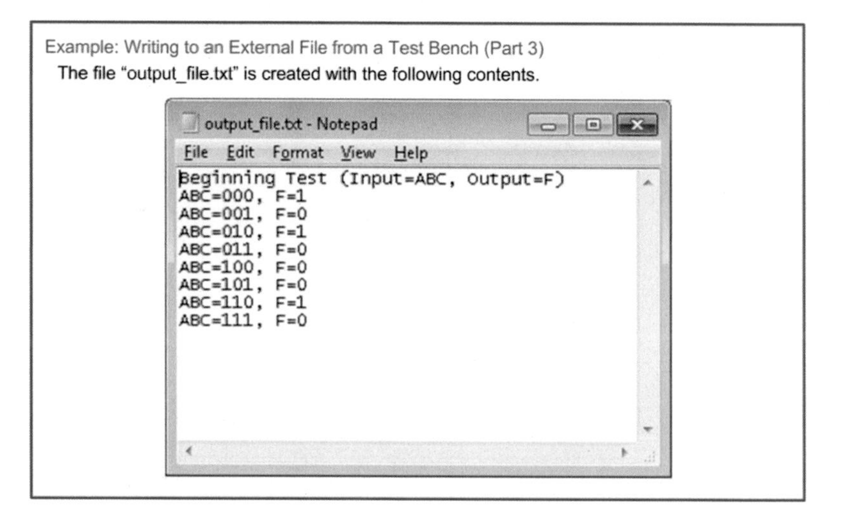


```
Beginning Test (Input=ABC, Output=F)
ABC=000, F=1
ABC=001, F=0
ABC=010, F=1
ABC=011, F=0
ABC=100, F=0
ABC=101, F=0
ABC=110, F=1
ABC=111, F=0
```

Example 7.7
Writing to an external file from a test bench (Part 3)

7.4.2 Writing to STD_OUTPUT from a Test Bench

The textio package also provides the ability to write to the standard output of the computer instead of to an external file. The standard output of the computer is typically routed to the transcript window of the simulator. This output mode is identical to how the *report* statement works, but using the textio package allows more functionality in the output text. The standard output of a computer is given a reserved file handle called **OUTPUT**. When using this file handle, a new file does not need to be declared in the test bench since it is already defined as part of the textio package. The reserved file handle name OUTPUT can be used directly in the writeline() procedure.

Let's look at an example of a test bench that outputs information about the test being conducted to STD_OUT. Example 7.8 shows this test bench approach. The test bench is identical as the one used in Example 7.6 with the exception that the writeline() procedure outputs are directed to the STD_OUTPUT of the computer using the reserved file handle name OUTPUT instead of to an external file.

Example: Writing to STD_OUTPUT from a Test Bench (Part 1)

This test bench directs the writeline() outputs to the STD_OUTPUT of the computer by using the reserved file handle "OUTPUT".

```
library IEEE;
use IEEE.std_logic_1164.all;
use IEEE.std_logic_textio.all;

library STD;
use STD.textio.all;

entity SystemX_TB is
end entity;

architecture SystemX_TB_arch of SystemX_TB is

  component SystemX
    port (ABC  : in  std_logic_vector(2 downto 0);
          F    : out std_logic);
  end component;

  signal   ABC_TB   : std_logic_vector(2 downto 0);
  signal   F_TB     : std_logic;

 begin

  DUT : SystemX port map (ABC => ABC_TB, F => F_TB);

  STIMULUS : process

    variable current_line : line;

    begin

      write(current_line, string'("Beginning Test (Input=ABC, Output=F)"));
      writeline(OUTPUT, current_line);

      ABC_TB <= "000"; wait for 125 ns;

      write(current_line, string'("ABC="));
      write(current_line, ABC_TB);
      write(current_line, string'(", F="));
      write(current_line, F_TB);
      writeline(OUTPUT, current_line);


      ABC_TB <= "001"; wait for t_wait;

      write(current_line, string'("ABC="));
      write(current_line, ABC_TB);
      write(current_line, string'(", F="));
      write(current_line, F_TB);
      writeline(OUTPUT, current_line);
                          :
      wait;

    end process;

end architecture;
```

The reserved file handle "OUTPUT" is used to direct the writeline() output to the computer STD_OUTPUT.

Repeat for all other possible inputs (not shown for brevity).

Example 7.8
Writing to STD_OUT from a test bench (Part 1)

Example 7.9 shows the output from the test bench. This output is displayed in the transcript window of the simulation tool.

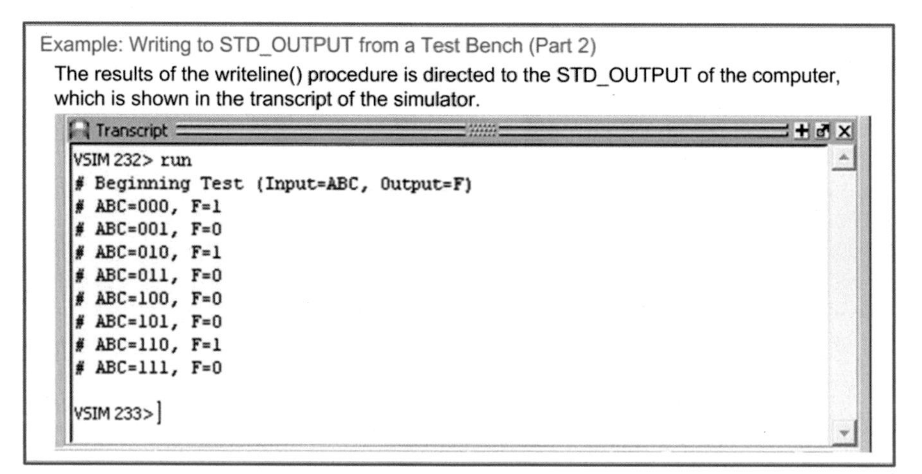

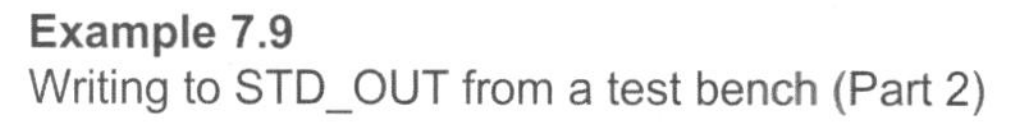
Example 7.9
Writing to STD_OUT from a test bench (Part 2)

7.4.3 Reading from an External File in a Test Bench

Let's now look at an example of reading test vectors from an external file using the textio package. Example 7.10 shows the test bench setup. In this example, the SystemX design from the prior example will be tested using vectors provided by an external file (input_file.txt). The test bench will read in each line of the file individually and sequentially. After reading a line, the test bench will drive the DUT with the input vector. In order to verify correct operation, the results will be written to the STD_OUTPUT of the computer.

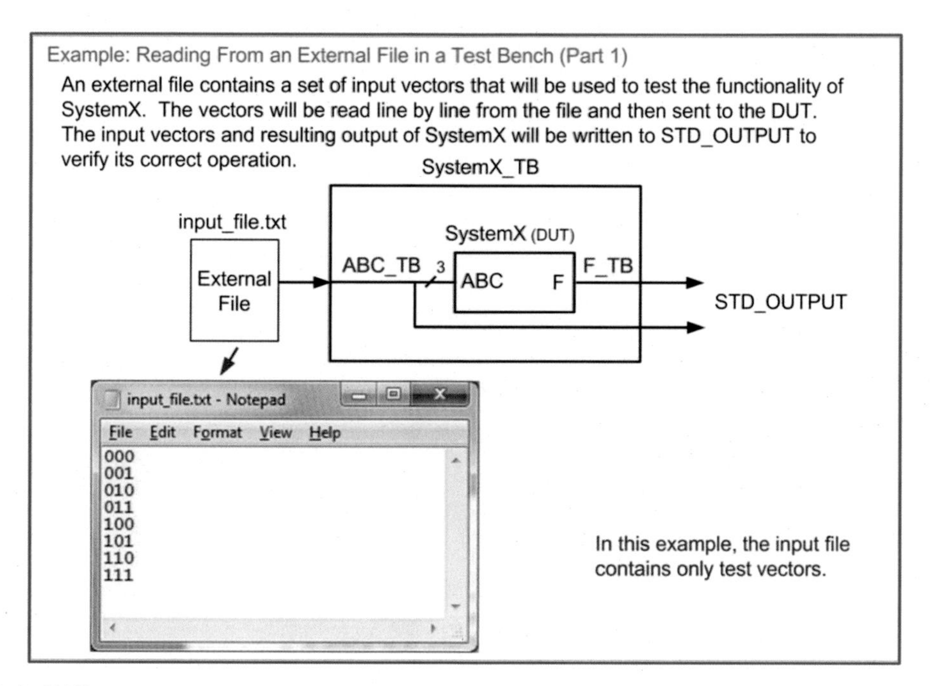

Example 7.10
Reading from an external file in a test bench (Part 1)

In order to read the external vectors, a file is declared in READ_MODE. This opens the external file and allows the VHDL test bench to access its lines. A variable is declared to hold the line that is read using the readline() procedure. In this example, the line variable for reading is called "current_read_line." A variable is also declared that will ultimately hold the vector that is extracted from current_read_line. This variable (called current_read_field) is declared to be of type std_logic_vector(2 downto 0) because the vectors in the file are 3-bit values. Once the line is read from the file using the readline() procedure, the vector can be read from the line variable using the read() procedure. Once the value resides in the *current_read_field* variable, it can be assigned to the DUT input signal vector *ABC_TB*. A set of messages are then written to the STD_OUTPUT of the computer using the reserved file handle OUTPUT. The messages contain descriptive text in addition to the values of the input vector and output value of the DUT. Example 7.11 shows the process to implement this behavior in the test bench.

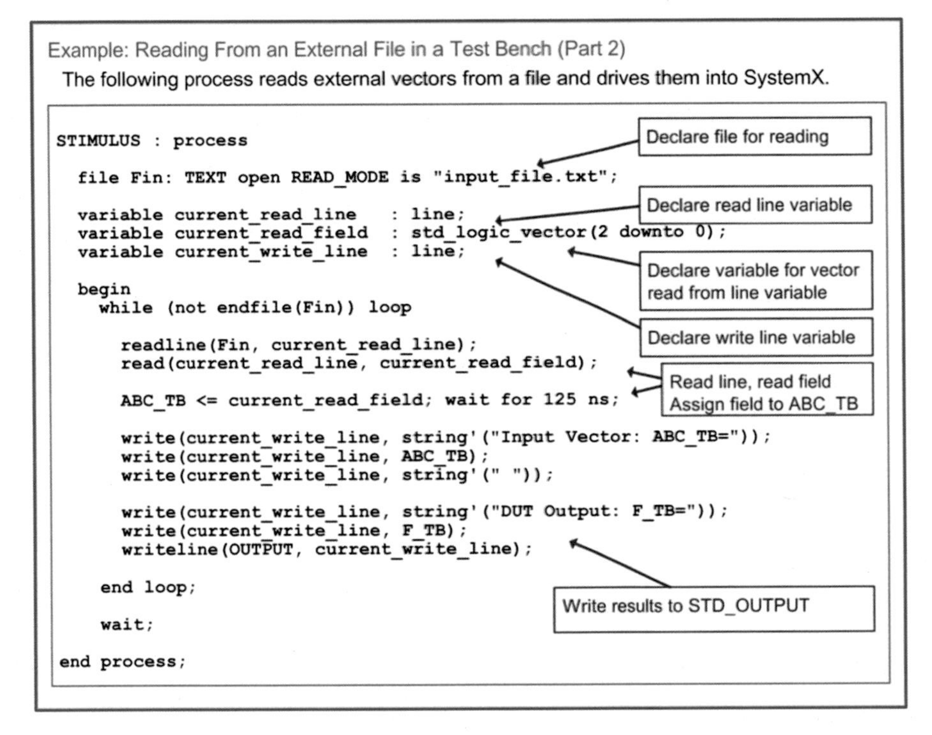

Example 7.11
Reading from an external file in a test bench (Part 2)

Example 7.12 shows the results of this test bench, which are written to STD_OUTPUT.

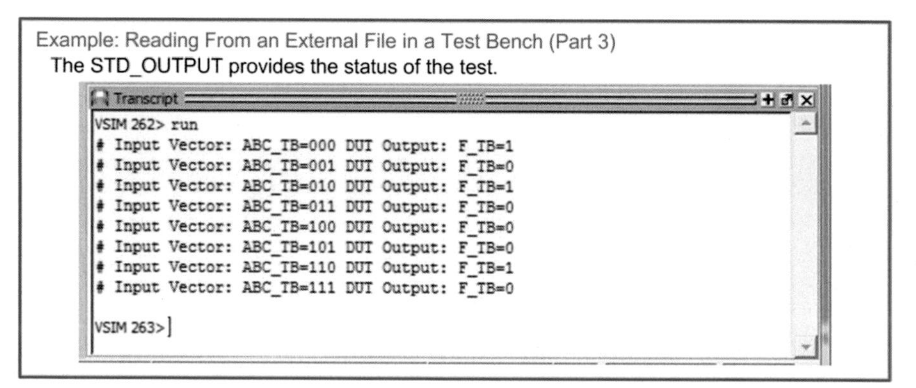

Example 7.12
Reading from an external file in a test bench (Part 3)

7.4.4 Reading Space-Delimited Data from an External File in a Test Bench

As mentioned earlier, information in a line variable is treated as white space-delimited by the read() procedure. This allows more information than just a single vector to be read from a file. When a read() procedure is performed on a line variable, it will extract information until it reaches either a white space or the end-of-line character. If a white space is encountered, the read() procedure will end. Let's look at an example of how to read information from a file when it contains both strings and vectors. Example 7.13 shows the test bench setup where an external file is to be read that contains both a text heading and test vector on each line. Since the header and the vector are separated with a white space character, two read() procedures need to be used to independently extract these distinct fields from the line variable.

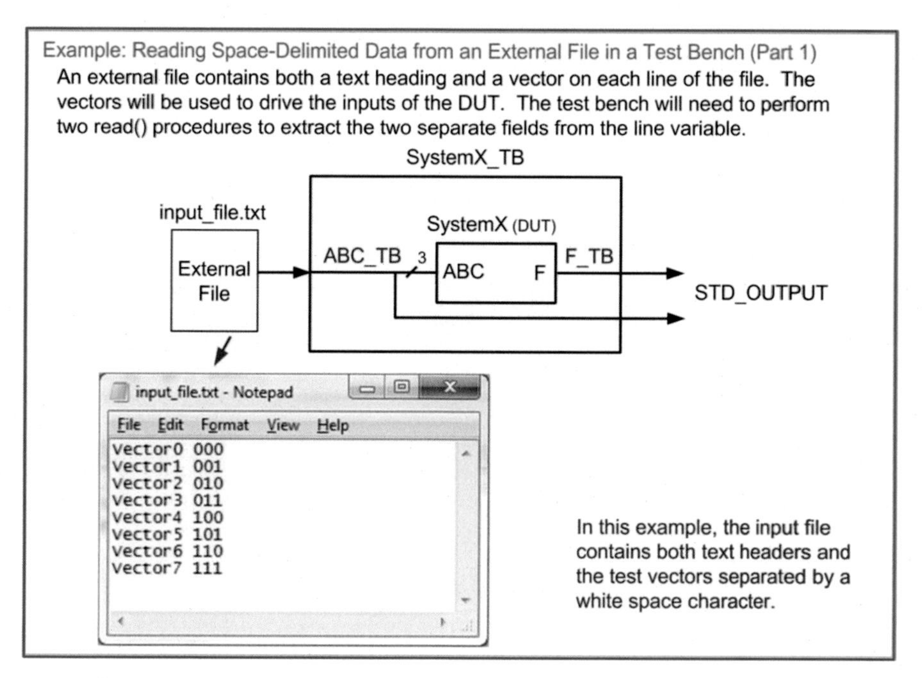

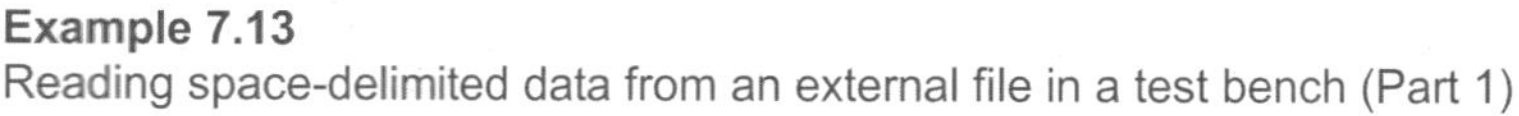

Example 7.13
Reading space-delimited data from an external file in a test bench (Part 1)

The test bench will transfer a line from the file into a line variable using the readline() procedure just as in the previous example; however, this time two different variables will need to be defined in order to read the two separate fields in the line. Each variable must be declared to be the proper type and size for the information in the field. For example, the first field in the file is a string of seven characters. As a result, the first variable declared (current_read_field1) will be of type *string(1 to 7)*. Recall that strings are typically indexed incrementally from left to right starting with the index 1. The second field in the file is a 3-bit vector, so the second variable declared (current_read_field2) will be of type std_logic_vector (2 downto 0). Each time a line is retrieved from the file using the readline() procedure, two subsequent read() procedures can be performed to extract the two fields from the line variable. The second field (i.e., the vector) can be used to drive the input of the DUT. In this example, both fields are written to STD_OUTPUT in addition to the output of the DUT to verify proper functionality. Example 7.14 shows the test bench process which models this behavior.

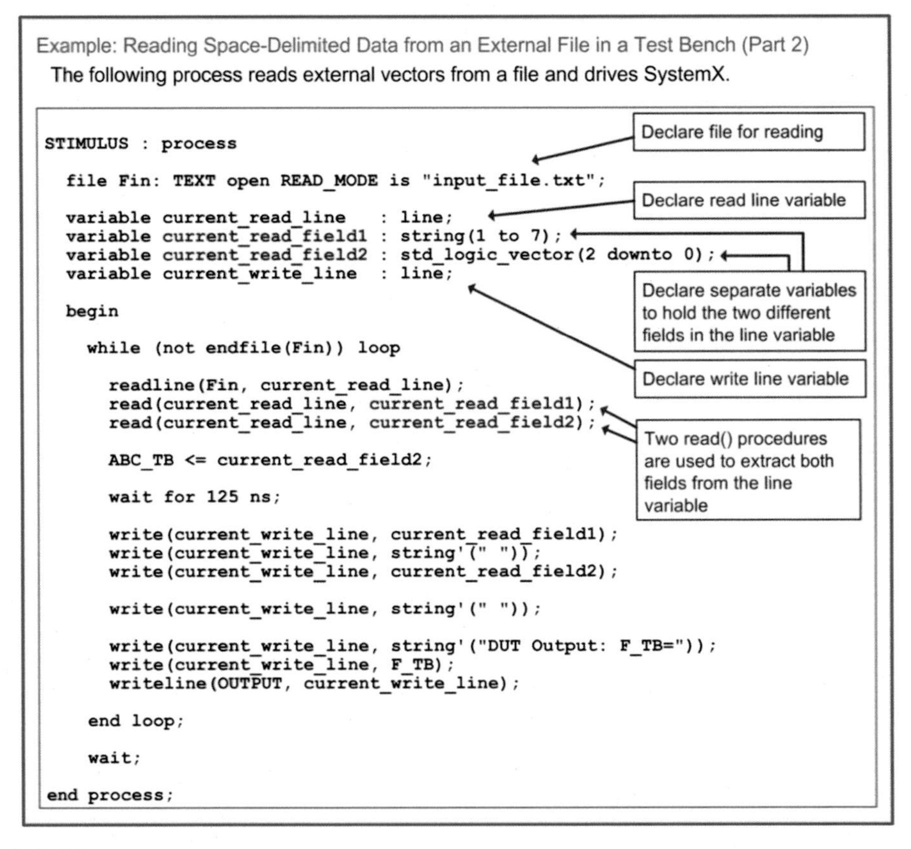

Example 7.14
Reading space-delimited data from an external file in a test bench (Part 2)

Example 7.15 shows the results of this test bench, which are written to STD_OUTPUT.

Example: Reading Space-Delimited Data from an External File in a Test Bench (Part 3)
The STD_OUTPUT provides the status of the test.

```
Transcript
VSIM 268> run
# Vector0 000 DUT Output: F_TB=1
# Vector1 001 DUT Output: F_TB=0
# Vector2 010 DUT Output: F_TB=1
# Vector3 011 DUT Output: F_TB=0
# Vector4 100 DUT Output: F_TB=0
# Vector5 101 DUT Output: F_TB=0
# Vector6 110 DUT Output: F_TB=1
# Vector7 111 DUT Output: F_TB=0

VSIM 269>
```

Example 7.15
Reading space-delimited data from an external file in a test bench (Part 3)

Concept Check

CC7.4 Can the TEXT_IO and STD_LOGIC_TEXTIO packages accomplish the same functionality as a report statement? If so, how?

A) Yes. If the line variable is simply written to the standard output of the computer, it will show up in the transcript window of the simulator just like the report statement does.

B) No. These packages only operate on external files.

Summary

- A *simulation test bench* is a VHDL file that drives stimulus into a device under test (DUT). Test benches do not have inputs or outputs and are not synthesizable.
- Stimulus patterns can be driven into a DUT within a process using a series of signal assignments with wait statements.
- Stimulus patterns can also be automatically generated using looping structures.
- The VHDL standard package supports the use of report and assert statement to provide a text-based output for tracking the status of a simulation.
- The TEXTIO and STD_LOGIC_TEXTIO allow the use of external files in test benches. This is useful for reading in more sophisticated input stimulus patterns and storing large output sets to files.

Exercise Problems

Section 7.1: Test Bench Overview

7.1.1 What is the purpose of a test bench?

7.1.2 Does a test bench have input and output ports?

7.1.3 Can a test bench be simulated?

7.1.4 Can a test bench be synthesized?

7.1.5 Design a VHDL test bench to verify the functional operation of the system in Fig. 7.1. Your test bench should drive in each input code for the vector ABCD in the order they would appear in a truth table (i.e., "0000," "0001," "0010," . . .). Your test bench should use a process and individual signal assignments for each pattern. Your test bench should change the input pattern every 10 ns using the *wait for* statement within your stimulus process.

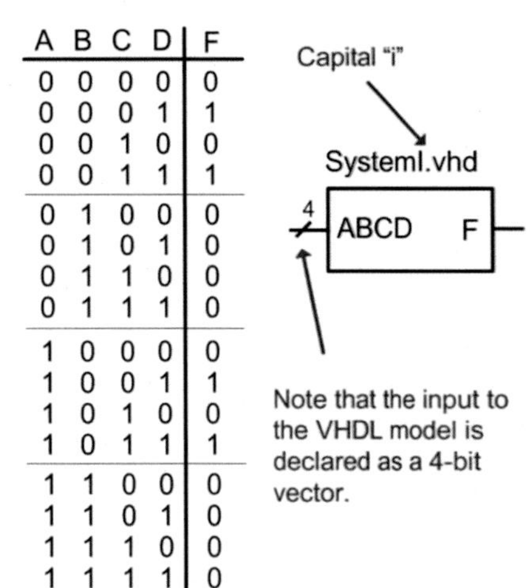

A	B	C	D	F
0	0	0	0	0
0	0	0	1	1
0	0	1	0	0
0	0	1	1	1
0	1	0	0	0
0	1	0	1	0
0	1	1	0	0
0	1	1	1	0
1	0	0	0	0
1	0	0	1	1
1	0	1	0	0
1	0	1	1	1
1	1	0	0	0
1	1	0	1	0
1	1	1	0	0
1	1	1	1	0

Fig. 7.1
System I functionality

7.1.6 Design a VHDL test bench to verify the functional operation of the system in Fig. 7.2. Your test bench should drive in each input code for the vector ABCD in the order they would appear in a truth table (i.e., "0000," "0001," "0010," . . .). Your test bench should use a process and individual signal assignments for each pattern. Your test bench should change the input pattern every 10 ns using the *wait for* statement within your stimulus process.

$F = \sum_{A,B,C,D}(4,5,7,12,13,15)$

SystemJ.vhd: 4 → ABCD F

Fig. 7.2
System J functionality

7.1.7 Design a VHDL test bench to verify the functional operation of the system in Fig. 7.3. Your test bench should drive in each input code for the vector ABCD in the order they would appear in a truth table (i.e., "0000," "0001," "0010," . . .). Your test bench should use a process and individual signal assignments for each pattern. Your test bench should change the input pattern every 10 ns using the *wait for* statement within your stimulus process.

$F = \prod_{A,B,C,D}(3,7,11,15)$

SystemK.vhd: 4 → ABCD F

Fig. 7.3
System K functionality

7.1.8 Design a VHDL test bench to verify the functional operation of the system in Fig. 7.4. Your test bench should drive in each input code for the vector ABCD in the order they would appear in a truth table (i.e., "0000," "0001," "0010," . . .). Your test bench should use a process and individual signal assignments for each pattern. Your test bench should change the input pattern every 10 ns using the *wait for* statement within your stimulus process.

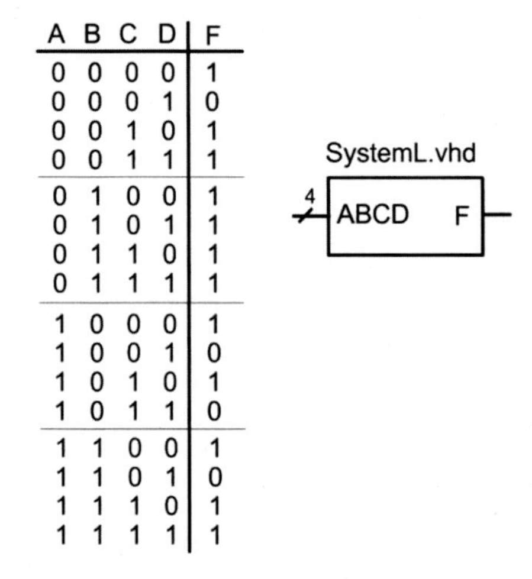

A	B	C	D	F
0	0	0	0	1
0	0	0	1	0
0	0	1	0	1
0	0	1	1	1
0	1	0	0	1
0	1	0	1	1
0	1	1	0	1
0	1	1	1	1
1	0	0	0	1
1	0	0	1	0
1	0	1	0	1
1	0	1	1	0
1	1	0	0	1
1	1	0	1	0
1	1	1	0	1
1	1	1	1	1

Fig. 7.4
System L functionality

Section 7.2: Generating Stimulus Vectors Using For Loops

7.2.1 Design a VHDL test bench to verify the functional operation of the system in Fig. 7.1. Your test bench should drive in each input code for the vector ABCD in the order they would appear in a truth table (i.e., "0000," "0001," "0010," . . .). Your test bench should use a single for loop within a process to generate all of the stimulus patterns automatically. Your test bench should change the input pattern every 10 ns using a *wait* statement.

7.2.2 Design a VHDL test bench to verify the functional operation of the system in Fig. 7.2. Your test bench should drive in each input code for the vector ABCD in the order they would appear in a truth table (i.e., "0000," "0001," "0010," . . .). Your test bench should use a single for loop within a process to generate all of the stimulus patterns automatically. Your test bench should change the input pattern every 10 ns using a *wait* statement.

7.2.3 Design a VHDL test bench to verify the functional operation of the system in Fig. 7.3. Your test bench should drive in each input code for the vector ABCD in the order they would appear in a truth table (i.e., "0000," "0001," "0010," . . .). Your test bench should use a single for loop within a process to generate all of the stimulus patterns automatically. Your test bench should change the input pattern every 10 ns using a *wait* statement.

7.2.4 Design a VHDL test bench to verify the functional operation of the system in Fig. 7.4. Your test bench should drive in each input code for the vector ABCD in the order they would appear in a truth table (i.e., "0000," "0001," "0010," . . .). Your test bench should use a single for loop within a process to generate all of the stimulus patterns automatically. Your test bench should change the input pattern every 10 ns using a *wait* statement.

7.2.5 Design a VHDL model for an 8-bit ripple carry adder (RCA) using a structural design approach. This involves creating a half adder (half_adder.vhd), full adder (full_adder.vhd), and then finally a top-level adder (rca.vhd) by instantiating eight full adder components. Model the ripple delay by inserting 1ns of gate delay for the XOR, AND, and OR operators using a delayed signal assignment. The general topology and entity definition for the design are shown in Example 4.6. Design a VHDL test bench to exhaustively verify this design under all input conditions. Your test bench should use two nested for loops within a process to generate all of the stimulus patterns automatically. Your test bench should change the input pattern every 30 ns using a *wait* statement in order to give sufficient time for the signals to ripple through the adder.

Section 7.3: Automated Checking Using Report and Assert Statements

7.3.1 Design a VHDL test bench to verify the functional operation of the system in Fig. 7.1. Your test bench should drive in each input code for the vector ABCD in the order they would appear in a truth table (i.e., "0000," "0001," "0010," . . .). Your test bench should change the input pattern every 10 ns using a *wait* statement. Use the report and assert statements to output a message on the status of each test to the simulation transcript window. For each input vector, create a message that indicates the current input vector being tested, the resulting output of your DUT, and whether the DUT output is correct.

7.3.2 Design a VHDL test bench to verify the functional operation of the system in Fig. 7.2. Your test bench should drive in each input code for the vector ABCD in the order they would appear in a truth table (i.e., "0000," "0001," "0010," . . .). Your test bench should change the input pattern every 10 ns using a *wait* statement. Use the report and assert statements to output a message on the status of each test to the simulation transcript window. For each input vector, create a message that indicates the current input vector being tested, the resulting output of your DUT, and whether the DUT output is correct.

7.3.3 Design a VHDL test bench to verify the functional operation of the system in Fig. 7.3. Your test bench should drive in each input code for the vector ABCD in the order they would appear in a truth table (i.e., "0000," "0001," "0010," . . .). Your test bench should change the input pattern every 10 ns using a *wait* statement. Use the report and assert statements to output a message on the status of each test to the simulation transcript window. For each input vector, create a message that indicates the current input vector being tested, the resulting output of your DUT, and whether the DUT output is correct.

7.3.4 Design a VHDL test bench to verify the functional operation of the system in Fig. 7.4. Your test bench should drive in each input code for the vector ABCD in the order they would appear in a truth table (i.e., "0000," "0001," "0010," . . .). Your test bench should change the input pattern every 10 ns using a *wait* statement. Use the report and assert statements to output a message on the status of each test to the simulation transcript window. For each input vector, create a message that indicates the current input vector being tested, the resulting output of your DUT, and whether the DUT output is correct.

Section 7.4: Using External I/O in Test Benches

7.4.1 Design a VHDL test bench to verify the functional operation of the system in Fig. 7.1. Your test bench should drive in each input code for the vector ABCD in the order they would appear in a truth table (i.e., "0000," "0001," "0010," . . .). Your test bench should use a process and individual signal assignments for each pattern. Your test bench should change the input pattern every 10 ns using the *wait for* statement within your stimulus process. Write the output results to an external file called "output_vectors.txt" using the TEXTIO and STD_LOGIC_TEXTIO packages.

7.4.2 Design a VHDL test bench to verify the functional operation of the system in Fig. 7.2. Your test bench should drive in each input code for the vector ABCD in the order they would appear in a truth table (i.e., "0000," "0001," "0010," . . .). Your test bench should use a process and individual signal assignments for each pattern. Your test bench should change the input pattern every 10 ns using the *wait for* statement within your stimulus process. Write the output results to the STD_OUTPUT of the simulator using the TEXTIO and STD_LOGIC_TEXTIO packages.

7.4.3 Design a VHDL test bench to verify the functional operation of the system in Fig. 7.3. Create an input text file called "input_vectors.txt" that contains each input code for the vector ABCD in the order they would appear in a truth table (i.e., "0000," "0001," "0010," . . .) on a separate line. Use the TEXTIO and STD_LOGIC_TEXTIO packages to read in each line of the file individually and use the corresponding input vector to drive the DUT. Write the output results to an external file called "output_vectors.txt."

7.4.4 Design a VHDL test bench to verify the functional operation of the system in Fig. 7.4. Create an input text file called "input_vectors.txt" that contains each input code for the vector ABCD in the order they would appear in a truth table (i.e., "0000," "0001," "0010," . . .) on a separate line. Use the TEXTIO and STD_LOGIC_TEXTIO packages to read in each line of the file individually and use the corresponding input vector to drive the DUT. Write the output results to the STD_OUTPUT of the simulator.

Chapter 8: Modeling Sequential Storage and Registers

In this chapter, we will look at modeling sequential storage devices. We begin by looking at modeling scalar storage devices such as D-latches and D-flip-flops and then move into multiple-bit storage models known as registers.

Learning Outcomes—After completing this chapter, you will be able to:

8.1 Design a VHDL model for a single-bit sequential logic storage device.
8.2 Design a VHDL model for a register.

8.1 Modeling Scalar Storage Devices

8.1.1 D-Latch

Let's begin with the model of a simple D-latch. Since the outputs of this sequential storage device are not updated continuously, its behavior is modeled using a process. Since we want to create a synthesizable model, we use a sensitivity list to trigger the process instead of wait statements. In the sensitivity list, we need to include the C input since it controls when the D-latch is in track or store mode. We also need to include the D input in the sensitivity list because during the track mode, the output Q will be continuously assigned the value of D, so any change on D needs to trigger the process. The use of an if/then statement is used to model the behavior during track mode (C = 1). Since the behavior is not explicitly stated for when C = 0, the outputs will hold their last value, which allows us to simply end the if/then statement to complete the model. Example 8.1 shows the behavioral model for a D-latch.

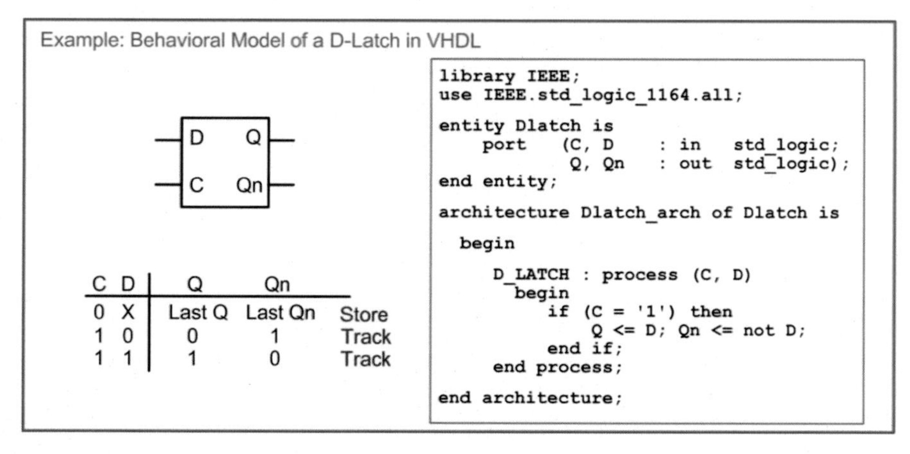

C	D	Q	Qn	
0	X	Last Q	Last Qn	Store
1	0	0	1	Track
1	1	1	0	Track

```
library IEEE;
use IEEE.std_logic_1164.all;

entity Dlatch is
    port    (C, D     : in   std_logic;
             Q, Qn    : out  std_logic);
end entity;

architecture Dlatch_arch of Dlatch is

  begin

    D_LATCH : process (C, D)
      begin
        if (C = '1') then
           Q <= D; Qn <= not D;
        end if;
    end process;

end architecture;
```

Example 8.1
Behavioral model of a D-latch in VHDL

B. J. LaMeres, *Quick Start Guide to VHDL*, https://doi.org/10.1007/978-3-031-42543-1_8

8.1.2 D-Flip-Flop

The rising edge behavior of a D-flip-flop is modeled using a (Clock'event and Clock = '1') Boolean condition within a process. The (rising_edge(Clock)) function can also be used for type std_logic. Example 8.2 shows the behavioral model for a rising edge-triggered D-flip-flop with both Q and Qn outputs.

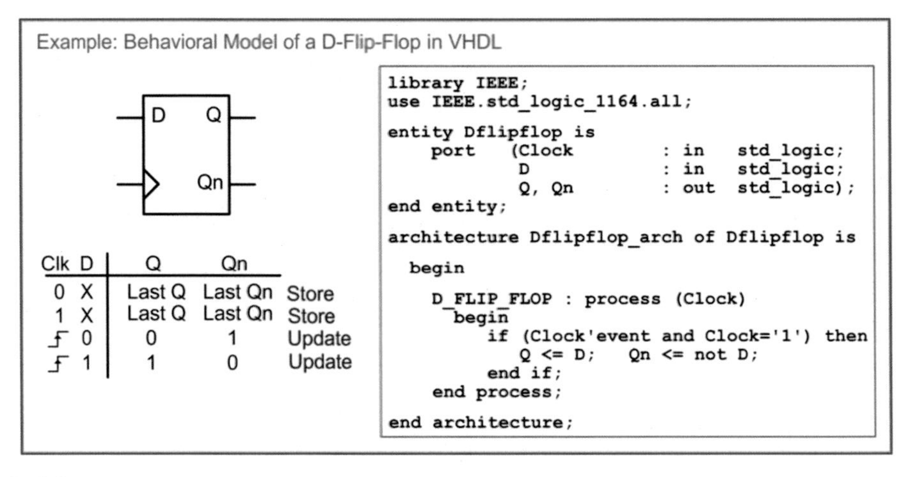

Example 8.2
Behavioral model of a D-flip-flop in VHDL

8.1.3 D-Flip-Flop with Asynchronous Resets

D-flip-flops typically have a reset line in order to initialize their outputs to a known state (e.g., Q = 0, Qn = 1). Resets are asynchronous, meaning that whenever they are asserted, assignments to the outputs take place immediately. If a reset was *synchronous*, the output assignments would only take place on the next rising edge of the clock. This behavior is undesirable because if there is a system failure, there is no guarantee that a clock edge will ever occur. Thus, the reset may never take place. Asynchronous resets are more desirable not only to put the D-flip-flops into a known state at startup but also to recover from a system failure that may have impacted the clock signal. In order to model this asynchronous behavior, the reset signal is placed in the sensitivity list. This allows both the clock and the reset inputs to trigger the process. Within the process, an if/then/elsif statement is used to determine whether the reset has been asserted or a rising edge of the clock has occurred. The if/then/elsif statement first checks whether the reset input has been asserted. If it has, it makes the appropriate assignments to the outputs (Q = 0, Qn = 1). If the reset has not been asserted, the *elsif* clause checks whether a rising edge of the clock has occurred using the (Clock'event and Clock = '1') Boolean condition. If it has, the outputs are updated accordingly (Q <= D, Qn <= not D). A final else statement is not included so that assignments to the outputs are not made under any other condition. This models the store behavior of the D-flip-flop. Example 8.3 shows the behavioral model for a rising edge-triggered D-flip-flop with an asynchronous, active LOW reset.

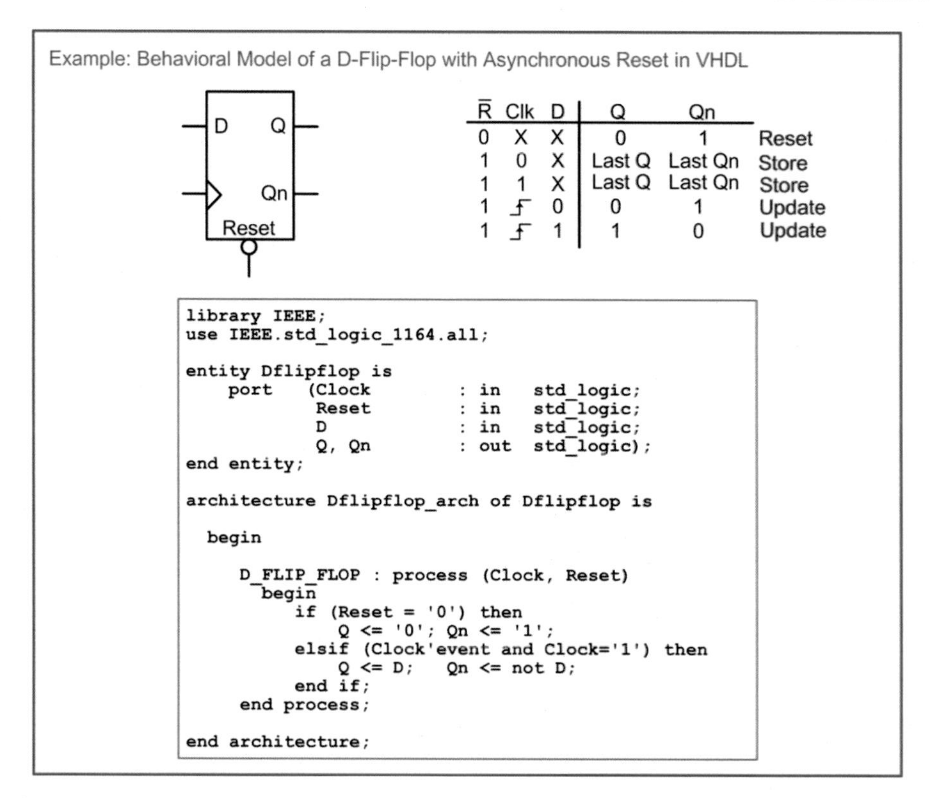

R̅	Clk	D	Q	Qn	
0	X	X	0	1	Reset
1	0	X	Last Q	Last Qn	Store
1	1	X	Last Q	Last Qn	Store
1	↑	0	0	1	Update
1	↑	1	1	0	Update

```
library IEEE;
use IEEE.std_logic_1164.all;

entity Dflipflop is
    port    (Clock        : in   std_logic;
             Reset        : in   std_logic;
             D            : in   std_logic;
             Q, Qn        : out  std_logic);
end entity;

architecture Dflipflop_arch of Dflipflop is

  begin

     D_FLIP_FLOP : process (Clock, Reset)
       begin
          if (Reset = '0') then
              Q <= '0'; Qn <= '1';
          elsif (Clock'event and Clock='1') then
              Q <= D;   Qn <= not D;
          end if;
     end process;

end architecture;
```

Example 8.3
Behavioral model of a D-flip-flop with asynchronous reset in VHDL

8.1.4 D-Flip-Flop with Asynchronous Reset and Preset

A D-flip-flop with both an asynchronous reset and asynchronous preset is handled in a similar manner as the D-flip-flop in the prior section. The preset input is included in the sensitivity list in order to trigger the process whenever a transition occurs on either the clock, reset, or preset inputs. An if/then/elsif statement is used to first check whether a reset has occurred; then whether a preset has occurred; and finally, whether a rising edge of the clock has occurred. Example 8.4 shows the model for a rising edge-triggered D-flip-flop with asynchronous, active LOW reset and preset.

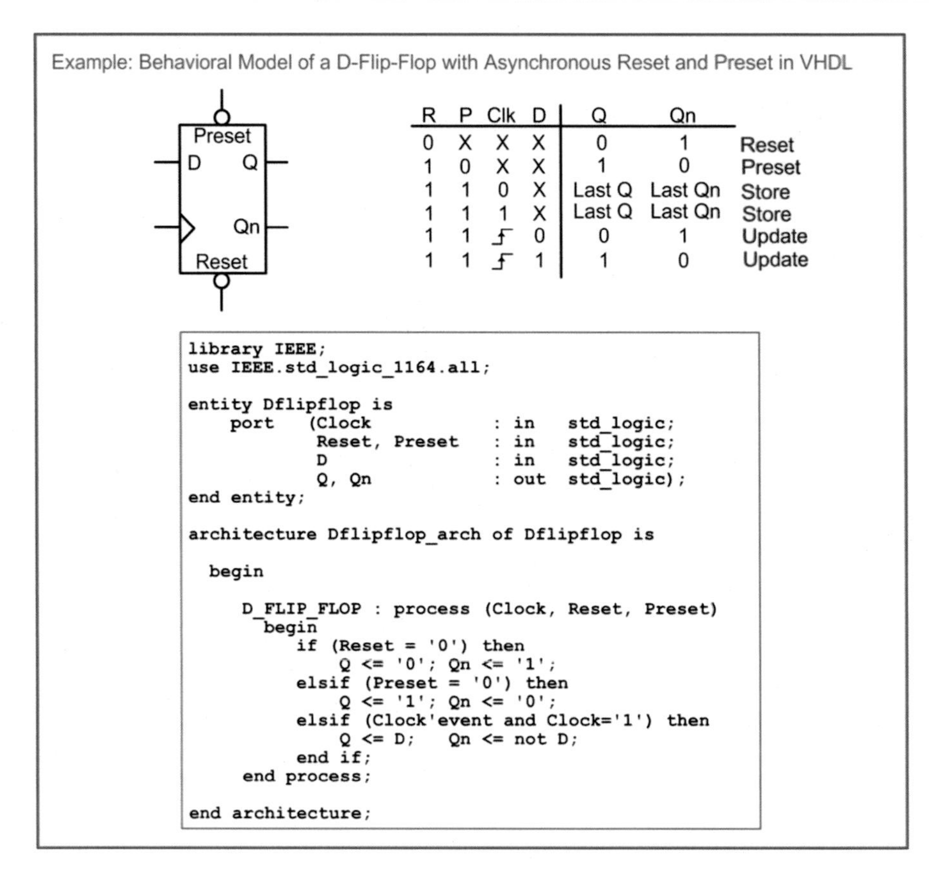

R	P	Clk	D	Q	Qn	
0	X	X	X	0	1	Reset
1	0	X	X	1	0	Preset
1	1	0	X	Last Q	Last Qn	Store
1	1	1	X	Last Q	Last Qn	Store
1	1	↑	0	0	1	Update
1	1	↑	1	1	0	Update

```
library IEEE;
use IEEE.std_logic_1164.all;

entity Dflipflop is
    port   (Clock             : in   std_logic;
            Reset, Preset     : in   std_logic;
            D                 : in   std_logic;
            Q, Qn             : out  std_logic);
end entity;

architecture Dflipflop_arch of Dflipflop is

  begin

     D_FLIP_FLOP : process (Clock, Reset, Preset)
       begin
          if (Reset = '0') then
              Q <= '0'; Qn <= '1';
          elsif (Preset = '0') then
              Q <= '1'; Qn <= '0';
          elsif (Clock'event and Clock='1') then
              Q <= D;    Qn <= not D;
          end if;
     end process;

end architecture;
```

Example 8.4
Behavioral model of a D-flip-flop with asynchronous reset and preset in VHDL

8.1.5 D-Flip-Flop with Synchronous Enable

An enable input is also a common feature of modern D-flip-flops. Enable inputs are synchronous, meaning that when they are asserted, action is only taken on the rising edge of the clock. This means that the enable input is not included in the sensitivity list of the process. Since action is only taken when there is a rising edge of the clock, a nested if/then statement is included beneath the *elsif (Clock'event and Clock = '1')* clause. Example 8.5 shows the model for a D-flip-flop with a synchronous enable (EN) input. When EN = 1, the D-flip-flop is enabled, and assignments are made to the outputs only on the rising edge of the clock. When EN = 0, the D-flip-flop is disabled, and assignments to the outputs are not made. When disabled, the D-flip-flop effectively ignores rising edges on the clock, and the outputs remain at their last values.

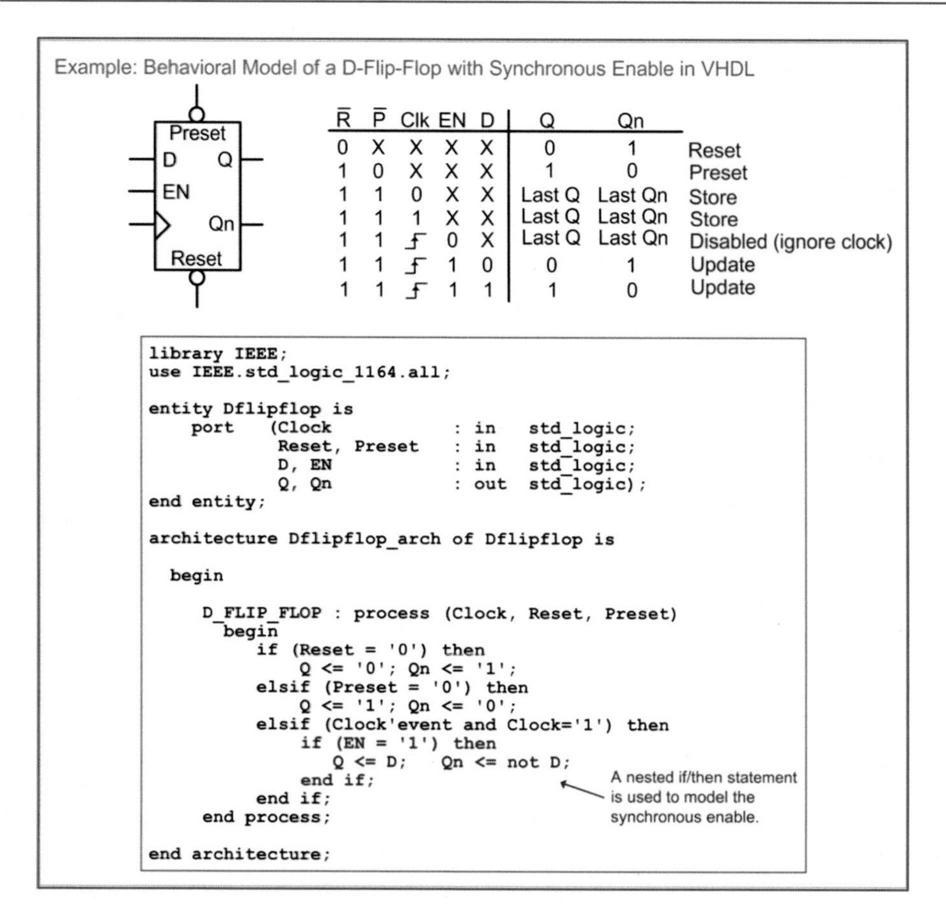

Example: Behavioral Model of a D-Flip-Flop with Synchronous Enable in VHDL

$\overline{R}$	$\overline{P}$	Clk	EN	D	Q	Qn	
0	X	X	X	X	0	1	Reset
1	0	X	X	X	1	0	Preset
1	1	0	X	X	Last Q	Last Qn	Store
1	1	1	X	X	Last Q	Last Qn	Store
1	1	↑	0	X	Last Q	Last Qn	Disabled (ignore clock)
1	1	↑	1	0	0	1	Update
1	1	↑	1	1	1	0	Update

```
library IEEE;
use IEEE.std_logic_1164.all;

entity Dflipflop is
    port   (Clock           : in   std_logic;
            Reset, Preset   : in   std_logic;
            D, EN           : in   std_logic;
            Q, Qn           : out  std_logic);
end entity;

architecture Dflipflop_arch of Dflipflop is

  begin

     D_FLIP_FLOP : process (Clock, Reset, Preset)
       begin
          if (Reset = '0') then
              Q <= '0'; Qn <= '1';
          elsif (Preset = '0') then
              Q <= '1'; Qn <= '0';
          elsif (Clock'event and Clock='1') then
              if (EN = '1') then
                 Q <= D;    Qn <= not D;
              end if;
          end if;
     end process;

end architecture;
```


Example 8.5
Behavioral model of a D-flip-flop with synchronous enable in VHDL

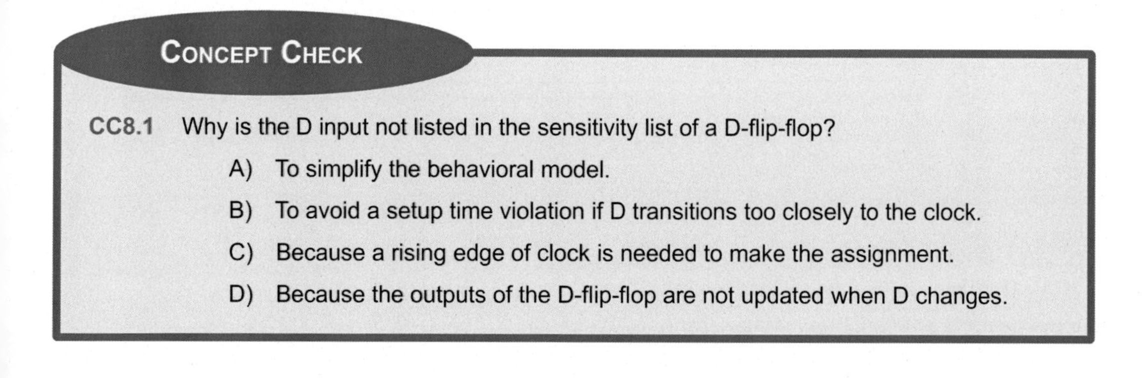

Concept Check

CC8.1 Why is the D input not listed in the sensitivity list of a D-flip-flop?

A) To simplify the behavioral model.

B) To avoid a setup time violation if D transitions too closely to the clock.

C) Because a rising edge of clock is needed to make the assignment.

D) Because the outputs of the D-flip-flop are not updated when D changes.

8.2 Modeling Registers

8.2.1 Registers with Enables

The term *register* describes a circuit that operates in a similar manner as a D-flip-flop with the exception that the input and output data are vectors. This circuit is implemented with a set of D-flip-flops all connected to the same clock, reset, and enable inputs. A register is a higher level of abstraction that allows vector data to be stored without getting into the details of the lower-level implementation of the D-flip-flop components. Register transfer level (RTL) modeling refers to a level of design abstraction in which vector data is moved and operated on in a synchronous manner. This design methodology is widely used in data path modeling and computer system design. Example 8.6 shows an RTL model of an 8-bit, synchronous register. This circuit has an active low, asynchronous reset that will cause the 8-bit output *Reg_Out* to go to 0 when it is asserted. When the reset is not asserted, the output will be updated with the 8-bit input *Reg_In* if the system is enabled (EN = 1) and there is a rising edge on the clock. If the register is disabled (EN = 0), the input clock is ignored. At all other times, the output holds its last value.

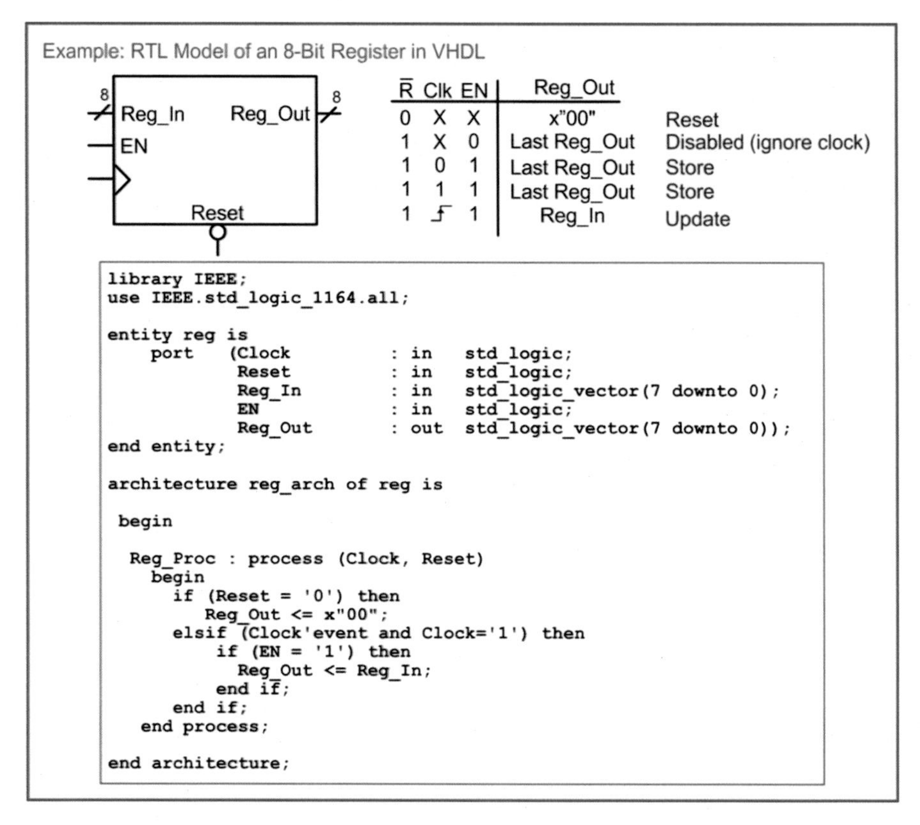

Example 8.6
RTL model of an 8-bit register in VHDL

8.2.2 Shift Registers

A shift register is a circuit which consists of multiple registers connected in series. Data is shifted from one register to another on the rising edge of the clock. This type of circuit is often used in serial-to-parallel data converters. Example 8.7 shows an RTL model for a 4-stage, 8-bit shift register. In the simulation waveform, the data is shown in hexadecimal format.

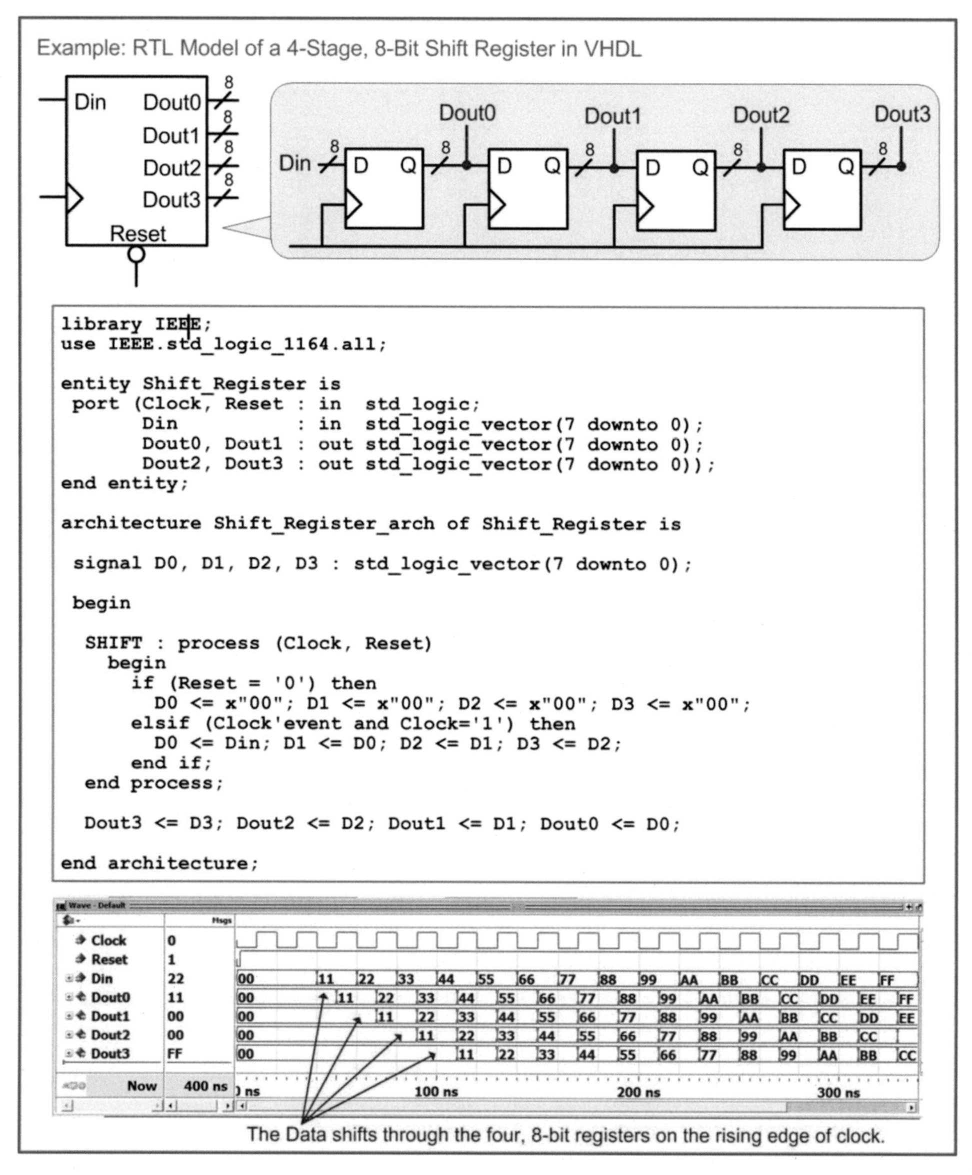


```
library IEEE;
use IEEE.std_logic_1164.all;

entity Shift_Register is
 port (Clock, Reset : in  std_logic;
       Din          : in  std_logic_vector(7 downto 0);
       Dout0, Dout1 : out std_logic_vector(7 downto 0);
       Dout2, Dout3 : out std_logic_vector(7 downto 0));
end entity;

architecture Shift_Register_arch of Shift_Register is

 signal D0, D1, D2, D3 : std_logic_vector(7 downto 0);

 begin

  SHIFT : process (Clock, Reset)
    begin
      if (Reset = '0') then
        D0 <= x"00"; D1 <= x"00"; D2 <= x"00"; D3 <= x"00";
      elsif (Clock'event and Clock='1') then
        D0 <= Din; D1 <= D0; D2 <= D1; D3 <= D2;
      end if;
  end process;

  Dout3 <= D3; Dout2 <= D2; Dout1 <= D1; Dout0 <= D0;

end architecture;
```

Example 8.7
RTL model of a 4-stage, 8-bit shift register in VHDL

8.2.3 Registers as Agents on a Data Bus

One of the powerful topologies that registers can easily model is a multi-drop bus. In this topology, multiple registers are connected to a data bus as receivers or agents. Each agent has an enable line that controls when it latches information from the data bus into its storage elements. This topology is synchronous, meaning that each agent and the driver of the data bus are connected to the same clock signal. Each agent has a dedicated, synchronous enable line that is provided by a system controller elsewhere in the design. Example 8.8 shows this multi-drop bus topology. In this example system, three registers (A, B, and C) are connected to a data bus as receivers. Each register is connected to the same clock and reset signals. Each register has its own dedicated enable line (A_EN, B_EN, and C_EN).

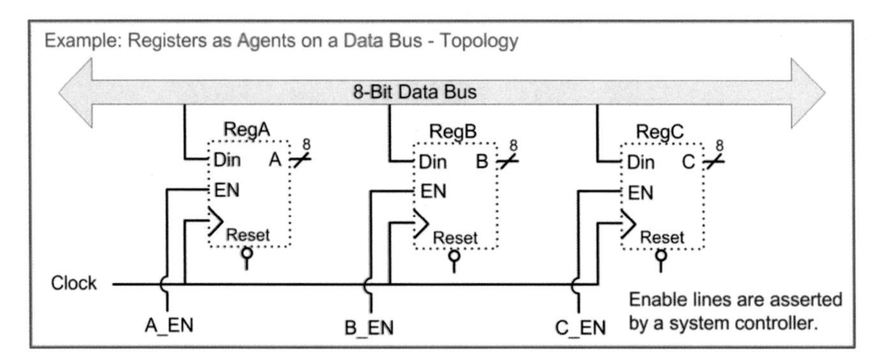

Example 8.8
Registers as agents on a data bus – system topology

This topology can be modeled using RTL abstraction by treating each register as a separate process. Example 8.9 shows how to describe this topology with an RTL model in VHDL. Notice that the three processes modeling the A, B, and C registers are nearly identical to each other with the exception of the signal names they use.

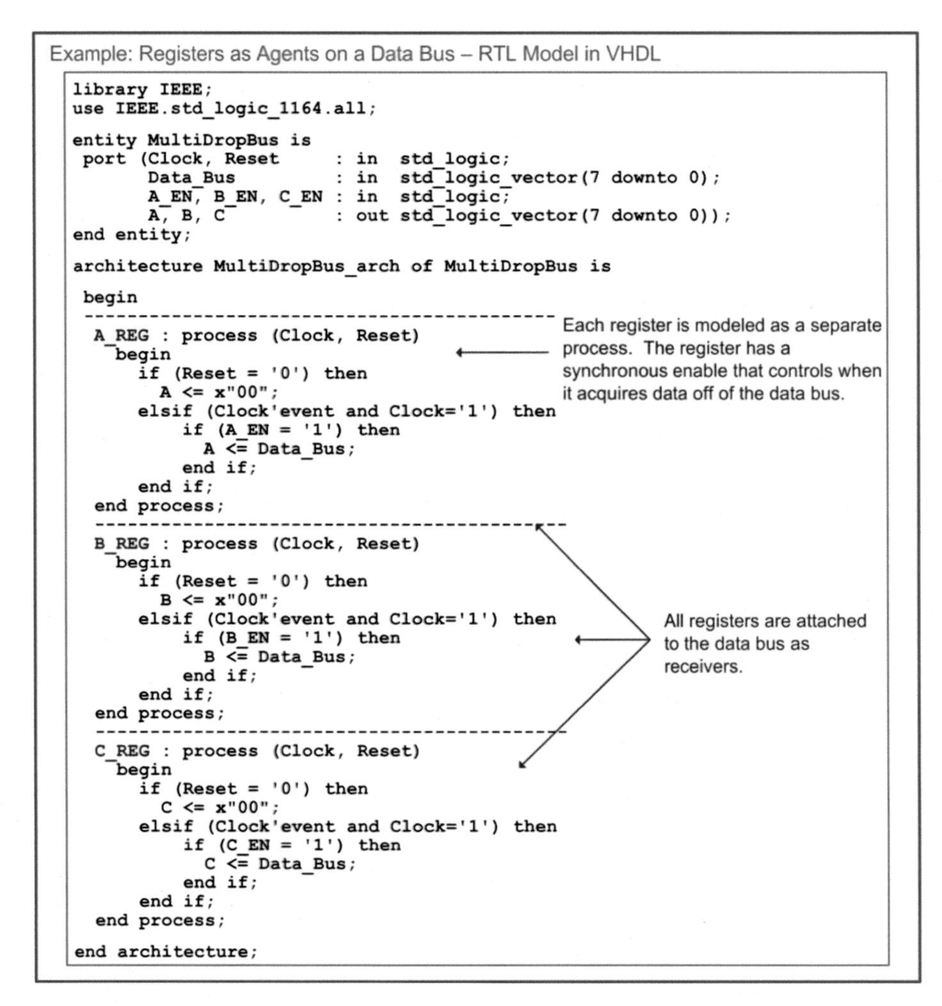
Example: Registers as Agents on a Data Bus – RTL Model in VHDL

```
library IEEE;
use IEEE.std_logic_1164.all;

entity MultiDropBus is
 port (Clock, Reset      : in  std_logic;
       Data_Bus          : in  std_logic_vector(7 downto 0);
       A_EN, B_EN, C_EN  : in  std_logic;
       A, B, C           : out std_logic_vector(7 downto 0));
end entity;

architecture MultiDropBus_arch of MultiDropBus is

 begin
 ------------------------------------------
  A_REG : process (Clock, Reset)
    begin
      if (Reset = '0') then
        A <= x"00";
      elsif (Clock'event and Clock='1') then
          if (A_EN = '1') then
            A <= Data_Bus;
          end if;
      end if;
  end process;
  ------------------------------------------
  B_REG : process (Clock, Reset)
    begin
      if (Reset = '0') then
        B <= x"00";
      elsif (Clock'event and Clock='1') then
          if (B_EN = '1') then
            B <= Data_Bus;
          end if;
      end if;
  end process;
  ------------------------------------------
  C_REG : process (Clock, Reset)
    begin
      if (Reset = '0') then
        C <= x"00";
      elsif (Clock'event and Clock='1') then
          if (C_EN = '1') then
            C <= Data_Bus;
          end if;
      end if;
  end process;

end architecture;
```


Example 8.9
Registers as agents on a data bus – RTL model in VHDL

Example 8.10 shows the resulting simulation waveform for this system. Each register is updated with the value on the data bus whenever its dedicated enable line is asserted.

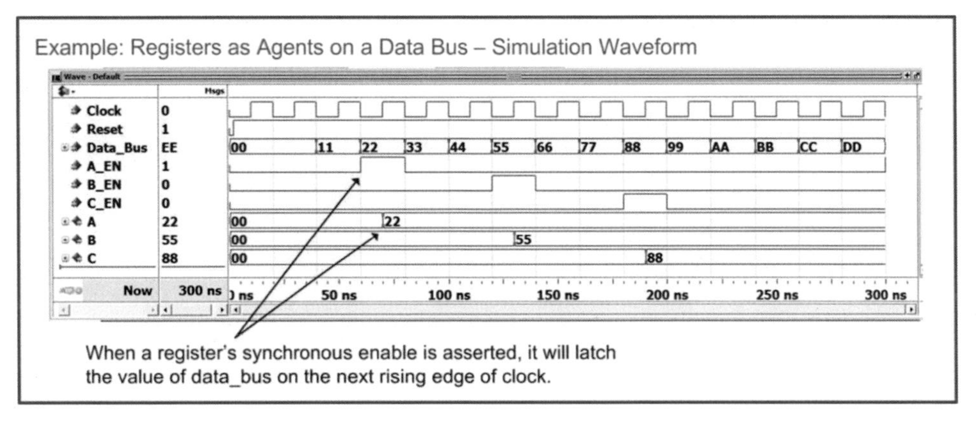

Example 8.10
Registers as agents on a data bus – simulation waveform

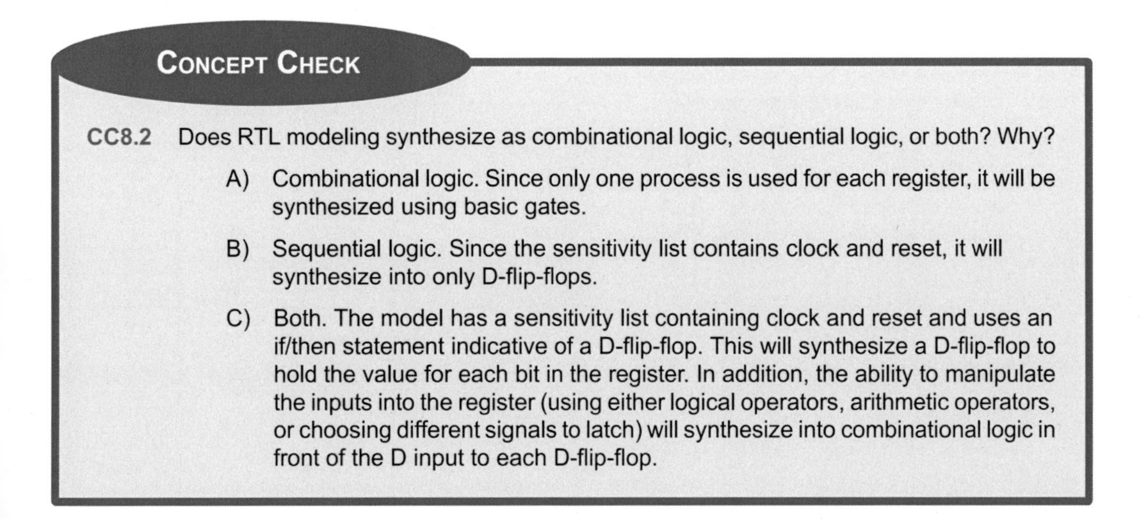

CONCEPT CHECK

CC8.2 Does RTL modeling synthesize as combinational logic, sequential logic, or both? Why?

A) Combinational logic. Since only one process is used for each register, it will be synthesized using basic gates.

B) Sequential logic. Since the sensitivity list contains clock and reset, it will synthesize into only D-flip-flops.

C) Both. The model has a sensitivity list containing clock and reset and uses an if/then statement indicative of a D-flip-flop. This will synthesize a D-flip-flop to hold the value for each bit in the register. In addition, the ability to manipulate the inputs into the register (using either logical operators, arithmetic operators, or choosing different signals to latch) will synthesize into combinational logic in front of the D input to each D-flip-flop.

Summary

- A synchronous system is modeled with a process and a sensitivity list. The clock and reset signals are always listed by themselves in the sensitivity list. Within the process is an if/then statement. The first clause of the if/then statement handles the asynchronous reset condition, while the second *elsif* clause handles the synchronous signal assignments.
- Edge sensitivity is modeled within a process using either the (*clock'event and clock = "1"*) syntax or an edge detection function provided by the STD_LOGIC_1164 package (i.e., *rising_edge()*).
- Most D-flip-flops and registers contain a synchronous *enable* line. This is modeled using a nested if/then statement within the main process if/then statement. The nested if/then goes beneath the clause for the synchronous signal assignments.
- Registers are modeled in VHDL in a similar manner to a D-flip-flop with a synchronous enable. The only difference is that the inputs and outputs are n-bit vectors.

Exercise Problems

Section 8.1: Modeling Scalar Storage Devices

8.1.1 How does a VHDL model for a D-flip-flop handle treating reset as the highest priority input?

8.1.2 For a VHDL model of a D-flip-flop with a synchronous enable (EN), why isn't EN listed in the sensitivity list?

8.1.3 For a VHDL model of a D-flip-flop with a synchronous enable (EN), what is the impact of listing EN in the sensitivity list?

8.1.4 For a VHDL model of a D-flip-flop with a synchronous enable (EN), why is the behavior of the enable modeled using a nested if/then statement under the clock edge clause rather than an additional elsif clause in the primary if/then statement?

Section 8.2: Modeling Registers

8.2.1 In *register transfer level* modeling, how does the width of the register relate to the number of D-flip-flops that will be synthesized?

8.2.2 In *register transfer level* modeling, how is the synchronous data movement managed if all registers are using the same clock?

8.2.3 Design a VHDL RTL model of a 32-bit, synchronous register. The block diagram for the entity definition is shown in Fig. 8.1. The register has a synchronous enable. The register should be modeled using a single process.

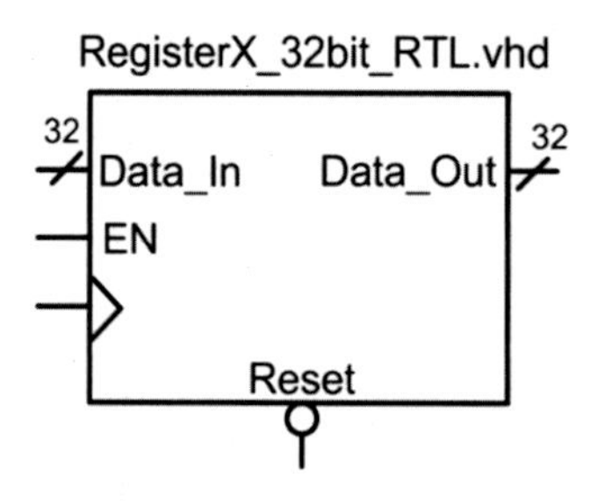

Fig. 8.1
32-bit register block diagram

8.2.4 Design a VHDL RTL model of an 8-stage, 16-bit shift register. The block diagram for the entity definition is shown in Fig. 8.2. Each stage of the shift register will be provided as an output of the system (A, B, C, D, E, F, G, and H). Use std_logic or std_logic_vector for all ports.

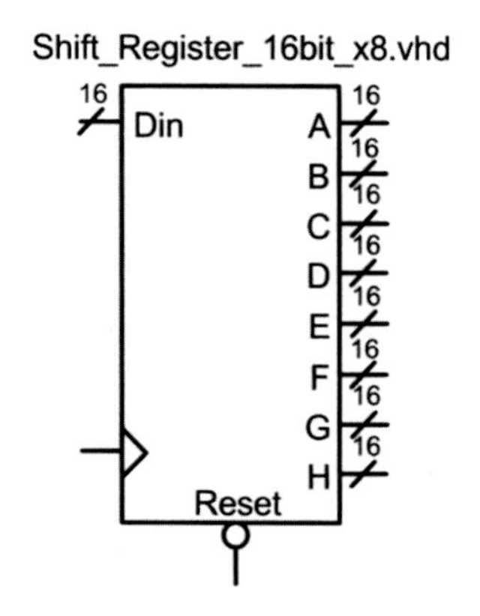

Fig. 8.2
16-bit shift register block diagram

8.2.5 Design a VHDL RTL model of the multi-drop bus topology in Fig. 8.3. Each of the 16-bit registers (RegA, RegB, RegC, and RegD) will latch the contents of the 16-bit data bus if their enable line is asserted. Each register should be modeled using an individual process.

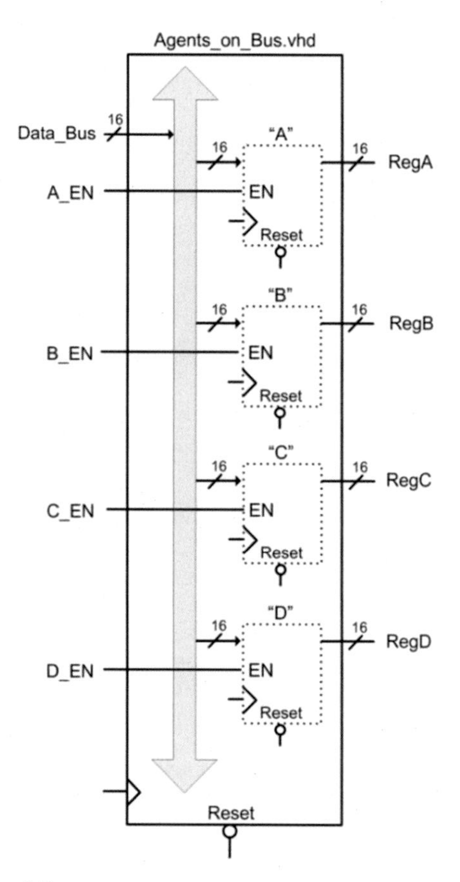

Fig. 8.3
Agents on a bus block diagram

Chapter 9: Modeling Finite-State Machines

In this chapter, we will look at modeling finite-state machines (FSMs). A FSM is one of the most powerful circuits in a digital system because it can make decisions about the next output based on both the current and past inputs. Finite-state machines are modeled using the constructs already covered in this book. In this chapter, we will look at the widely accepted three-process model for designing a FSM.

Learning Outcomes—After completing this chapter, you will be able to:

9.1 Describe the three-process modeling approach for FSM design.
9.2 Design a VHDL model for a FSM from a state diagram.

9.1 The FSM Design Process and a Push-Button Window Controller Example

The most common modeling practice for FSMs is to create a new *user-defined* type that can take on the descriptive state names from the state diagram. Two signals are then created of this type, *current_state* and *next_state*. Once these signals are created, all of the functional blocks in the state machine can use the descriptive state names in their conditional signal assignments. The synthesizer will automatically assign the state codes based on the most effective use of the target technology (e.g., binary, gray code, one-hot). Within the VHDL state machine model, three processes are used to describe each of the functional blocks, *state memory*, *next-state logic*, and *output logic*. In order to examine how to model a finite-state machine using this approach, let's use a push-button window controller example. Example 9.1 gives the overview of the design objectives for this example and the state diagram describing the behavior to be modeled in VHDL.

B. J. LaMeres, *Quick Start Guide to VHDL*, https://doi.org/10.1007/978-3-031-42543-1_9

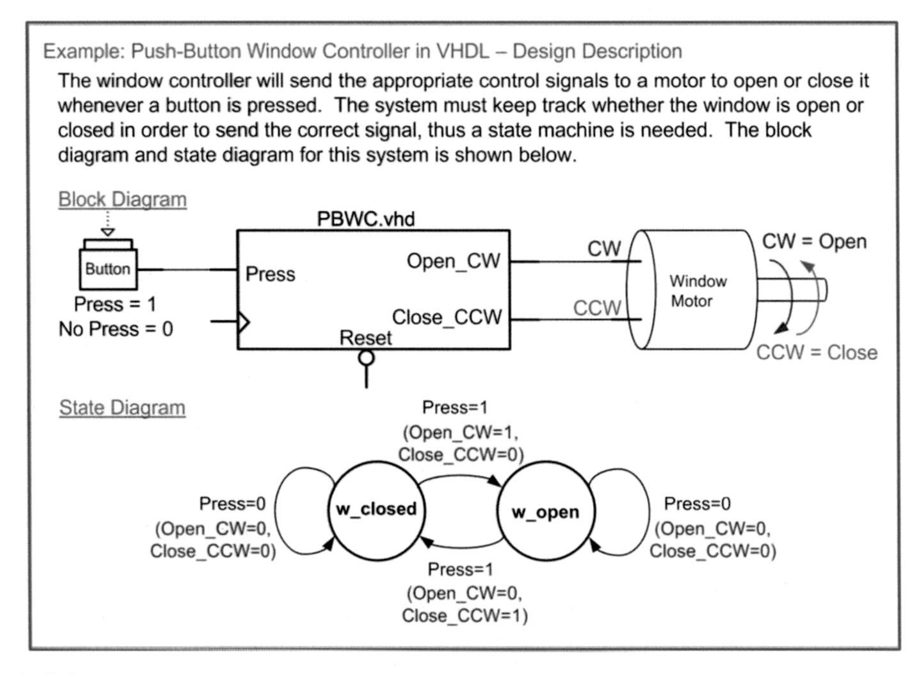

Example 9.1
Push-button window controller in VHDL – design description

Let's begin by defining the entity. The system has an input called *Press* and two outputs called *Open_CW* and *Close_CCW*. The system also has clock and reset inputs. We will design the system to update on the rising edge of the clock and have an asynchronous, active LOW, reset. Example 9.2 shows the VHDL entity definition for this example.

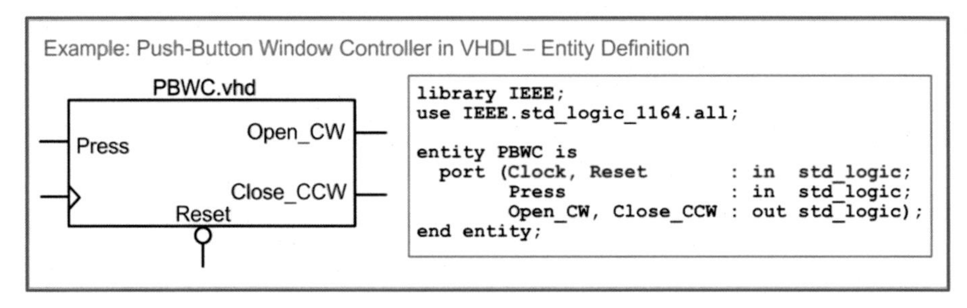

Example 9.2
Push-button window controller in VHDL – entity definition

9.1.1 Modeling the States with User-Defined, Enumerated Data Types

Now we begin designing the finite-state machine in VHDL using behavioral modeling constructs. The first step is to create a new user-defined, enumerated data type that can take on values that match the descriptive state names we've chosen in the state diagram (i.e., w_closed and w_open). This is accomplished by declaring a new type before the begin statement in the architecture with the keyword *type*. For this example, we will create a new type called *State_Type* and explicitly enumerate the values that it can take on. This type can now be used in future signal declarations. We then create two new signals called *current_state* and *next_state* of type State_Type. These two signals will be used throughout the VHDL model in order to provide a high-level, readable description of the FSM behavior. The following syntax shows how to declare the new type and declare the current_state and next_state signals:

```
type  State_Type is (w_closed, w_open);
signal current_state, next_state : State_Type;
```

9.1.2 The State Memory Process

Now we model the state memory of the FSM using a process. This process models the behavior of the D-flip-flops in the FSM that are holding the current state on their Q outputs. Each time there is a rising edge of the clock, the current state is updated with the next-state value present on the D inputs of the D-flip-flops. This process must also model the reset condition. For this example, we will have the state machine go to the *w_closed* state when Reset is asserted. At all other times, the process will simply update current_state with next_state on every rising edge of the clock. The process model is very similar to the model of a D-flip-flop. This is as expected since this process will synthesize into one or more D-flip-flops to hold the current state. The sensitivity list contains only Clock and Reset, and assignments are only made to the signal current_state. The following syntax shows how to model the state memory of this FSM example:

```
STATE_MEMORY : process (Clock, Reset)
 begin
   if (Reset = '0') then
      current_state <= w_closed;
   elsif (Clock'event and Clock='1') then
      current_state <= next_state;
   end if;
end process;
```

9.1.3 The Next-State Logic Process

Now we model the next-state logic of the FSM using a second process. Recall that the next-state logic is combinational logic; thus, we need to include all of the input signals that the circuit considers in the next-state calculation in the sensitivity list. The current_state signal will always be included in the sensitivity list of the next-state logic process in addition to any inputs to the system. For this example, the system has one other input called *Press*. This process makes assignments to the next_state signal. It is common to use a case statement to separate out the assignments that occur at each state. At each state within the case statement, an if/then statement is used to model the assignments for different input conditions on Press. The following syntax shows how to model the next-state logic of this FSM example. Notice that we include a *when others* clause to ensure that the state machine has a path back to the reset state in the case of an unexpected fault.

```
NEXT_STATE_LOGIC : process (current_state, Press)
     begin
       case (current_state) is
          when w_closed => if (Press = '1') then
                                next_state <= w_open;
                           else
                                next_state <= w_closed;
                           end if;
          when w_open   => if (Press = '1') then
                                next_state <= w_closed;
                           else
                                next_state <= w_open;
                           end if;
          when others   => next_state <= w_closed;
     end case;
 end process;
```

9.1.4 The Output Logic Process

Now we model the output logic of the FSM using a third process. Recall that output logic is combinational logic; thus, we need to include all of the input signals that this circuit considers in the output assignments. The current_state will always be included in the sensitivity list. If the FSM is a Mealy machine, then the system inputs will also be included in the sensitivity list. If the machine is a Moore machine, then only the current_state will be present in the sensitivity list. For this example, the FSM is a Mealy machine, so the input Press needs to be included in the sensitivity list. Note that this process only makes assignments to the outputs of the machine (Open_CW and Close_CCW). The following syntax shows how to model the output logic of this FSM example. Again, we include a *when others* clause to ensure that the state machine has explicit output behavior in the case of a fault.

```
OUTPUT_LOGIC : process (current_state, Press)
  begin
    case (current_state) is
      when w_closed => if (Press = '1') then
                           Open_CW <= '1'; Close_CCW <= '0';
                       else
                           Open_CW <= '0'; Close_CCW <= '0';
                       end if;

      when w_open   => if (Press = '1') then
                          Open_CW <= '0'; Close_CCW <= '1';
                       else
                          Open_CW <= '0'; Close_CCW <= '0';
                       end if;

      when others   => Open_CW <= '0'; Close_CCW <= '0';
  end case;
end process;
```

Putting this all together in the VHDL architecture yields a functional model for the FSM that can be simulated and synthesized. Once again, it is important to keep in mind that since we did not explicitly assign the state codes, the synthesizer will automatically assign the codes based on the most efficient use of the target technology. Example 9.3 shows the entire architecture for this example.

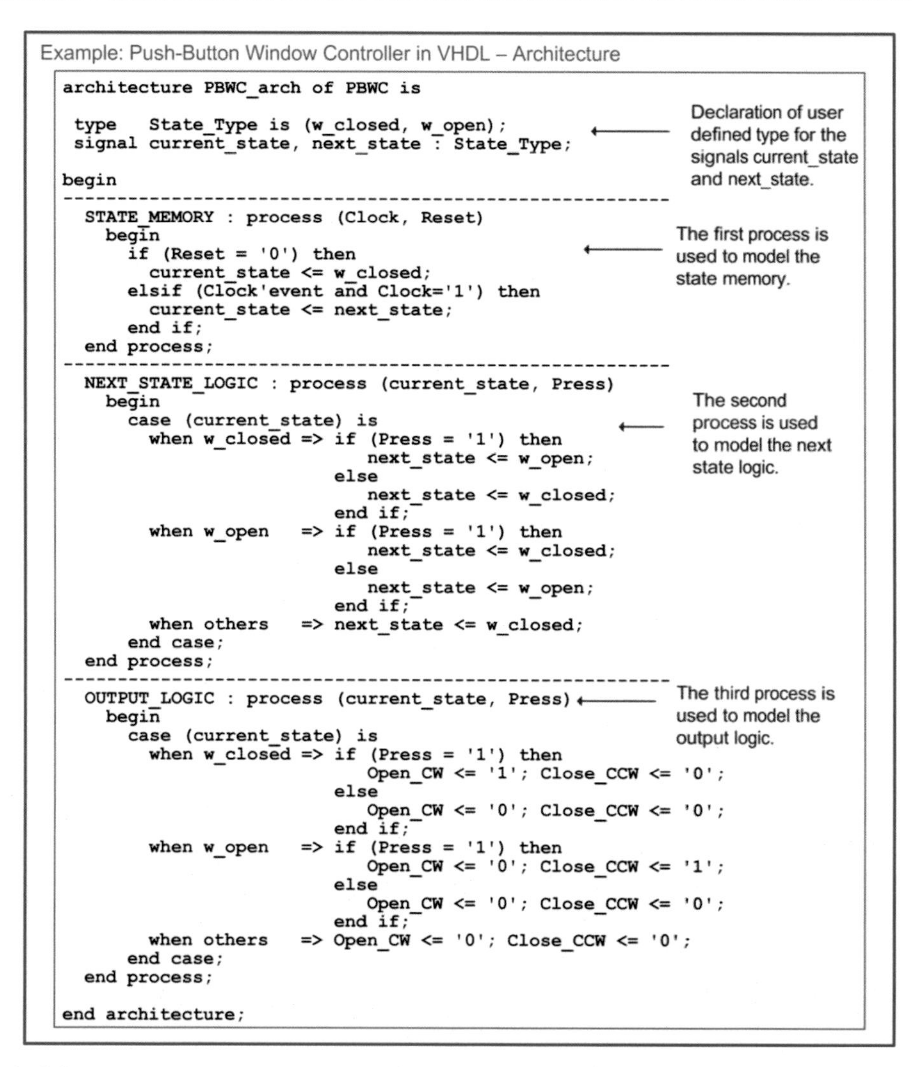
Example: Push-Button Window Controller in VHDL – Architecture

```
architecture PBWC_arch of PBWC is

 type   State_Type is (w_closed, w_open);
 signal current_state, next_state : State_Type;

begin
------------------------------------------------------
  STATE_MEMORY : process (Clock, Reset)
    begin
      if (Reset = '0') then
        current_state <= w_closed;
      elsif (Clock'event and Clock='1') then
        current_state <= next_state;
      end if;
  end process;
------------------------------------------------------
  NEXT_STATE_LOGIC : process (current_state, Press)
    begin
      case (current_state) is
        when w_closed => if (Press = '1') then
                            next_state <= w_open;
                         else
                            next_state <= w_closed;
                         end if;
        when w_open   => if (Press = '1') then
                            next_state <= w_closed;
                         else
                            next_state <= w_open;
                         end if;
        when others   => next_state <= w_closed;
      end case;
  end process;
------------------------------------------------------
  OUTPUT_LOGIC : process (current_state, Press)
    begin
      case (current_state) is
        when w_closed => if (Press = '1') then
                            Open_CW <= '1'; Close_CCW <= '0';
                         else
                            Open_CW <= '0'; Close_CCW <= '0';
                         end if;
        when w_open   => if (Press = '1') then
                            Open_CW <= '0'; Close_CCW <= '1';
                         else
                            Open_CW <= '0'; Close_CCW <= '0';
                         end if;
        when others   => Open_CW <= '0'; Close_CCW <= '0';
      end case;
  end process;

end architecture;
```


Example 9.3
Push-button window controller in VHDL – architecture

Example 9.4 shows the simulation waveform for this state machine. This functional simulation was performed using ModelSim-Altera Starter Edition 10.1d.

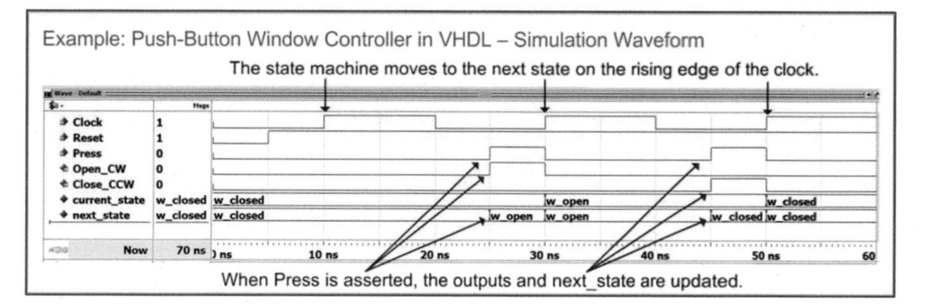

Example 9.4
Push-button window controller in VHDL – simulation waveform

9.1.5 Explicitly Defining State Codes with Subtypes

In the prior example, we did not have control over the state variable encoding. While the previous example is the most common way to model FSMs, there are situations where we would like to assign the state variable codes manually. This is accomplished using a *subtype* and constants. A subtype is simply a constrained type, meaning that it defines a subset of values that an existing type can take on. For example, we could create a subtype to constrain the std_logic data type to only allow values of 0 and 1 and *not* the values of U, X, Z, W, L, H, and -. This would not be considered a new type since it is simply a constraint put upon the existing std_logic type. A subtype defines the constraint and has a unique name that can be used to declare other signals. To use this approach for manually encoding the states of a FSM, we first declare a new subtype called *State_Type* that is simply a version of the existing type std_logic. We then create constants to represent the descriptive state names in the state diagram. These constants are given the type State_Type and a specific value. The value given is the state code we wish to assign to the particular state name. Finally, the current_state and next_state signals are declared of type State_Type. In this way, we can use the same VHDL processes as in the previous example that use the descriptive state names from the state diagram. The following is the VHDL syntax for manually assigning the state codes using subtypes. This syntax would replace the State_Type declaration in the previous example. Example 9.5 shows the resulting simulation waveforms.

```
subtype State_Type is std_logic;
constant w_open   : State_Type := '0';
constant w_closed : State_Type := '1';
signal  current_state, next_state : State_Type;
```

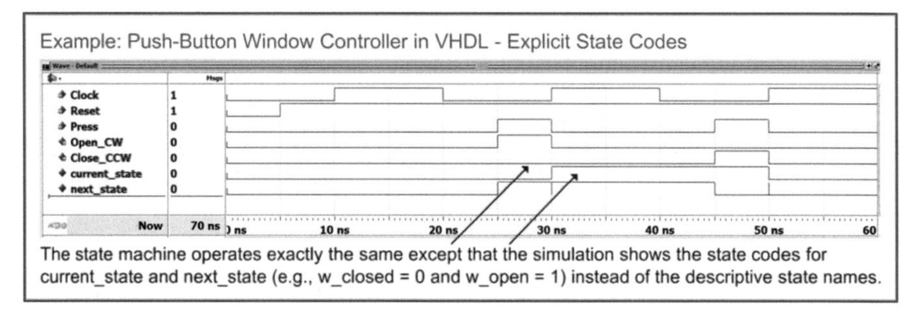

The state machine operates exactly the same except that the simulation shows the state codes for current_state and next_state (e.g., w_closed = 0 and w_open = 1) instead of the descriptive state names.

Example 9.5
Push-button window controller in VHDL – explicit state codes

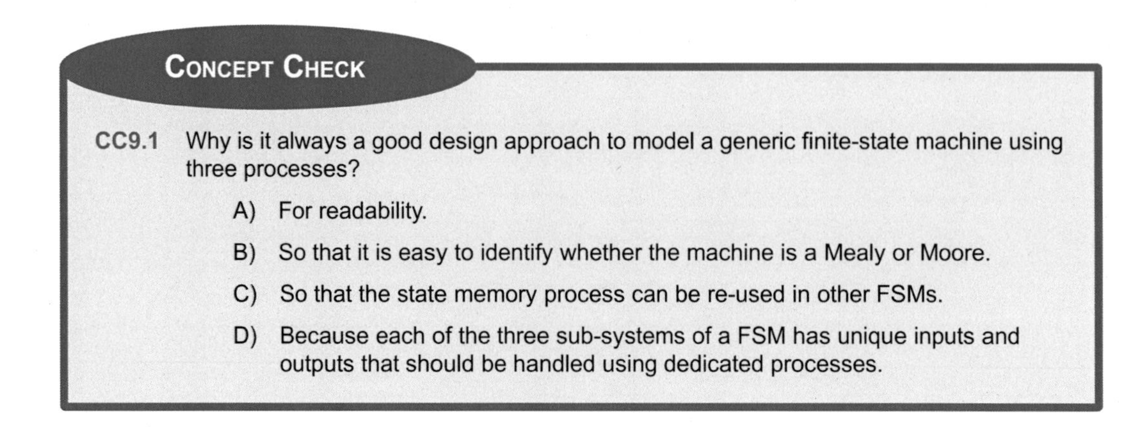

Concept Check

CC9.1 Why is it always a good design approach to model a generic finite-state machine using three processes?

A) For readability.

B) So that it is easy to identify whether the machine is a Mealy or Moore.

C) So that the state memory process can be re-used in other FSMs.

D) Because each of the three sub-systems of a FSM has unique inputs and outputs that should be handled using dedicated processes.

9.2 FSM Design Examples

9.2.1 Serial Bit Sequence Detector in VHDL

Let's look at the design of a serial bit sequence detector finite-state machine using the behavioral modeling constructs of VHDL. Example 9.6 shows the design description and entity definition for this state machine.

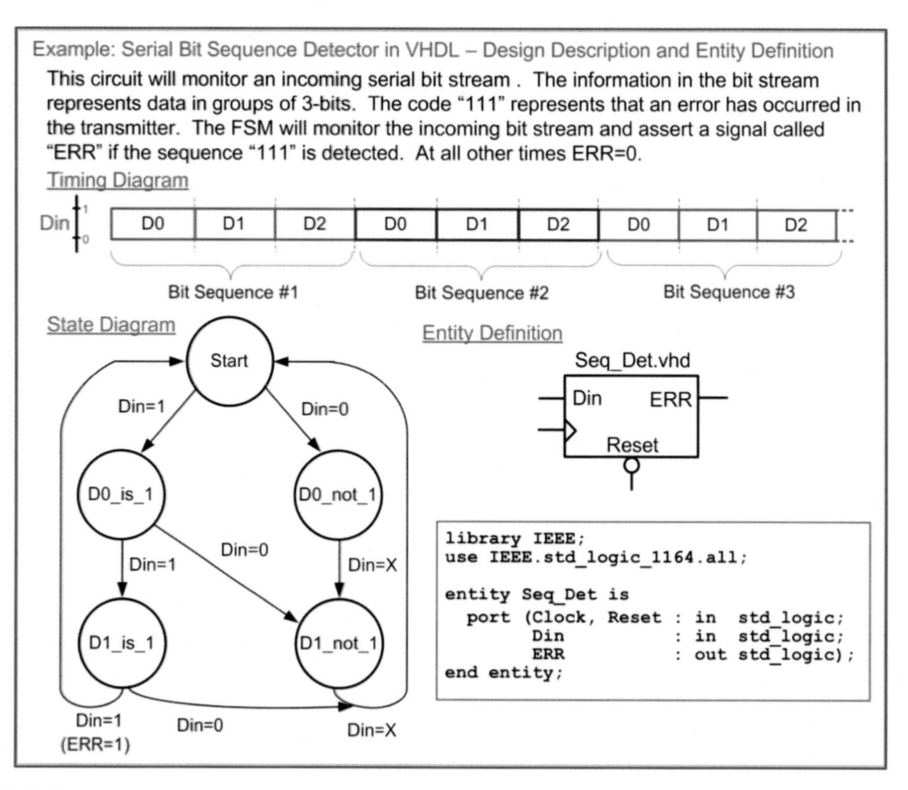


```
library IEEE;
use IEEE.std_logic_1164.all;

entity Seq_Det is
  port (Clock, Reset : in  std_logic;
        Din          : in  std_logic;
        ERR          : out std_logic);
end entity;
```

Example 9.6
Serial bit sequence detector in VHDL – design description and entity definition

Example 9.7 shows the architecture for the serial bit sequence detector. In this example, a user-defined type is created to model the descriptive state names in the state diagram.

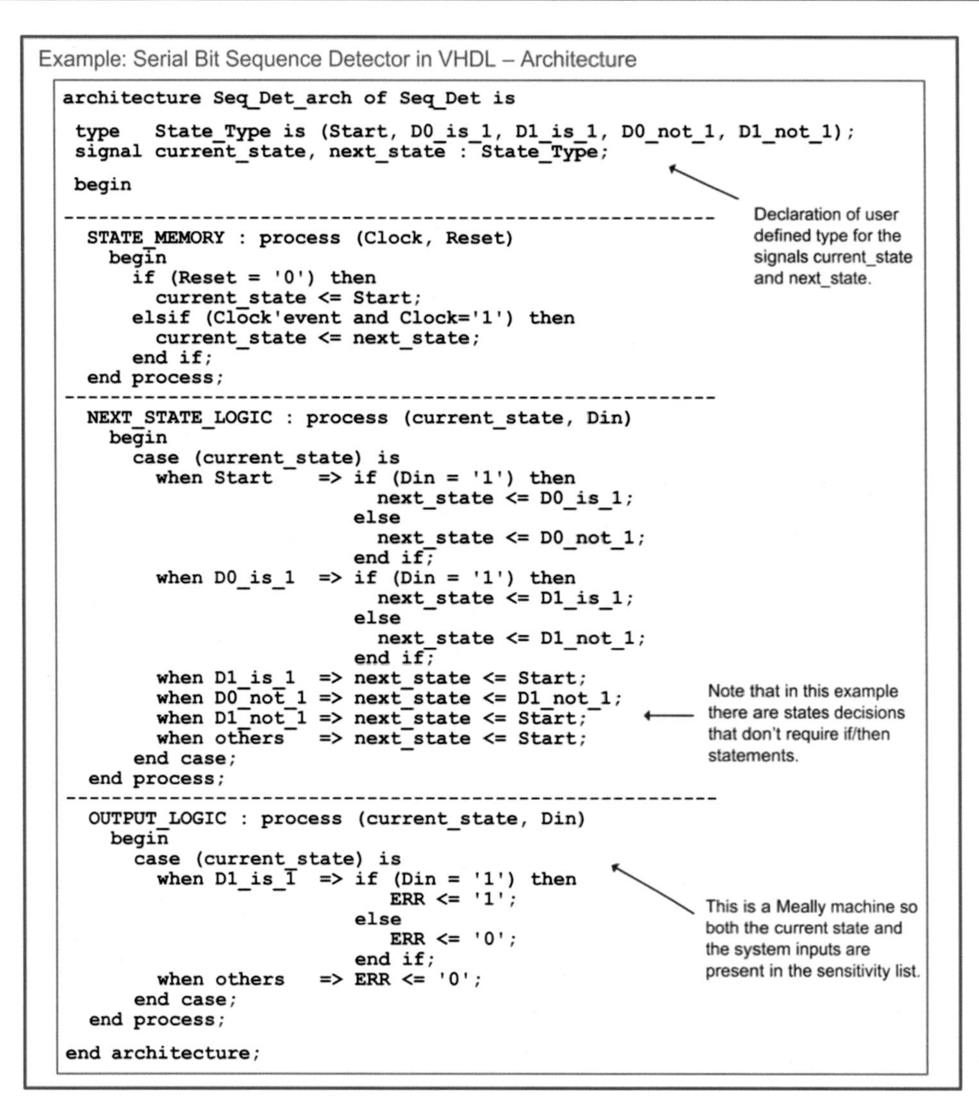

Example: Serial Bit Sequence Detector in VHDL – Architecture

```
architecture Seq_Det_arch of Seq_Det is

 type   State_Type is (Start, D0_is_1, D1_is_1, D0_not_1, D1_not_1);
 signal current_state, next_state : State_Type;

 begin
--------------------------------------------------------
  STATE_MEMORY : process (Clock, Reset)
    begin
      if (Reset = '0') then
        current_state <= Start;
      elsif (Clock'event and Clock='1') then
        current_state <= next_state;
      end if;
  end process;
--------------------------------------------------------
  NEXT_STATE_LOGIC : process (current_state, Din)
    begin
      case (current_state) is
        when Start    => if (Din = '1') then
                           next_state <= D0_is_1;
                         else
                           next_state <= D0_not_1;
                         end if;
        when D0_is_1  => if (Din = '1') then
                           next_state <= D1_is_1;
                         else
                           next_state <= D1_not_1;
                         end if;
        when D1_is_1  => next_state <= Start;
        when D0_not_1 => next_state <= D1_not_1;
        when D1_not_1 => next_state <= Start;
        when others   => next_state <= Start;
      end case;
  end process;
--------------------------------------------------------
  OUTPUT_LOGIC : process (current_state, Din)
    begin
      case (current_state) is
        when D1_is_1  => if (Din = '1') then
                           ERR <= '1';
                         else
                           ERR <= '0';
                         end if;
        when others   => ERR <= '0';
      end case;
  end process;

end architecture;
```


Example 9.7
Serial bit sequence detector in VHDL – architecture

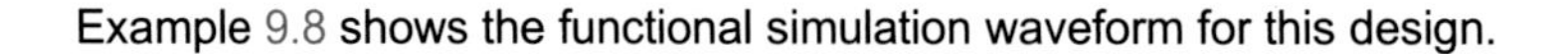

Example 9.8 shows the functional simulation waveform for this design.

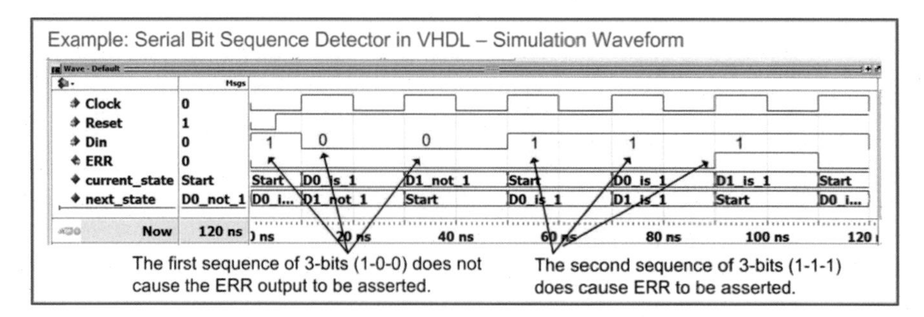

Example 9.8
Serial bit sequence detector in VHDL – simulation waveform

9.2.2 Vending Machine Controller in VHDL

Let's now look at the design of a vending machine controller using the behavioral modeling constructs of VHDL. Example 9.9 shows the design description and entity definition.

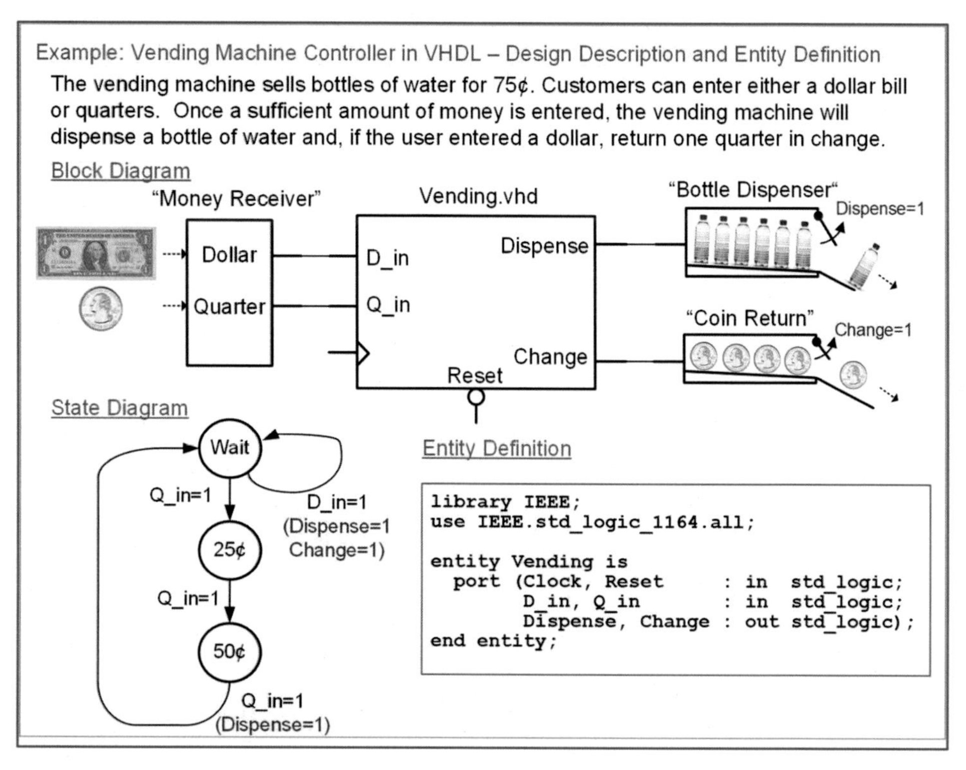

Example 9.9
Vending machine controller in VHDL – design description and entity definition

Example 9.10 shows the VHDL architecture for the vending machine controller. In this model, the descriptive state names Wait, 25¢, and 50¢ cannot be used directly. This is because Wait is a VHDL keyword and user-defined names cannot begin with a number. Instead, the letter "s" is placed in front of the state names in order to make them legal VHDL names (i.e., sWait, s25, s50).

Example: Vending Machine Controller in VHDL – Architecture

```
architecture Vending_arch of Vending is

 type   State_Type is (sWait, s25, s50);
 signal current_state, next_state : State_Type;

 begin
--------------------------------------------------------
  STATE_MEMORY : process (Clock, Reset)
    begin
      if (Reset = '0') then
        current_state <= sWait;
      elsif (Clock'event and Clock='1') then
        current_state <= next_state;
      end if;
  end process;

--------------------------------------------------------
  NEXT_STATE_LOGIC : process (current_state, D_in, Q_in)
    begin
      case (current_state) is
        when sWait  => if (Q_in = '1') then
                         next_state <= s25;
                       else
                         next_state <= sWait;
                       end if;
        when s25    => if (Q_in = '1') then
                         next_state <= s50;
                       else
                         next_state <= s25;
                       end if;
        when s50    => if (Q_in = '1') then
                         next_state <= sWait;
                       else
                         next_state <= s50;
                       end if;
        when others => next_state <= sWait;
      end case;
  end process;

--------------------------------------------------------
  OUTPUT_LOGIC : process (current_state, D_in, Q_in)
    begin
      case (current_state) is
        when sWait  => if (D_in = '1') then
                          Dispense <= '1'; Change <='1';
                       else
                          Dispense <= '0'; Change <='0';
                       end if;
        when s25    => Dispense <= '0'; Change <='0';
        when s50    => if (Q_in = '1') then
                          Dispense <= '1'; Change <='0';
                       else
                          Dispense <= '0'; Change <='0';
                       end if;
        when others => Dispense <= '0'; Change <='0';
      end case;
  end process;

end architecture;
```

Note that an "s" is added to the beginning of the state names since "Wait" is a VHDL keyword and names cannot start with a number.

This is a Meally machine so both the current state and the system inputs are present in the sensitivity list.

Example 9.10
Vending machine controller in VHDL – architecture

Example 9.11 shows the resulting simulation waveform for this design.

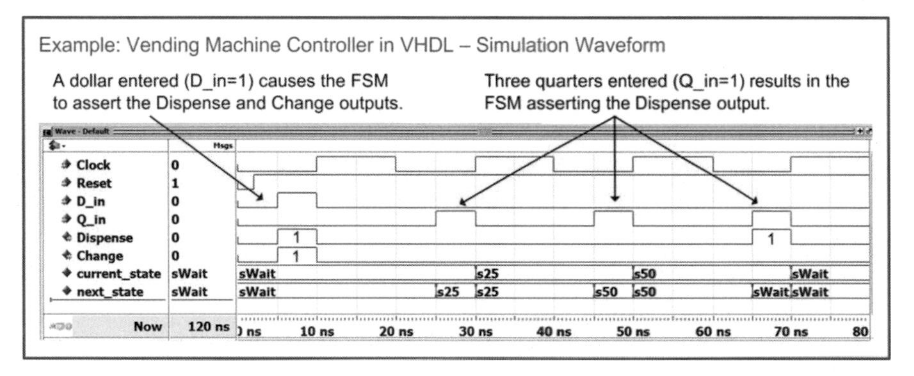

Example 9.11
Vending machine controller in VHDL – simulation waveform

9.2.3 2-Bit Binary Up/Down Counter in VHDL

Let's now look at how a simple counter can be implemented using the three-process behavioral modeling approach in VHDL. Example 9.12 shows the design description and entity definition for a 2-bit binary up/down counter FSM.

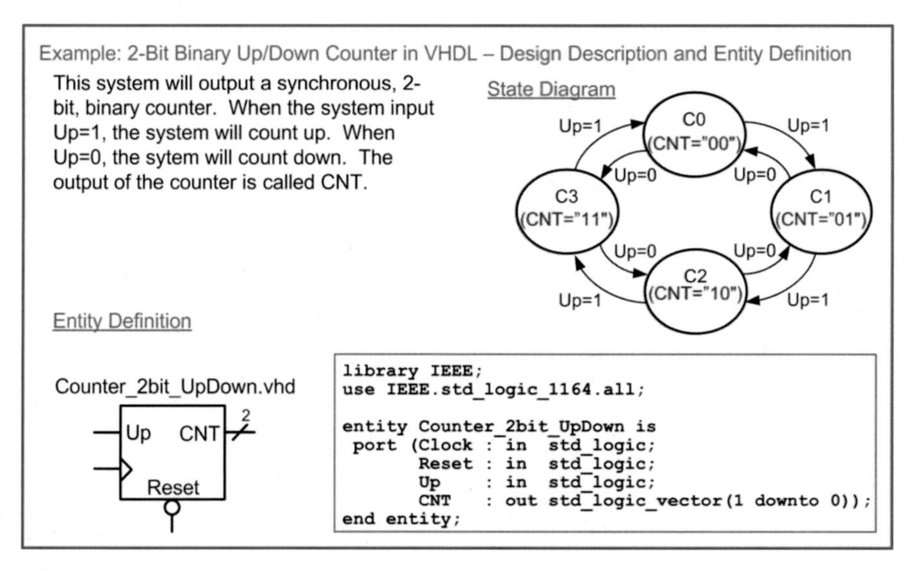

Example 9.12
2-bit binary up/down counter in VHDL – design description and entity definition

Example 9.13 shows the architecture for the 2-bit up/down counter using the three-process modeling approach. Since a counter's outputs only depend on the current state, counters are Moore machines. This simplifies the output logic process since it only needs to contain the current state in its sensitivity list.

Example: 2-Bit Binary Up/Down Counter in VHDL – Architecture (Three Process Model)

```
library IEEE;
use IEEE.std_logic_1164.all;

entity Counter_2bit_UpDown is
 port (Clock, Reset : in  std_logic;
       Up           : in  std_logic;
       CNT          : out std_logic_vector(1 downto 0));
end entity;

architecture Counter_2bit_UpDown_arch of Counter_2bit_UpDown is

 type   State_Type is (C0, C1, C2, C3);
 signal current_state, next_state : State_Type;

 begin
-------------------------------------------------------
  STATE_MEMORY : process (Clock, Reset)
    begin
      if (Reset = '0') then
        current_state <= C0;
      elsif (Clock'event and Clock='1') then
        current_state <= next_state;
      end if;
  end process;
-------------------------------------------------------
  NEXT_STATE_LOGIC : process (current_state, Up)
    begin
      case (current_state) is
        when C0      => if (Up = '1') then
                          next_state <= C1;
                        else
                          next_state <= C3;
                        end if;
        when C1      => if (Up = '1') then
                          next_state <= C2;
                        else
                          next_state <= C0;
                        end if;
        when C2      => if (Up = '1') then
                          next_state <= C3;
                        else
                          next_state <= C1;
                        end if;
        when C3      => if (Up = '1') then
                          next_state <= C0;
                        else
                          next_state <= C2;
                        end if;
        when others => next_state <= C0;
      end case;
  end process;
-------------------------------------------------------
  OUTPUT_LOGIC : process (current_state)
    begin
      case (current_state) is
        when C0      => CNT <= "00";
        when C1      => CNT <= "01";
        when C2      => CNT <= "10";
        when C3      => CNT <= "11";
        when others => CNT <= "00";
      end case;
  end process;

end architecture;
```

A counter is a Moore machine so the output only depends on the current state.

Example 9.13
2-bit binary up/down counter in VHDL – architecture (three-process model)

Example 9.14 shows the resulting simulation waveform for this counter finite-state machine.

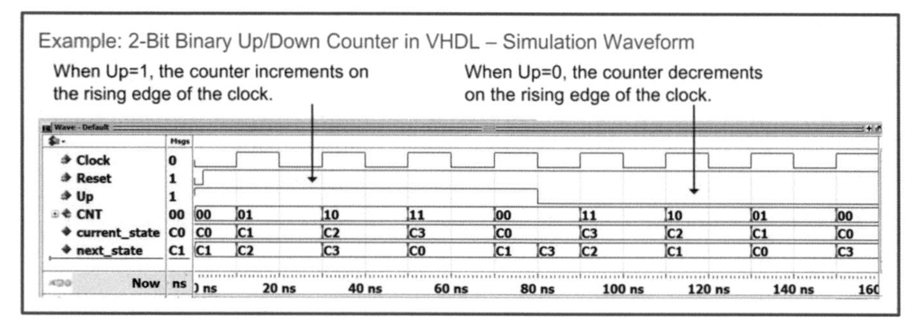

Example 9.14
2-bit binary up/down counter in VHDL – simulation waveform

Concept Check

CC9.2 The state memory process is nearly identical for all finite-state machines with one exception. What is it?

A) The sensitivity list may need to include a preset signal.

B) Sometimes it is modeled using an SR latch storage approach instead of with D-flip-flop behavior.

C) The name of the reset state will be different.

D) The current_state and next_state signals are often swapped.

Summary

- Generic finite-state machines are modeled using three separate processes that describe the behavior of the next-state logic, the state memory, and the output logic. Separate processes are used because each of the three functions in a FSM is dependent on different input signals.
- In VHDL, descriptive state names can be created for a FSM with a user-defined, enumerated data type. The new type is first declared, and each of the descriptive state names is provided that the new data type can take on. Two signals are then created called *current_state* and *next_state* using the new data type. These two signals can then be assigned the descriptive state names of the FSM directly. This approach allows the synthesizer to assign the state codes arbitrarily.
- A subtype can be used when defining the state names if it is desired to explicitly define the state codes.

Exercise Problems

Section 9.1: The FSM Design Process

9.1.1 What is the advantage of using *user-defined, enumerated data types* for the states when modeling a finite-state machine?

9.1.2 What is the advantage of using *subtypes* for the states when modeling a finite-state machine?

9.1.3 When using the three-process behavioral modeling approach for finite-state machines, does the next-state logic process model combinational or sequential logic?

9.1.4 When using the three-process behavioral modeling approach for finite-state machines, does the state memory process model combinational or sequential logic?

9.1.5 When using the three-process behavioral modeling approach for finite-state machines, does the output logic process model combinational or sequential logic?

9.1.6 When using the three-process behavioral modeling approach for finite-state machines, what inputs are listed in the sensitivity list of the next-state logic process?

9.1.7 When using the three-process behavioral modeling approach for finite-state machines, what inputs are listed in the sensitivity list of the state memory process?

9.1.8 When using the three-process behavioral modeling approach for finite-state machines, what inputs are listed in the sensitivity list of the output logic process?

9.1.9 When using the three-process behavioral modeling approach for finite-state machines, how can the signals listed in the sensitivity list of the output logic process immediately tell whether the FSM is a Mealy or a Moore machine?

9.1.10 Why is it not a good design approach to combine the next-state logic and output logic behavior into a single process?

Section 9.2: FSM Design Examples

9.2.1 Design a VHDL behavioral model to implement the finite-state machine described by the state diagram in Fig. 9.1. Use the entity definition provided in this figure for your design. Use the three-process approach to modeling FSMs described in this chapter for your design. Model the states in this machine with a user-defined enumerated type.

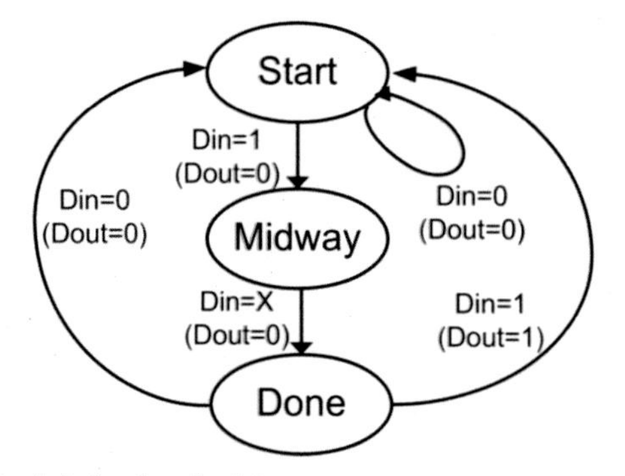

fsm1_behavioral.vhd

```
entity fsm1_behavioral is
    port   (Clock, Reset : in   std_logic;
            Din          : in   std_logic;
            Dout         : out  std_logic);
end entity;
```

Fig. 9.1
FSM 1 state diagram and entity

9.2.2 Design a VHDL behavioral model to implement the finite-state machine described by the state diagram in Fig. 9.1. Use the entity definition provided in this figure for your design. Use the three-process approach to modeling FSMs described in this chapter for your design. Explicitly assign binary state codes using VHDL subtypes. Use the following state codes: Start = "00," Midway = "01," Done = "10."

9.2.3 Design a VHDL behavioral model to implement the finite-state machine described by the state diagram in Fig. 9.2. Use the entity definition provided in this figure for your design. Use the three-process approach to modeling FSMs described in this chapter for your design. Model the states in this machine with a user-defined enumerated type.

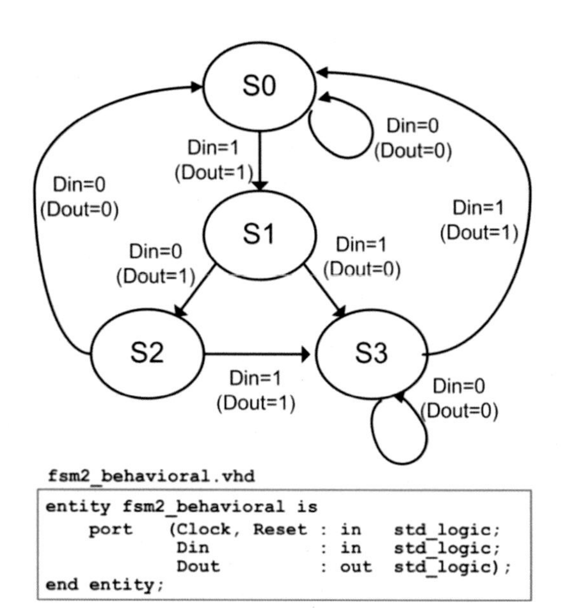

fsm2_behavioral.vhd

```
entity fsm2_behavioral is
    port   (Clock, Reset : in   std_logic;
            Din          : in   std_logic;
            Dout         : out  std_logic);
end entity;
```

Fig. 9.2
FSM 2 state diagram and entity

9.2.4 Design a VHDL behavioral model to implement the finite-state machine described by the state diagram in Fig. 9.2. Use the entity definition provided in this figure for your design. Use the three-process approach to modeling FSMs described in this chapter for your design. Assign one-hot state codes using VHDL subtypes. Use the following state codes: S0 = "0001," S1 = "0010," S2 = "0100," S3 = "1000."

9.2.5 Design a VHDL behavioral model for a 4-bit serial bit sequence detector similar to Example 9.6. Use the entity definition provided in Fig. 9.3. Use the three-process approach to modeling FSMs described in this chapter for your design. The input to your sequence detector is called *DIN*, and the output is called *FOUND*. Your detector will assert FOUND anytime there is a 4-bit sequence of "0101." For all other input sequences, the output is not asserted. Model the states in your machine with a user-defined enumerated type.

Seq_Det_behavioral.vhd

```
entity Seq_Det_behavioral is
  port (Clock, Reset : in  std_logic;
        DIN          : in  std_logic;
        FOUND        : out std_logic);
end entity;
```

Fig. 9.3
Sequence detector entity

9.2.6 Design a VHDL behavioral model for a 20-cent vending machine controller similar to Example 9.9. Use the entity definition provided in Fig. 9.4. Use the three-process approach to modeling FSMs described in this chapter for your design. Your controller will take in nickels and dimes and dispense a product anytime the customer has entered 20 cents. Your FSM has two inputs, *Nin* and *Din*. Nin is asserted whenever the customer enters a nickel, while Din is asserted anytime the customer enters a dime. Your FSM has two outputs, *Dispense* and *Change*. Dispense is asserted anytime the customer has entered at least 20 cents, and Change is asserted anytime the customer has entered more than 20 cents and needs a nickel in change. Model the states in this machine with a user-defined enumerated type.

Vending_behavioral.vhd

```
entity Vending_behavioral is
  port (Clock, Reset      : in  std_logic;
        Nin, Din          : in  std_logic;
        Dispense, Change  : out std_logic);
end entity;
```

Fig. 9.4
Vending machine entity

9.2.7 Design a VHDL behavioral model for a finite-state machine for a traffic light controller at the intersection of a busy highway and a seldom used side road. Use the entity definition provided in Fig. 9.5. You will be designing the control signals for just the red, yellow, and green lights facing the highway. Under normal conditions, the highway has a green light. The side road has car detector that indicates when car pulls up by asserting a signal called *CAR*. When CAR is asserted, you will change the highway traffic light from green to yellow and then from yellow to red. Once in the red position, a built-in timer will begin a countdown and provide your controller a signal called *TIMEOUT* when 15 seconds has passed. Once TIMEOUT is asserted, you will change the highway traffic light back to green. Your system will have three outputs, *GRN*, *YLW*, and *RED*, which control when the highway-facing traffic lights are on (1 = ON, 0 = OFF). Model the states in this machine with a user-defined enumerated type.

tlc_behavioral.vhd

```
entity tlc_behavioral is
  port (Clock, Reset    : in  std_logic;
        CAR, TIMEOUT    : in  std_logic;
        GRN, YLW, RED   : out std_logic);
end entity;
```

Fig. 9.5
Traffic light controller entity

Chapter 10: Modeling Counters

Counters are a special case of finite-state machines because they move linearly through their discrete states (either forward or backward) and typically are implemented with state-encoded outputs. Due to this simplified structure and wide spread use in digital systems, VHDL allows counters to be modeled using a single process and with arithmetic operators (i.e., + and −). This enables a more compact model and allows much wider counters to be implemented. This chapter will cover some of the most common techniques for modeling counters.

Learning Outcomes—After completing this chapter, you will be able to:

10.1 Design a behavioral model for a counter using a single process.
10.2 Design a behavioral model for a counter with enable and load capability.

10.1 Modeling Counters with a Single Process

10.1.1 Counters in VHDL Using the Type UNSIGNED

Let's look at how we can model a 4-bit binary up counter with an output called *CNT*. First, we want to model this counter using the "+" operator. Recall that the "+" operator is not defined in the std_logic_1164 package. We need to include the numeric_std package in order to add this capability. Within the numeric_std package, the "+" operator is only defined for types signed and unsigned (and not for std_logic_vector), so the output CNT will be declared as type unsigned. Next, we want to implement the counter using a signal assignment in the form *CNT <= CNT + 1*; however, since CNT is an output port, it cannot be used as an argument (right hand side) in an operation. We will need to create an internal signal to implement the counter functionality (i.e., CNT_tmp). Since a signal does not contain directionality, it can be used as both the target and an argument of an operation. Outside of the counter process, a concurrent signal assignment is used to continuously assign CNT_tmp to CNT in order to drive the output of the system. This means that we need to create the internal signal CNT_tmp of type unsigned also to support this assignment. Example 10.1 shows the VHDL model and simulation waveform for this counter. When the counter reaches its maximum value of "1111," it rolls over to "0000" and continues counting because it is defined to only contain 4 bits.

B. J. LaMeres, *Quick Start Guide to VHDL*, https://doi.org/10.1007/978-3-031-42543-1_10

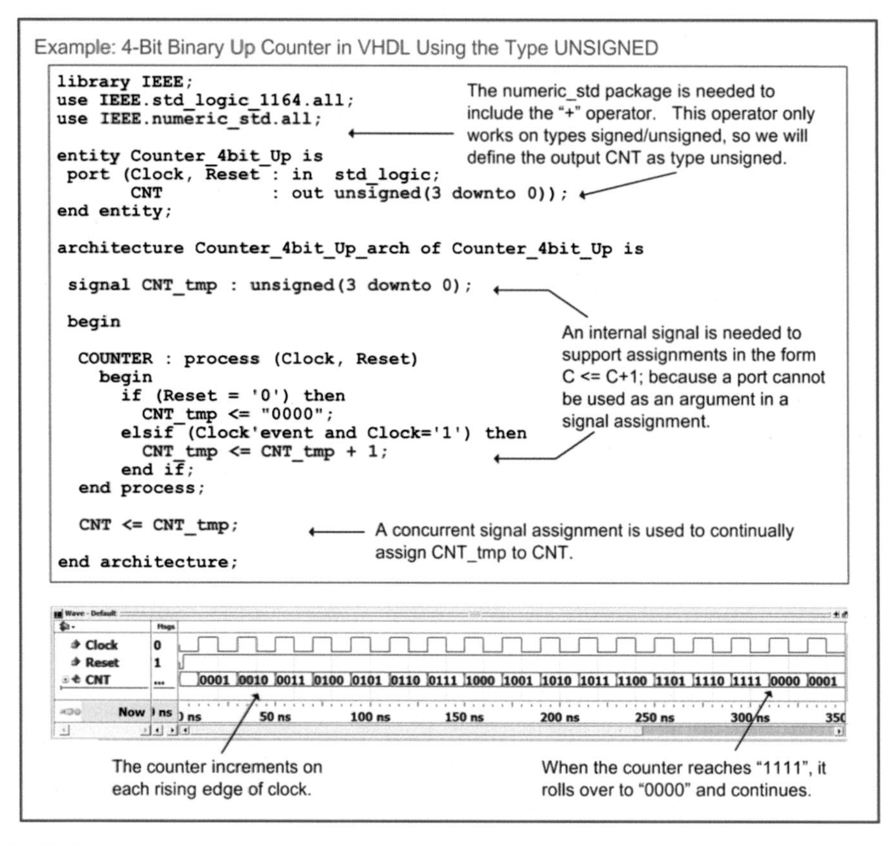

Example 10.1
4-bit binary up counter in VHDL using the type UNSIGNED

10.1.2 Counters in VHDL Using the Type INTEGER

Another common technique to model counters with a single process is to use the type integer. The numeric_std package supports the "+" operator for type integer. It also contains a conversion between the types integer and unsigned/signed. This means a process can be created to model the counter functionality with integers, and then the result can be converted and assigned to the output of the system of type unsigned. One thing that must be considered when using integers is that they are defined as 32-bit, two's complement numbers. This means that if a counter is defined to use integers and the maximum range of the counter is not explicitly controlled, the counter will increment through the entire range of 32-bit values it can take on. There are a variety of ways to explicitly bound the size of an integer counter. The first is to use an if/then clause within the process to check for the upper limit desired in the counter. For example, if we wish to create a 4-bit binary counter, we will check if the integer counter has reached 15 each time through the process. If it has, we will reset it to zero. Synthesizers will recognize that the integer counter is never allowed to exceed 15 (or "1111" for an unsigned counter) and remove the unused bits of the integer type during implementation (i.e., the remaining 28 bits). Example 10.2 shows the VHDL model and simulation waveform for this implementation of the 4-bit counter using integers.

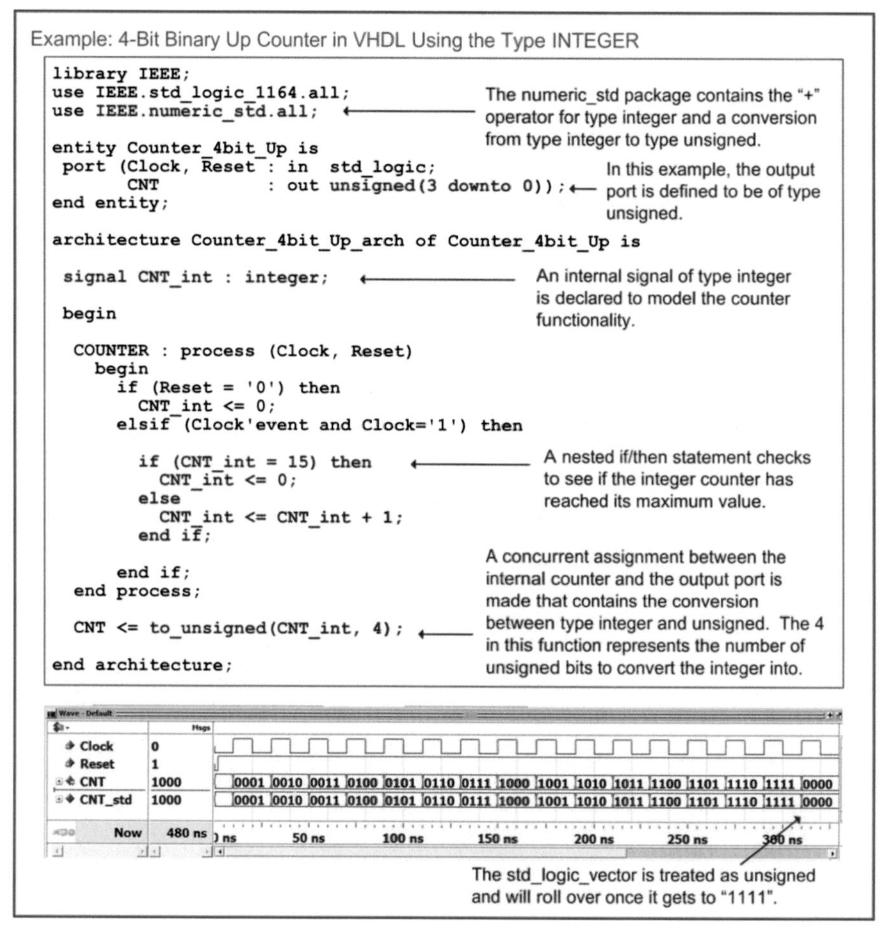

Example: 4-Bit Binary Up Counter in VHDL Using the Type INTEGER

```
library IEEE;
use IEEE.std_logic_1164.all;
use IEEE.numeric_std.all;

entity Counter_4bit_Up is
 port (Clock, Reset : in  std_logic;
       CNT          : out unsigned(3 downto 0));
end entity;

architecture Counter_4bit_Up_arch of Counter_4bit_Up is

 signal CNT_int : integer;

 begin

  COUNTER : process (Clock, Reset)
    begin
      if (Reset = '0') then
        CNT_int <= 0;
      elsif (Clock'event and Clock='1') then

        if (CNT_int = 15) then
          CNT_int <= 0;
        else
          CNT_int <= CNT_int + 1;
        end if;

      end if;
  end process;

  CNT <= to_unsigned(CNT_int, 4);

end architecture;
```


Example 10.2
4-bit binary up counter in VHDL using the type INTEGER

10.1.3 Counters in VHDL Using the Type STD_LOGIC_VECTOR

It is often desired to have the ports of a system be defined of type std_logic/std_logic_vector for compatibility with other systems. One technique to accomplish this and also model the counter behavior internally using std_logic_vector is through inclusion of the numeric_std_unsigned package. This package allows the use of std_logic_vector when declaring the ports and signals within the design and treats them as unsigned when performing arithmetic and comparison functions. Example 10.3 shows the VHDL model and simulation waveform for this alternative implementation of the 4-bit counter.

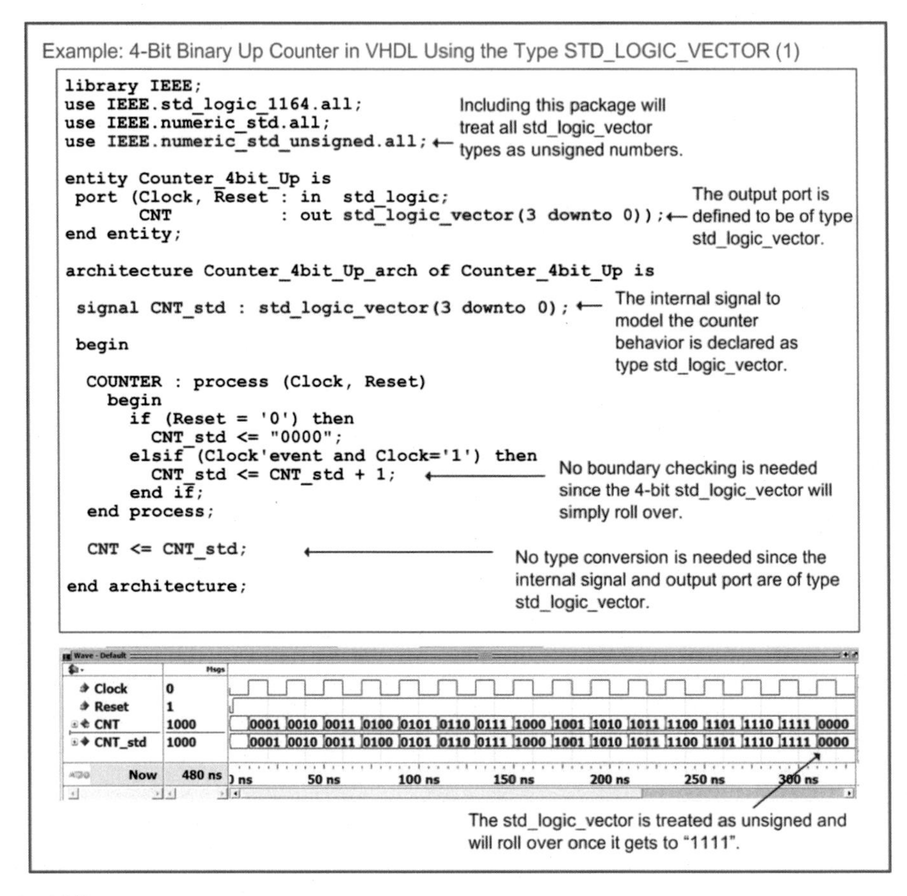

Example 10.3
4-bit binary up counter in VHDL using the type STD_LOGIC_VECTOR (1)

If it is designed to have an output type of std_logic_vector and use an integer in modeling the behavior of the counter, then a double conversion can be used. In the following example, the counter behavior is modeled using an integer type with range checking. A concurrent signal assignment is used at the end of the architecture in order to convert the integer to type std_logic_vector. This is accomplished by first converting the type integer to unsigned and then converting the type unsigned to std_logic_vector. Example 10.4 shows the VHDL model and simulation waveform for this alternative implementation of the 4-bit counter.

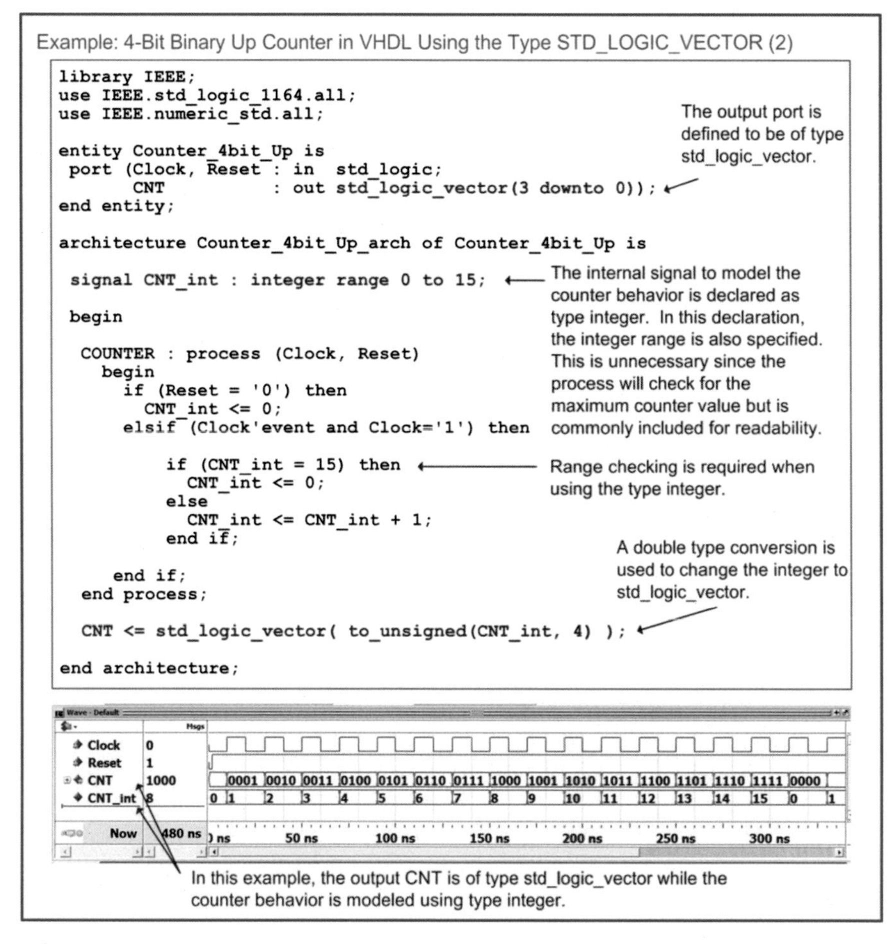

Example: 4-Bit Binary Up Counter in VHDL Using the Type STD_LOGIC_VECTOR (2)

```
library IEEE;
use IEEE.std_logic_1164.all;
use IEEE.numeric_std.all;

entity Counter_4bit_Up is
 port (Clock, Reset : in  std_logic;
       CNT          : out std_logic_vector(3 downto 0));
end entity;

architecture Counter_4bit_Up_arch of Counter_4bit_Up is

 signal CNT_int : integer range 0 to 15;

 begin

  COUNTER : process (Clock, Reset)
    begin
      if (Reset = '0') then
        CNT_int <= 0;
      elsif (Clock'event and Clock='1') then

          if (CNT_int = 15) then
            CNT_int <= 0;
          else
            CNT_int <= CNT_int + 1;
          end if;

     end if;
  end process;

  CNT <= std_logic_vector( to_unsigned(CNT_int, 4) );

end architecture;
```


Example 10.4
4-bit binary up counter in VHDL using the type STD_LOGIC_VECTOR (2)

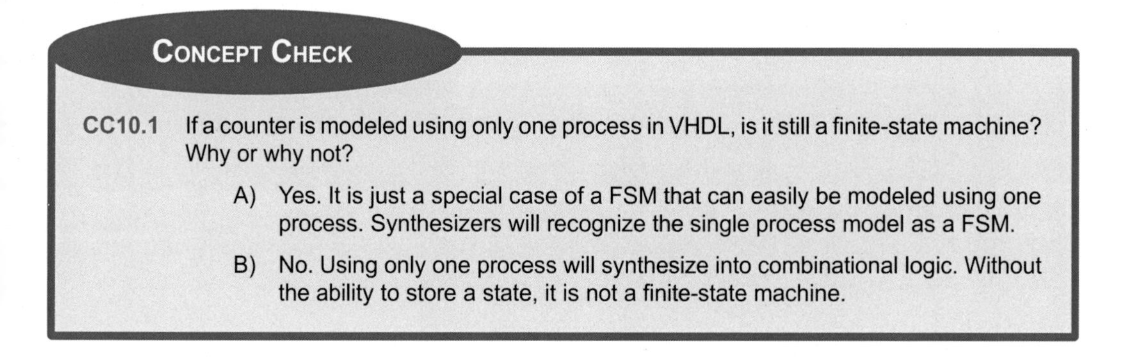

Concept Check

CC10.1 If a counter is modeled using only one process in VHDL, is it still a finite-state machine? Why or why not?

A) Yes. It is just a special case of a FSM that can easily be modeled using one process. Synthesizers will recognize the single process model as a FSM.

B) No. Using only one process will synthesize into combinational logic. Without the ability to store a state, it is not a finite-state machine.

10.2 Counters with Enables and Loads

10.2.1 Modeling Counters with Enables

Including an enable in a counter is a common technique to prevent the counter from running continuously. When the enable is asserted, the counter will increment on the rising edge of the clock as usual. When the enable is de-asserted, the counter will simply hold its last value. Enable lines are synchronous, meaning that they are only evaluated on the rising edge of the clock. As such, they are modeled using a nested if/then statement within the if/then statement checking for a rising edge of the clock. Example 10.5 shows an example model for a 4-bit counter with enable.

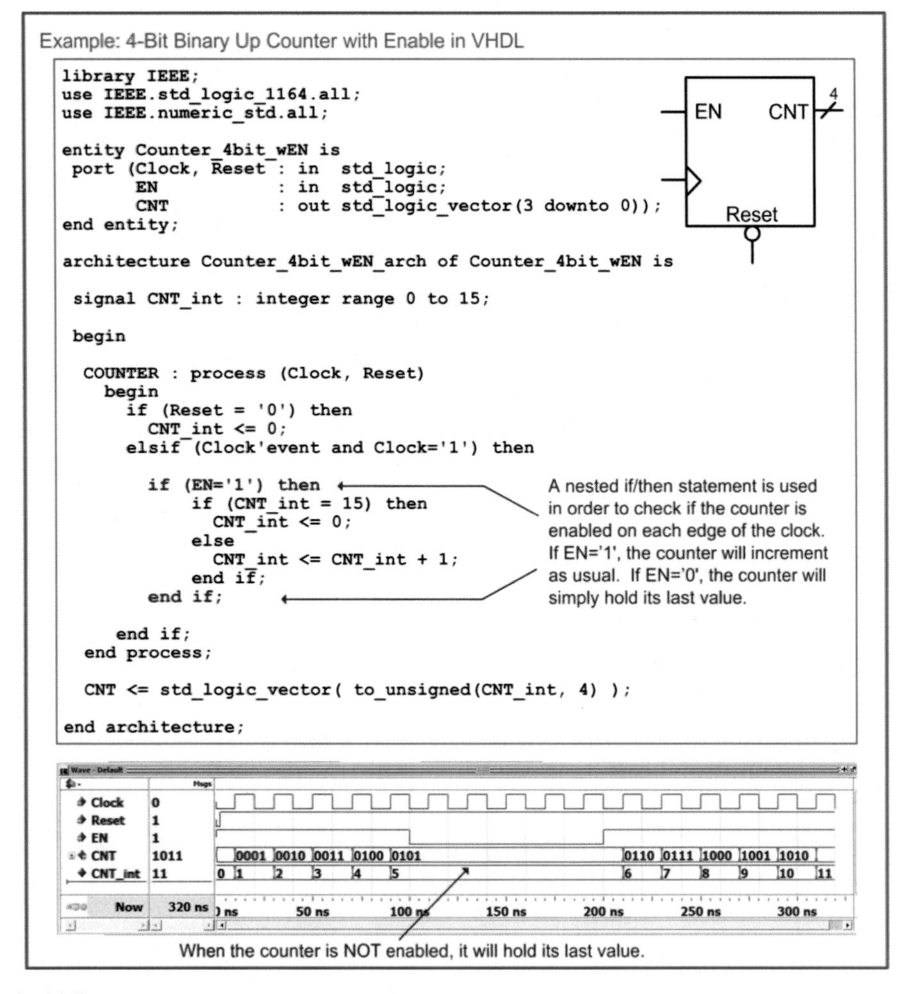

Example 10.5
4-bit binary up counter with enable in VHDL

10.2.2 Modeling Counters with Loads

A counter with a *load* has the ability to set the counter to a specified value. The specified value is provided on an input port (i.e., CNT_in) with the same width as the counter output (CNT). A synchronous load input signal (i.e., Load) is used to indicate when the counter should set its value to the value present on CNT_in. Example 10.6 shows an example model for a 4-bit counter with load capability.

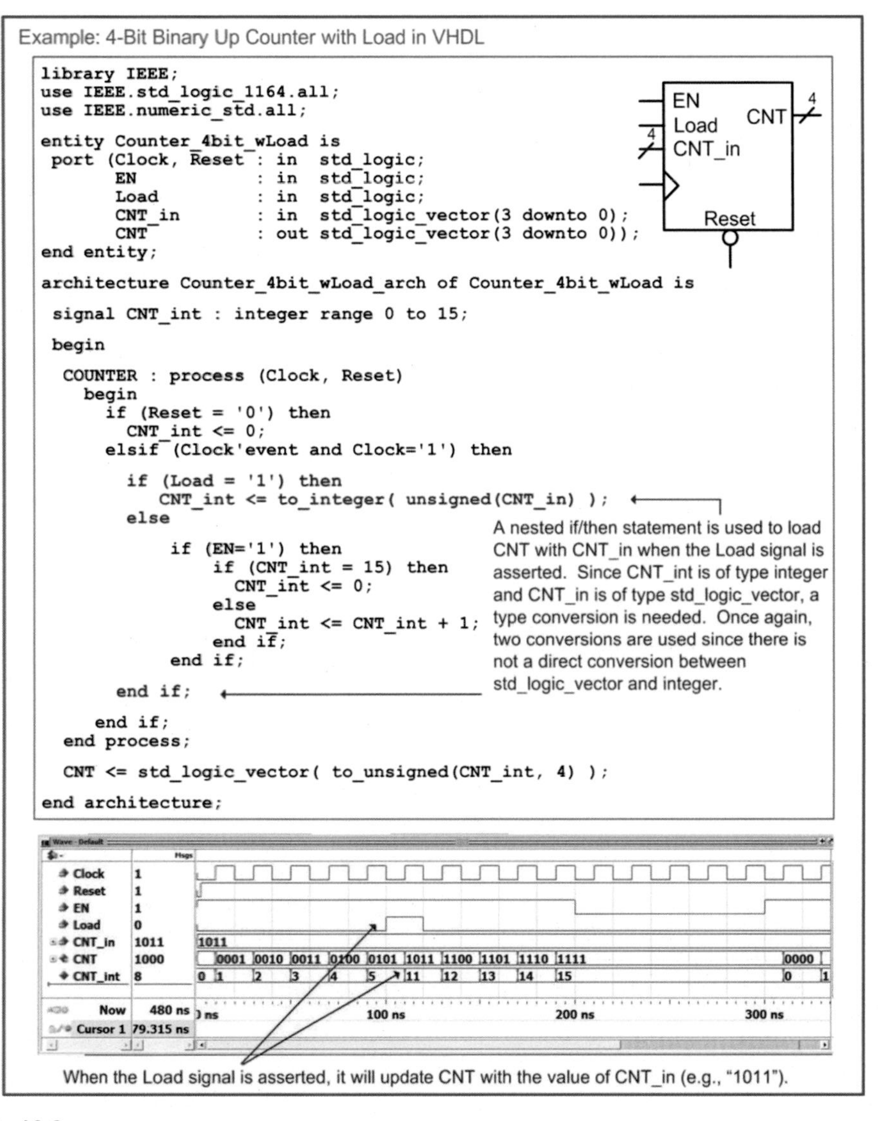

Example 10.6
4-bit binary up counter with load in VHDL

Concept Check

CC10.2 If the counter has other inputs such as loads and enables, shouldn't they be listed in the sensitivity list along with clock and reset?

A) Yes. All inputs should go in the sensitivity list.

B) No. Only signals that trigger an assignment are listed in the sensitivity list. The only two signals that have this behavior are clock and reset.

Summary

- A counter is a special type of finite-state machine in which the states are traversed linearly. The linear progression of states allows the next-state logic to be simplified. The complexity of the output logic in a counter can also be reduced by encoding the states with the desired counter output for that state. This technique, known as *state-encoded outputs*, allows the system outputs to simply be the current state of the FSM.
- Counters are a special type of finite-state machine that can be modeled using a single process in VHDL. Only the clock and reset signals are listed in the sensitivity list of the counter process because they are the only signals that trigger signal assignments.
- Within the process of a counter, arithmetic operators (i.e., + or −) can be used to modify the counter value. Since these operators aren't defined for the type std_logic_vector, type casting is usually required.

Exercise Problems

Section 10.1: Modeling Counters with a Single Process

10.1.1 Design a VHDL behavioral model for a 16-bit binary up counter using a single process. The block diagram for the entity definition is shown in Fig. 10.1. In your model, declare Count_Out to be of type unsigned, and implement the internal counter functionality with a signal of type unsigned.

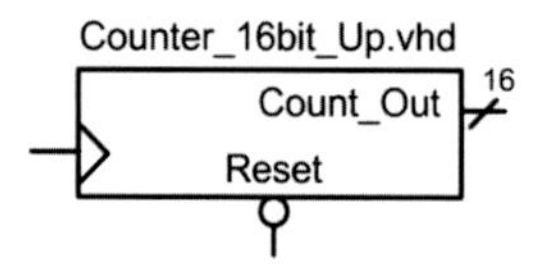

Fig. 10.1
16-bit binary up counter block diagram

10.1.2 Design a VHDL behavioral model for a 16-bit binary up counter using a single process. The block diagram for the entity definition is shown in Fig. 10.1. In your model, declare Count_Out to be of type unsigned, and implement the internal counter functionality with a signal of type integer.

10.1.3 Design a VHDL behavioral model for a 16-bit binary up counter using a single process. The block diagram for the entity definition is shown in Fig. 10.1. In your model, declare Count_Out to be of type std_logic_vector, and implement the internal counter functionality with a signal of type integer.

Section 10.2: Counters with Enables and Loads

10.2.1 Design a VHDL behavioral model for a 16-bit binary up counter with enable using a single process. The block diagram for the entity definition is shown in Fig. 10.2. In your model, declare Count_Out to be of type unsigned, and implement the internal counter functionality with a signal of type unsigned.

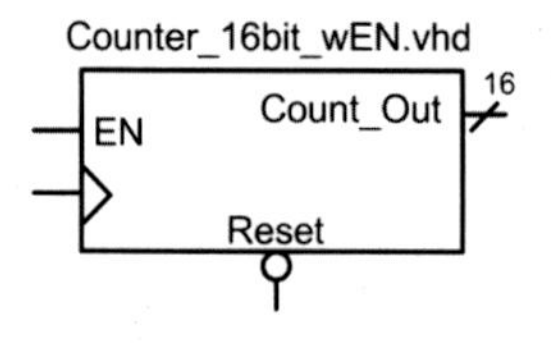

Fig. 10.2
16-bit binary up counter with enable block diagram

10.2.2 Design a VHDL behavioral model for a 16-bit binary up counter with enable using a single process. The block diagram for the entity definition is shown in Fig. 10.2. In your model, declare Count_Out to be of type unsigned, and implement the internal counter functionality with a signal of type integer.

10.2.3 Design a VHDL behavioral model for a 16-bit binary up counter with enable using a single process. The block diagram for the entity definition is shown in Fig. 10.2. In your model, declare Count_Out to be of type std_logic_vector, and implement the internal counter functionality with a signal of type integer.

10.2.4 Design a VHDL behavioral model for a 16-bit binary up counter with enable and load using a single process. The block diagram for the entity definition is shown in Fig. 10.3. In your model, declare Count_Out to be of type unsigned, and implement the internal counter functionality with a signal of type unsigned.

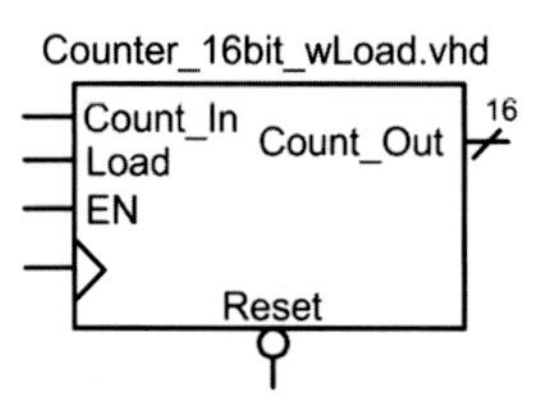

Fig. 10.3
16-bit binary up counter with load block diagram

10.2.5 Design a VHDL behavioral model for a 16-bit binary up counter with enable and load using a single process. The block diagram for the entity definition is shown in Fig. 10.3. In your model, declare Count_Out to be of type unsigned, and implement the internal counter functionality with a signal of type integer.

10.2.6 Design a VHDL behavioral model for a 16-bit binary up counter with enable and load using a single process. The block diagram for the entity definition is shown in Fig. 10.3. In your model, declare Count_Out to be of type std_logic_vector, and implement the internal counter functionality with a signal of type integer.

10.2.7 Design a VHDL behavioral model for a 16-bit binary up/down counter using a single process. The block diagram for the entity definition is shown in Fig. 10.4. When Up = 1, the counter will increment. When Up = 0, the counter will decrement. In your model, declare Count_Out to be of type unsigned, and implement the internal counter functionality with a signal of type unsigned.

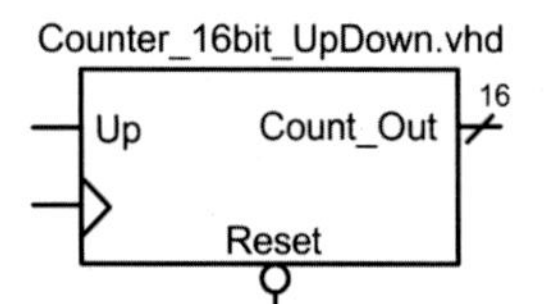

Fig. 10.4
16-bit binary up/down counter block diagram

10.2.8 Design a VHDL behavioral model for a 16-bit binary up/down counter using a single process. The block diagram for the entity definition is shown in Fig. 10.4. When Up = 1, the counter will increment. When Up = 0, the counter will decrement. In your model, declare Count_Out to be of type unsigned, and implement the internal counter functionality with a signal of type integer.

10.2.9 Design a VHDL behavioral model for a 16-bit binary up/down counter using a single process. The block diagram for the entity definition is shown in Fig. 10.4. When Up = 1, the counter will increment. When Up = 0, the counter will decrement. In your model, declare Count_Out to be of type std_logic_vector, and implement the internal counter functionality with a signal of type integer.

Chapter 11: Modeling Memory

This chapter covers how to model memory arrays in VHDL. These models are technology independent, meaning that they can be ultimately synthesized into a wide range of semiconductor memory devices.

Learning Outcomes—After completing this chapter, you will be able to:

11.1 Describe the basic architecture and terminology for semiconductor-based memory systems.
11.2 Design a VHDL model for a read-only memory array.
11.3 Design a VHDL model for a read/write memory array.

11.1 Memory Architecture and Terminology

The term *memory* is used to describe a system with the ability to store digital information. The term *semiconductor memory* refers to systems that are implemented using integrated circuit technology. These types of systems store the digital information using transistors, fuses, and/or capacitors on a single semiconductor substrate. Memory can also be implemented using technology other than semiconductors. Disk drives store information by altering the polarity of magnetic fields on a circular substrate. The two magnetic polarities (north and south) are used to represent different logic values (i.e., 0 or 1). Optical disks use lasers to burn pits into reflective substrates. The binary information is represented by light either being reflected (no pit) or not reflected (pit present). Semiconductor memory does not have any moving parts, so it is called *solid-state memory* and can hold more information per unit area than disk memory. Regardless of the technology used to store the binary data, all memory has common attributes and terminologies that are discussed in this chapter.

11.1.1 Memory Map Model

The information stored in memory is called the **data**. When information is placed into memory, it is called a **write**. When information is retrieved from memory, it is called a **read**. In order to access data in memory, an **address** is used. While data can be accessed as individual bits, in order to reduce the number of address locations needed, data is typically grouped into *N-bit words*. If a memory system has $N = 8$, this means that 8 bits of data are stored at each address. The number of address locations is described using the variable *M*. The overall size of the memory is typically stated by saying "MxN." For example, if we had a 16×8 memory system, that means that there are 16 address locations, each capable of storing a byte of data. This memory would have a **capacity** of $16 \times 8 = 128$ bits. Since the address is implemented as a binary code, the number of lines in the address bus (n) will dictate the number of address locations that the memory system will have ($M = 2^n$). Figure 11.1 shows a graphical depiction of how data resides in memory. This type of graphic is called a *memory map model*.

B. J. LaMeres, *Quick Start Guide to VHDL*, https://doi.org/10.1007/978-3-031-42543-1_11

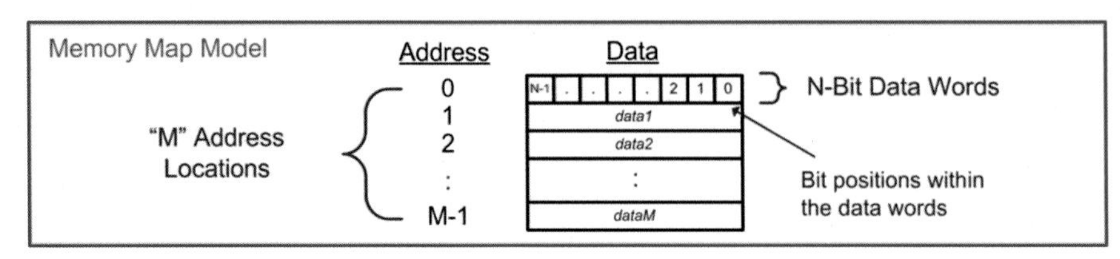

Fig. 11.1
Memory map model

11.1.2 Volatile vs. Non-volatile Memory

Memory is classified into two categories depending on whether it can store information when power is removed or not. The term **non-volatile** is used to describe memory that *holds* information when the power is removed, while the term **volatile** is used to describe memory that loses its information when power is removed. Historically, volatile memory is able to run at faster speeds compared to non-volatile memory, so it is used as the primary storage mechanism while a digital system is running. Non-volatile memory is necessary in order to hold critical operation information for a digital system such as startup instructions, operations systems, and applications.

11.1.3 Read-Only Memory vs. Read/Write Memory

Memory can also be classified into two categories with respect to how data is accessed. **Read-only memory (ROM)** is a device that cannot be written to during normal operation. This type of memory is useful for holding critical system information or programs that should not be altered while the system is running. **Read/write** memory refers to memory that can be read and written to during normal operation and is used to hold temporary data and variables.

11.1.4 Random Access vs. Sequential Access

Random-access memory (RAM) describes memory in which any location in the system can be accessed at any time. The opposite of this is **sequential access** memory, in which not all address locations are immediately available. An example of a sequential access memory system is a tape drive. In order to access the desired address in this system, the tape spool must be spun until the address is in a position that can be observed. Most semiconductor memory in modern systems is random access. The terms RAM and ROM have been adopted, somewhat inaccurately, to also describe groups of memory with particular behavior. While the term ROM technically describes a system that cannot be written to, it has taken on the additional association of being the term to describe non-volatile memory. While the term RAM technically describes how data is accessed, it has taken on the additional association of being the term to describe volatile memory. When describing modern memory systems, the terms RAM and ROM are used most commonly to describe the characteristics of the memory being used; however, modern memory systems can be both read/write and non-volatile, and the majority of memory is random access.

Concept Check

CC11.1 An 8-bit-wide memory has 8 address lines. What is its capacity in bits?

A) 64 B) 256 C) 1024 D) 2048

11.2 Modeling Read-Only Memory

Modeling of memory in VHDL is accomplished using the *array* data type. Recall the syntax for declaring a new array type below:

```
type name is array (<range>) of <element_type>;
```

To create the ROM array, a new type is declared (e.g., ROM_type) that is an array. The *range* represents the addressing of the memory array and is provided as an integer. The *element_type* of the array specifies the data type to be stored at each address and represents the data in the memory array. The type of the element should be std_logic_vector with a width of *N*. To define a 4 × 4 array of memory, we would use the following syntax:

Example:

```
type ROM_type is array (0 to 3) of std_logic_vector(3 downto 0);
```

Notice that the address is provided as an integer (0 to 3). This will require two address bits. Also notice that this defines 4-bit data words. Next, we define a new constant of type ROM_type. When defining a constant, we provide the contents at each address.

Example:

```
constant ROM : ROM_type := (0 => "1110",
                            1 => "0010",
                            2 => "1111",
                            3 => "0100");
```

At this point, the ROM array is declared and initialized. In order to model the read behavior, a concurrent signal assignment is used. The assignment will be made to the output data_out based on the incoming address. The assignment to data_out will be the contents of the constant ROM at a particular address. Since the index of a VHDL array needs to be provided as an integer (e.g., 0,1,2,3) and the address of the memory system is provided as a std_logic_vector, a type conversion is required. Since there is not a direct conversion from type std_logic_vector to integer, two conversions are required. The first step is to convert the address from std_logic_vector to unsigned using the *unsigned* type conversion. This conversion exists within the numeric_std package. The second step is to convert the address from unsigned to integer using the *to_integer* conversion. The final assignment is as follows:

Example:

```
data_out <= ROM(to_integer(unsigned(address)));
```

Example 11.1 shows the entire VHDL model for this memory system and the simulation waveform. In the simulation, each possible address is provided (i.e., "00," "01," "10," and "11"). For each address, the corresponding information appears on the data_out port. Since this is an asynchronous memory system, the data appears immediately upon receiving a new address.

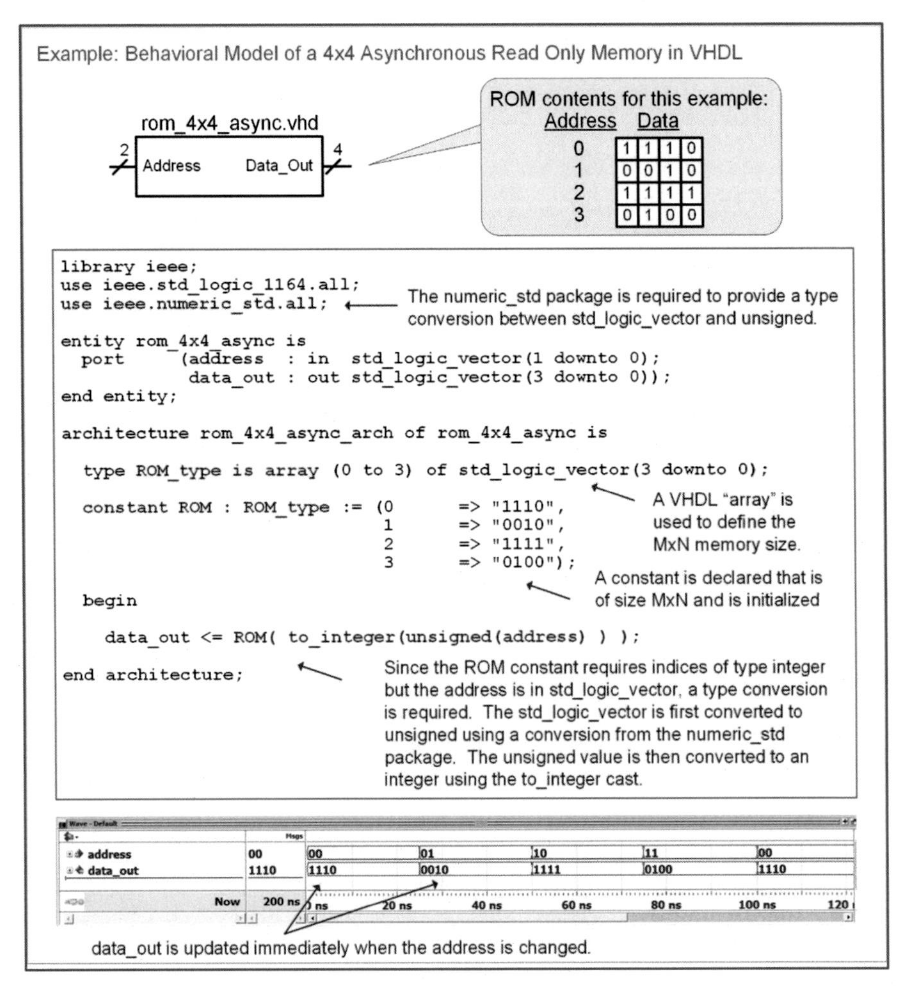

Example 11.1
Behavioral model of a 4 × 4 asynchronous read-only memory in VHDL

Latency can be modeled in memory systems by using delayed signal assignments. In the above example, if the memory system had a latency of 5 ns, this could be modeled using the following approach:

Example:

```
data_out <= ROM(to_integer(unsigned(address))) after 5 ns;
```

A synchronous ROM can be created in a similar manner. In this approach, a clock edge is used to trigger when the data_out port is updated. A sensitivity list is used that contains only the signal clock to trigger the assignment. A rising edge condition is then used in an if/then statement to make the assignment only on a rising edge. Example 11.2 shows the VHDL model and simulation waveform for this system. Notice that prior to the first clock edge, the simulator does not know what to assign to data_out, so it lists the value as *uninitialized*.

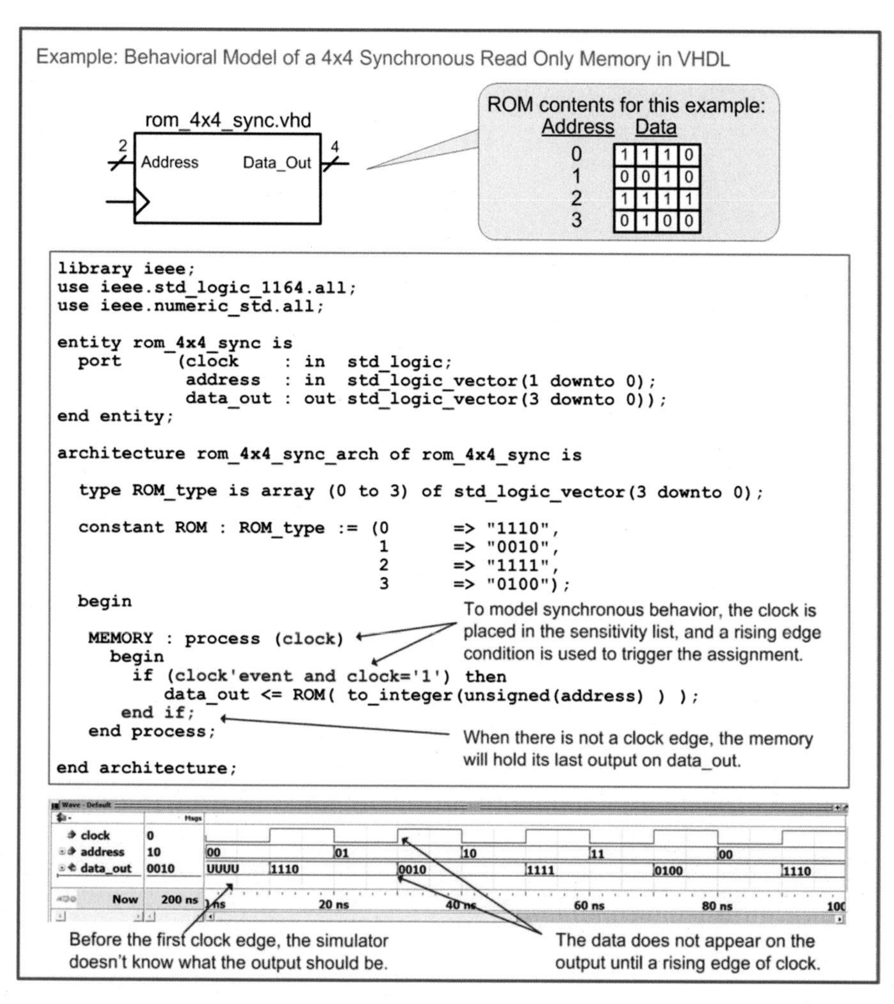

Example 11.2
Behavioral model of a 4 × 4 synchronous read-only memory in VHDL

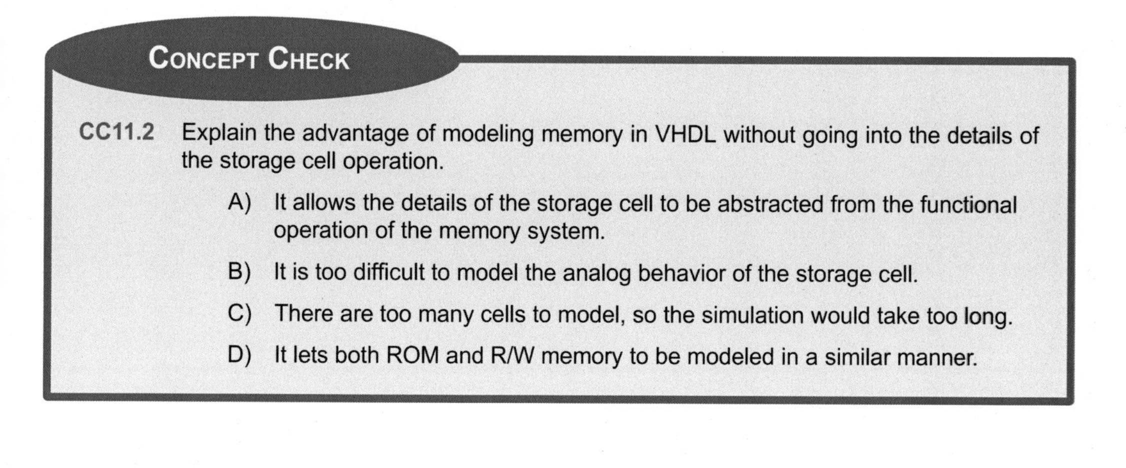

CONCEPT CHECK

CC11.2 Explain the advantage of modeling memory in VHDL without going into the details of the storage cell operation.

A) It allows the details of the storage cell to be abstracted from the functional operation of the memory system.

B) It is too difficult to model the analog behavior of the storage cell.

C) There are too many cells to model, so the simulation would take too long.

D) It lets both ROM and R/W memory to be modeled in a similar manner.

11.3 Modeling Read/Write Memory

In a read/write memory model, a new type is created using a VHDL *array* (e.g., RW_type) that defines the size of the storage system. To create the memory, a new signal is declared with the array type.

Example:

```
type RW_type is array (0 to 3) std_logic_vector(3 downto 0);
signal RW : RW_type;
```

Note that a signal is used in a read/write system as opposed to a constant as in the read-only memory system. This is because a read/write system is uninitialized until it is written to. A process is then used to model the behavior of the memory system. Since this is an asynchronous system, all inputs are listed in the sensitivity list (i.e., address, WE, and data_in). The process first checks whether the write enable line is asserted (WE = 1), which indicates a write cycle is being performed. If it is, then it makes an assignment to the RW signal at the location provided by the address input with the data provided by the data_in input. Since the RW array is indexed using integers, type conversions are required to convert the address from std_logic_vector to integer. When WE is not asserted (WE = 0), a read cycle is being performed. In this case, the process makes an assignment to data_out with the contents stored at the provided address. This assignment also requires type conversions to change the address from std_logic_vector to integer. The following syntax implements this behavior:

Example:

```
MEMORY: process (address, WE, data_in)
 begin
   if (WE = '1') then
    RW(to_integer(unsigned(address))) <= data_in;
   else
    data_out <= RW(to_integer(unsigned(address)));
   end if;
end process;
```

A read/write memory does not contain information until its storage locations are written to. As a result, if the memory is read from before it has been written to, the simulation will return *uninitialized*. Example 11.3 shows the entire VHDL model for an asynchronous read/write memory and the simulation waveform showing read/write cycles.

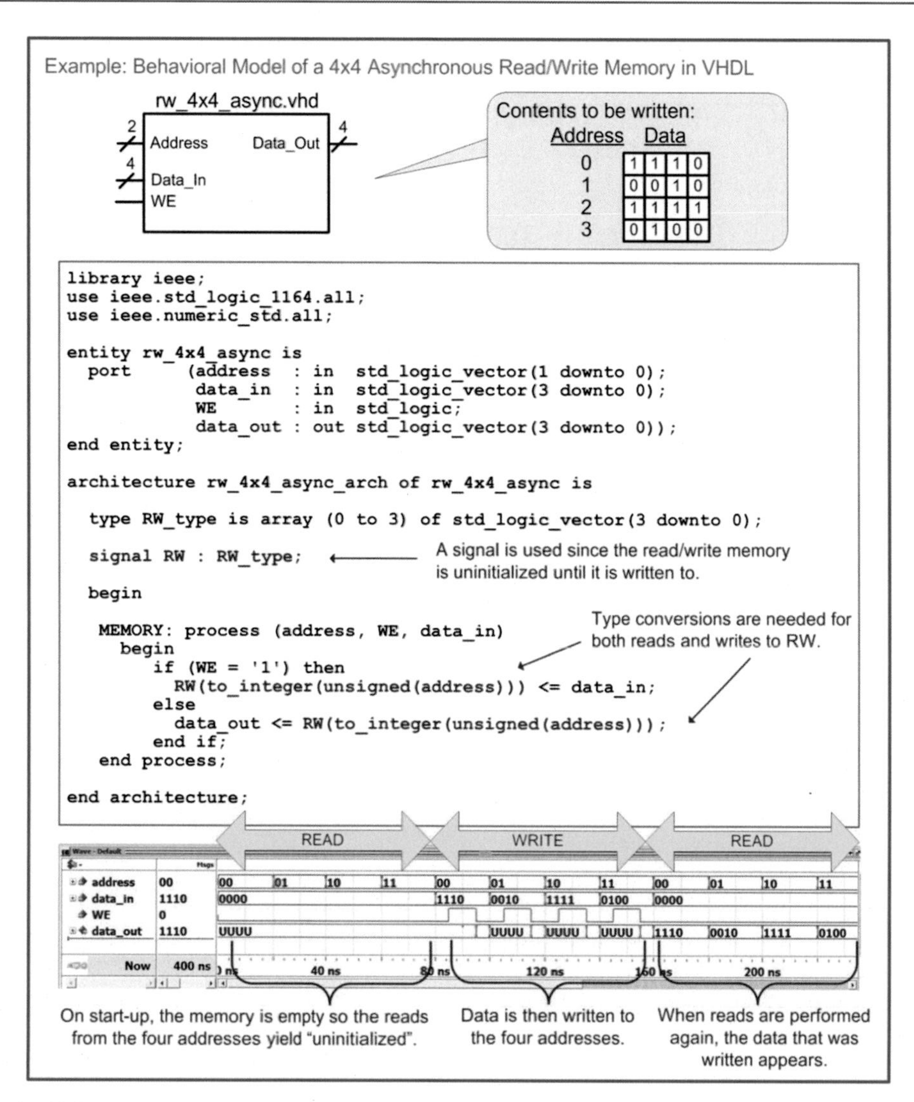

Example 11.3
Behavioral model of a 4 × 4 asynchronous read/write memory in VHDL

A synchronous read/write memory is made in a similar manner with the exception that a clock is used to trigger the signal assignments in the sensitivity list. The WE signal acts as a synchronous control signal indicating whether assignments are read from or written to the RW array. Example 11.4 shows the entire VHDL model for a synchronous read/write memory and the simulation waveform showing both read and write cycles.

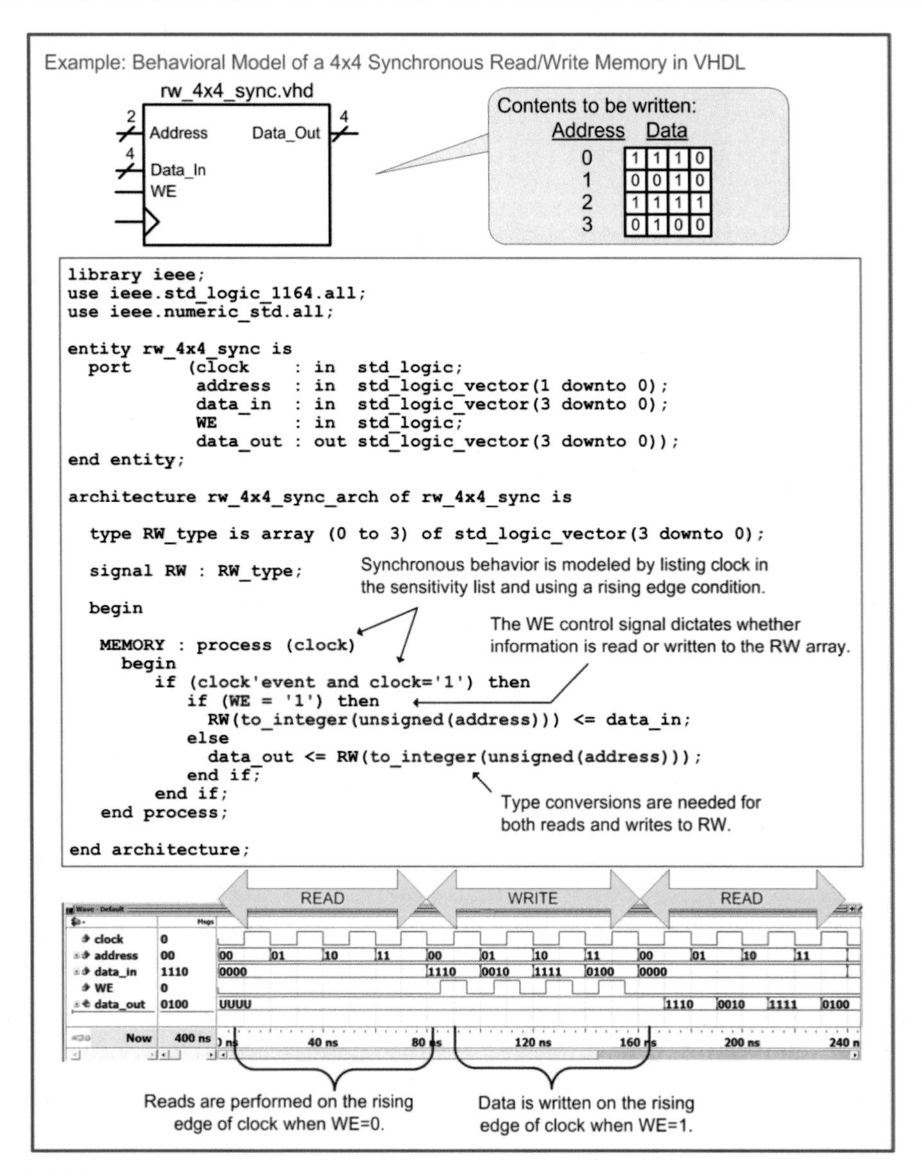

Example 11.4
Behavioral model of a 4 × 4 synchronous read/write memory in VHDL

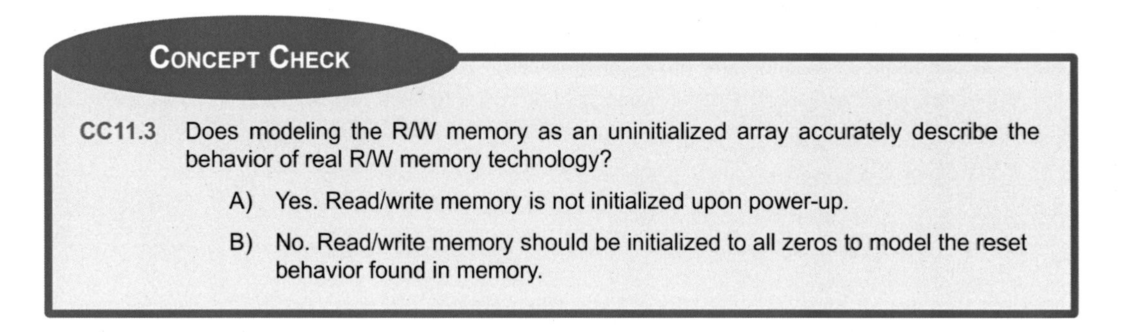

Concept Check

CC11.3 Does modeling the R/W memory as an uninitialized array accurately describe the behavior of real R/W memory technology?

A) Yes. Read/write memory is not initialized upon power-up.

B) No. Read/write memory should be initialized to all zeros to model the reset behavior found in memory.

Summary

- The term memory refers to large arrays of digital storage. The technology used in memory is typically optimized for storage density at the expense of control capability. This is different from a D-flip-flop, which is optimized for complete control at the bit level.
- A memory device always contains an address bus input. The number of bits in the address bus dictates how many storage locations can be accessed. An n-bit address bus can access 2^n (or M) storage locations.
- The width of each storage location (N) allows the density of the memory array to be increased by reading and writing vectors of data instead of individual bits.
- A memory map is a graphical depiction of a memory array. A memory map is useful to give an overview of the capacity of the array and how different address ranges of the array are used.
- A read is an operation in which data is retrieved from memory. A write is an operation in which data is stored to memory.
- An asynchronous memory array responds immediately to its control inputs. A synchronous memory array only responds on the triggering edge of clock.
- Volatile memory will lose its data when the power is removed. Non-volatile memory will retain its data when the power is removed.
- Read-only memory (ROM) is a memory type that cannot be written to during normal operation. Read/write (R/W) memory is a memory type that can be written to during normal operation. Both ROM and R/W memory can be read from during normal operation.
- Random-access memory (RAM) is a memory type in which any location in memory can be accessed at any time. In sequential access memory, the data can only be retrieved in a linear sequence. This means that in sequential memory the data cannot be accessed arbitrarily.
- Memory can be modeled in VHDL using the array data type.
- Read-only memory in VHDL is implemented as an array of constants.
- Read/write memory in VHDL is implemented an as array of signal vectors.

Exercise Problems

Section 11.1: Memory Architecture and Terminology

11.1.1 For a 512k × 32 memory system, how many unique address locations are there? Give the exact number.

11.1.2 For a 512k × 32 memory system, what is the data width at each address location?

11.1.3 For a 512k × 32 memory system, what is the *capacity* in bits?

11.1.4 For a 512k × 32-bit memory system, what is the *capacity* in bytes?

11.1.5 For a 512k × 32 memory system, how wide does the incoming address bus need to be in order to access every unique address location?

Section 11.2: Modeling Read-Only Memory

11.2.1 Design a VHDL model for the 16 × 8, asynchronous, read-only memory system shown in Fig. 11.2. The system should contain the information provided in the memory map. Create a test bench to simulate your model by reading from each of the 16 unique addresses and observing Data_Out to verify that it contains the information in the memory map.

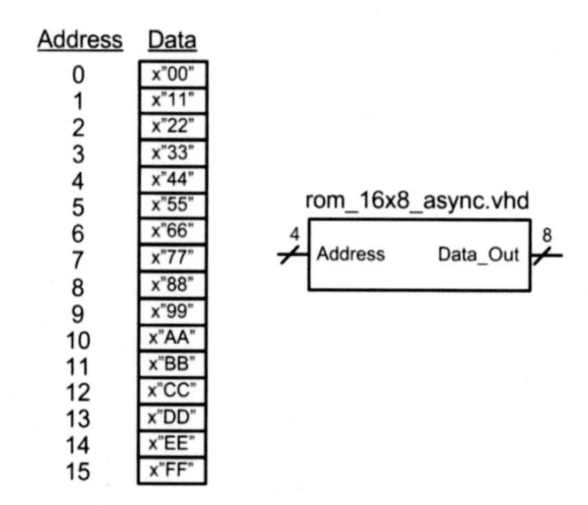

Fig. 11.2
16 × 8 asynchronous ROM block diagram

11.2.2 Design a VHDL model for the 16 × 8, synchronous, read-only memory system shown in Fig. 11.3. The system should contain the information provided in the memory map. Create a test bench to simulate your model by reading from each of the 16 unique addresses and observing Data_Out to verify that it contains the information in the memory map.

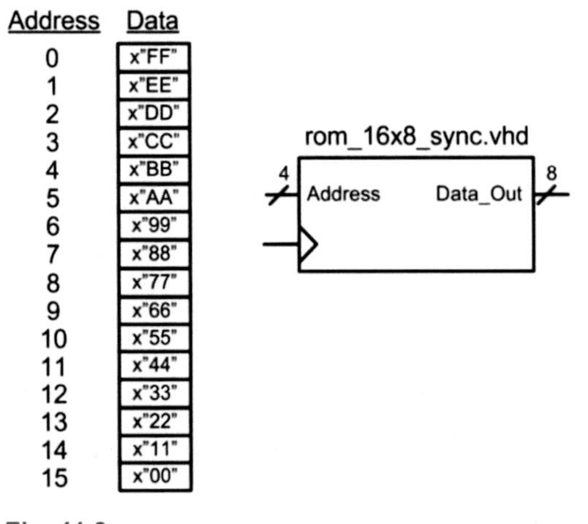

Fig. 11.3
16 × 8 synchronous ROM block diagram

Section 11.3: Modeling Read/Write Memory

11.3.1 Design a VHDL model for the 16 × 8, asynchronous, read/write memory system shown in Fig. 11.4. Create a test bench to simulate your model. Your test bench should first read from all of the address locations to verify that they are uninitialized. Next, your test bench should write unique information to each of the address locations. Finally, your test bench should read from each address location to verify that the information that was written was stored and can be successfully retrieved.

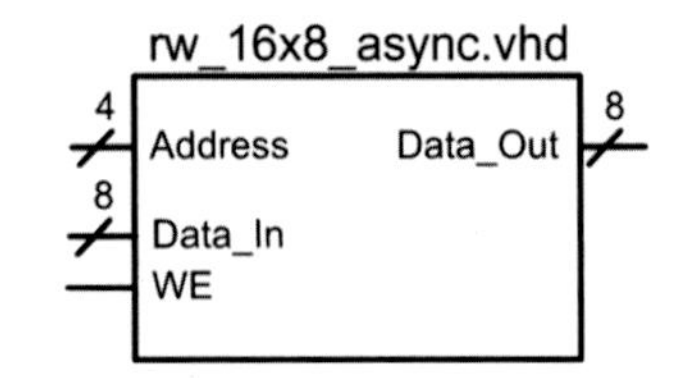

Fig. 11.4
16 × 8 asynchronous R/W block diagram

11.3.2 Design a VHDL model for the 16 × 8, synchronous, read/write memory system shown in Fig. 11.5. Create a test bench to simulate your model. Your test bench should first read from all of the address locations to verify that they are uninitialized. Next, your test bench should write unique information to each of the address locations. Finally, your test bench should read from each address location to verify that the information that was written was stored and can be successfully retrieved.

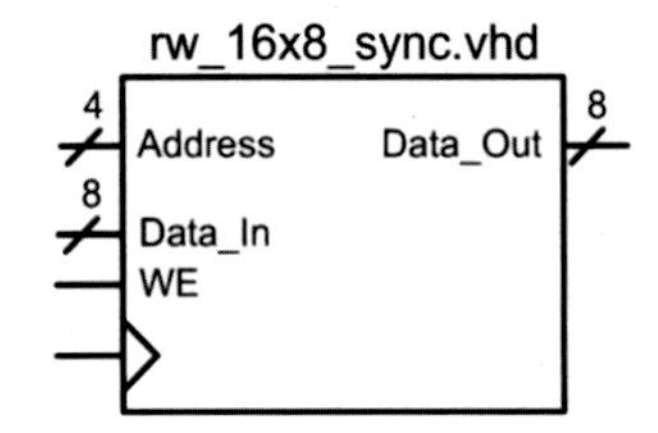

Fig. 11.5
16 × 8 synchronous R/W block diagram

Chapter 12: Computer System Design

This chapter presents the design of a simple computer system that will illustrate the use of many of the VHDL modeling techniques covered in this book. The goal of this chapter is not to provide an in-depth coverage of modern computer architecture, but rather to present a simple operational computer that can be implemented in VHDL to show how to use many of the modeling techniques covered thus far. This chapter begins with some architectural definitions so that consistent terminology can be used throughout the computer design example.

Learning Outcomes—After completing this chapter, you will be able to:

12.1 Describe the basic components and operation of computer hardware.
12.2 Describe the basic components and operation of computer software.
12.3 Design a fully operational computer system using VHDL.

12.1 Computer Hardware

A computer accomplishes tasks through an architecture that uses both *hardware* and *software*. The hardware in a computer consists of many of the elements that we have covered so far. These include registers, arithmetic and logic circuits, finite-state machines, and memory. What makes a computer so useful is that the hardware is designed to accomplish a predetermined set of **instructions**. These instructions are relatively simple, such as moving data between memory and a register or performing arithmetic on two numbers. The instructions are comprised of binary codes that are stored in a memory device and represent the sequence of operations that the hardware will perform to accomplish a task. This sequence of instructions is called a computer **program**. What makes this architecture so useful is that the preexisting hardware can be *programmed* to perform an almost unlimited number of tasks by simply defining the sequence of instructions to be executed. The process of designing the sequence of instructions, or program, is called *software development* or *software engineering*.

The idea of a general-purpose computing machine dates back to the nineteenth century. The first computing machines were implemented with mechanical systems and were typically analog in nature. As technology advanced, computer hardware evolved from electromechanical switches to vacuum tubes and ultimately to integrated circuits. These newer technologies enabled switching circuits and provided the capability to build binary computers. Today's computers are built exclusively with semiconductor materials and integrated circuit technology. The term *microcomputer* is used to describe a computer that has its processing hardware implemented with integrated circuitry. Nearly all modern computers are binary. Binary computers are designed to operate on a fixed set of bits. For example, an 8-bit computer would perform operations on 8 bits at a time. This means it moves data between registers and memory and performs arithmetic and logic operations in groups of 8 bits.

Computer hardware refers to all of the physical components within the system. This hardware includes all circuit components in a computer such as the memory devices, registers, and finite-state machines. Figure 12.1 shows a block diagram of the basic hardware components in a computer.

B. J. LaMeres, *Quick Start Guide to VHDL*, https://doi.org/10.1007/978-3-031-42543-1_12

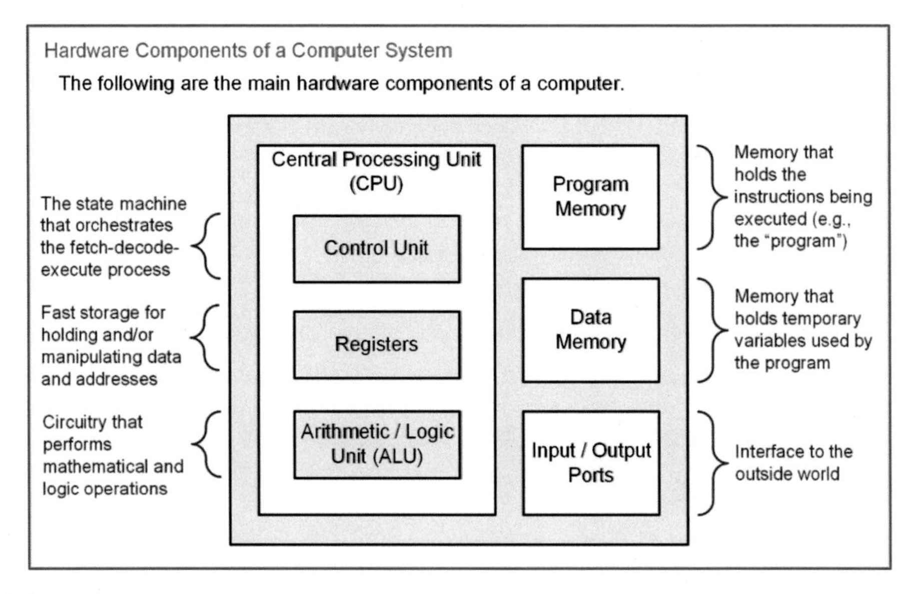

Fig. 12.1
Hardware components of a computer system

12.1.1 Program Memory

The instructions that are executed by a computer are held in *program memory*. Program memory is treated as read only during execution in order to prevent the instructions from being overwritten by the computer. Programs are typically held in non-volatile memory so that the computer system does not lose its program when power is removed. Modern computers will often copy a program from non-volatile memory (e.g., a hard disk drive) to volatile memory (i.e., SRAM or DRAM) after startup in order to speed up instruction execution as volatile memory is often a faster technology.

12.1.2 Data Memory

Computers also require *data memory*, which can be written to and read from during normal operation. This memory is used to hold temporary variables that are created by the software program. This memory expands the capability of the computer system by allowing large amounts of information to be created and stored by the program. Additionally, computations can be performed that are larger than the width of the computer system by holding interim portions of the calculation (e.g., performing a 128-bit addition on a 32-bit computer). Data memory is typically implemented with volatile memory as it is often faster than read-only memory technology.

12.1.3 Input/Output Ports

The term *port* is used to describe the mechanism to get information from the output world into or out of the computer. Ports can be input, output, or bidirectional. I/O ports can be designed to pass information in a serial or parallel format.

12.1.4 Central Processing Unit

The *central processing unit* (CPU) is considered the *brains* of the computer. The CPU handles reading instructions from memory, decoding them to understand which instruction is being performed, and executing the necessary steps to complete the instruction. The CPU also contains a set of registers that are used for general-purpose data storage, operational information, and system status. Finally, the CPU contains circuitry to perform arithmetic and logic operations on data.

12.1.4.1 Control Unit

The *control unit* is a finite-state machine that controls the operation of the computer. This FSM has states that perform fetching the instruction (i.e., reading it from program memory), decoding the instruction, and executing the appropriate steps to accomplish the instruction. This process is known as *fetch, decode, and execute* and is repeated each time an instruction is performed by the CPU. As the control unit state machine traverses through its states, it asserts control signals that move and manipulate data in order to achieve the desired functionality of the instruction.

12.1.4.2 Data Path: Registers

The CPU groups its registers and ALU into a sub-system called the *data path*. The data path refers to the fast storage and data manipulations within the CPU. All of these operations are initiated and managed by the control unit state machine. The CPU contains a variety of registers that are necessary to execute instructions and hold status information about the system. Basic computers have the following registers in their CPU:

- **Instruction register (IR)** – The instruction register holds the current binary code of the instruction being executed. This code is read from program memory as the first part of instruction execution. The IR is used by the control unit to decide which states in its FSM to traverse in order to execute the instruction.
- **Memory address register (MAR)** – The memory address register is used to hold the current address being used to access memory. The MAR can be loaded with addresses in order to fetch instructions from program memory or with addresses to access data memory and/or I/O ports.
- **Program counter (PC)** – The program counter holds the address of the current instruction being executed in program memory. The program counter will increment sequentially through the program memory reading instructions until a dedicated instruction is used to set it to a new location.
- **General-purpose registers** – These registers are available for temporary storage by the program. Instructions exist to move information from memory into these registers and to move information from these registers into memory. Instructions also exist to perform arithmetic and logic operations on the information held in these registers.
- **Condition code register (CCR)** – The condition code register holds status flags that provide information about the arithmetic and logic operations performed in the CPU. The most common flags are *negative* (N), zero (Z), two's complement overflow (V), and carry (C). This register can also contain flags that indicate the status of the computer, such as if an interrupt has occurred or if the computer has been put into a low-power mode.

12.1.4.3 Data Path: Arithmetic Logic Unit (ALU)

The *arithmetic logic unit* is the system that performs all mathematical (i.e., addition, subtraction, multiplication, and division) and logic operations (i.e., and, or, not, shifts, etc.). This system operates on data being held in CPU registers. The ALU has a unique symbol associated with it to distinguish it from other functional units in the CPU.

Figure 12.2 shows the typical organization of a CPU. The registers and ALU are grouped into the data path. In this example, the computer system has two general-purpose registers called A and B. This CPU organization will be used throughout this chapter to illustrate the detailed execution of instructions.

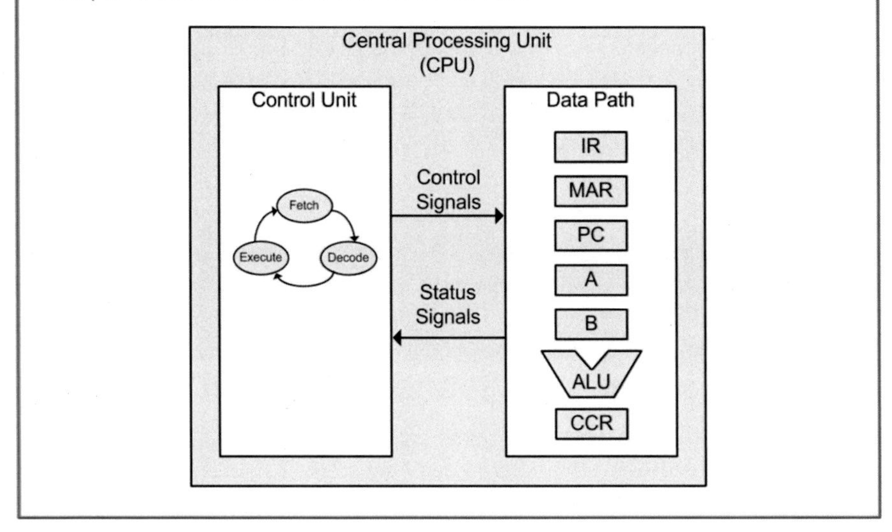

Fig. 12.2
Typical CPU organization

12.1.5 A Memory-Mapped System

A common way to simplify moving data in or out of the CPU is to assign a unique address to all hardware components in the memory system. Each input/output port and each location in both program and data memory are assigned a unique address. This allows the CPU to access everything in the memory system with a dedicated address. This reduces the number of lines that must pass into the CPU. A *bus system* facilitates transferring information within the computer system. An address bus is driven by the CPU to identify which location in the memory system is being accessed. A data bus is used to transfer information to/from the CPU and the memory system. Finally, a control bus is used to provide other required information about the transactions such as *read* or *write* lines. Figure 12.3 shows the computer hardware in a memory-mapped architecture.

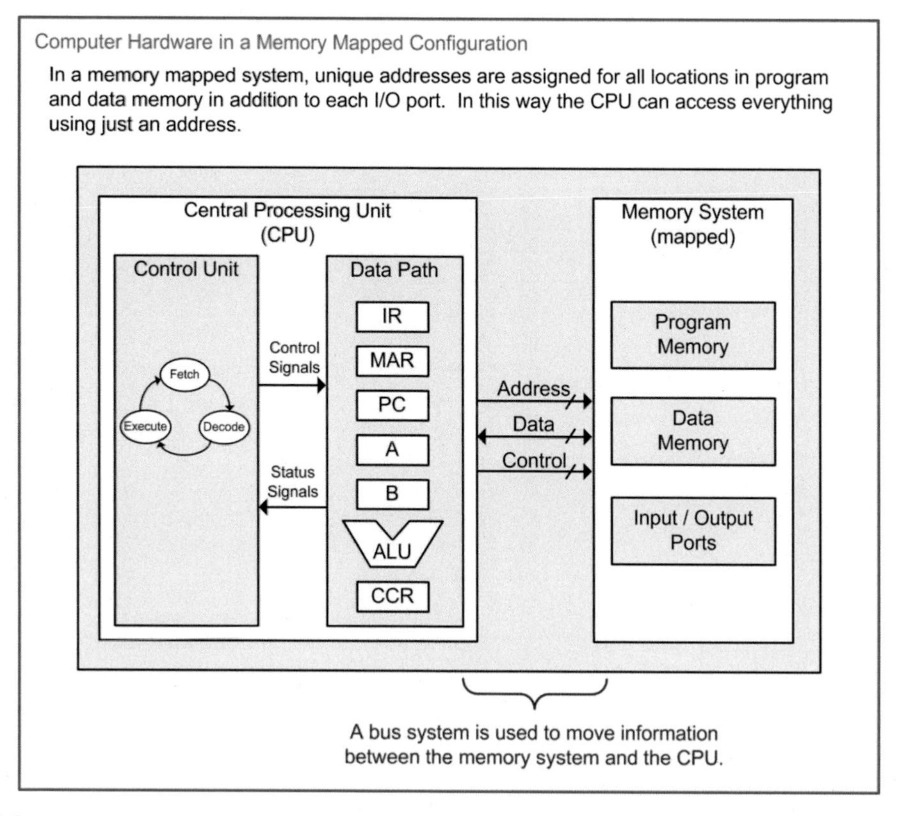

Fig. 12.3
Computer hardware in a memory-mapped configuration

To help visualize how the memory addresses are assigned, a *memory map* is used. This is a graphical depiction of the memory system. In the memory map, the ranges of addresses are provided for each of the main subsections of memory. This gives the programmer a quick overview of the available resources in the computer system. Example 12.1 shows a representative memory map for a computer system with an address bus with a width of 8 bits. This address bus can provide 256 unique locations. For this example, the memory system is also 8-bits wide; thus, the entire memory system is 256×8 in size. In this example 128 bytes are allocated for program memory; 96 bytes are allocated for data memory; 16 bytes are allocated for output ports; and 16 bytes are allocated for input ports.

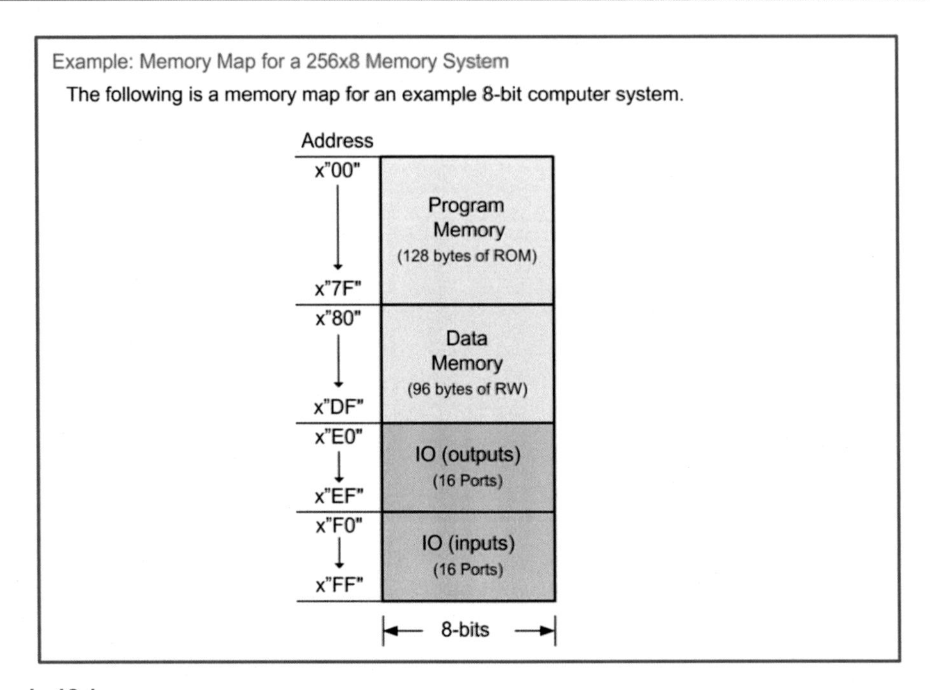

Example 12.1
Memory map for a 256×8 memory system

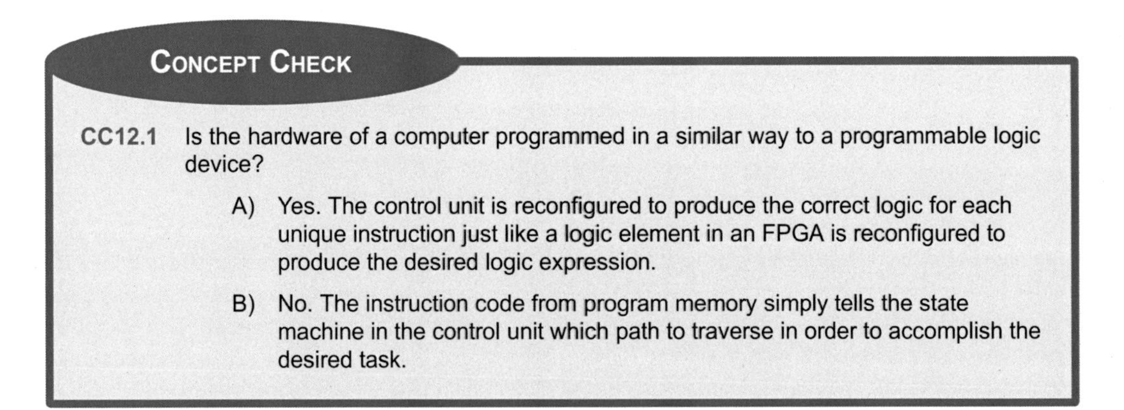

CONCEPT CHECK

CC12.1 Is the hardware of a computer programmed in a similar way to a programmable logic device?

A) Yes. The control unit is reconfigured to produce the correct logic for each unique instruction just like a logic element in an FPGA is reconfigured to produce the desired logic expression.

B) No. The instruction code from program memory simply tells the state machine in the control unit which path to traverse in order to accomplish the desired task.

12.2 Computer Software

Computer software refers to the instructions that the computer can execute and how they are designed to accomplish various tasks. The specific group of instructions that a computer can execute is known as its **instruction set**. The instruction set of a computer needs to be defined first before the computer hardware can be implemented. Some computer systems have a very small number of instructions in order to reduce the physical size of the circuitry needed in the CPU. This allows the CPU to execute the instructions very quickly but requires a large number of operations to accomplish a given task. This architectural approach is called a **reduced instruction set computer** (RISC). The alternative to this approach is to make an instruction set with a large number of dedicated instructions that can accomplish a given task in fewer CPU operations. The drawback of this approach is that the physical size of the CPU must be larger in order to accommodate the various instructions. This architectural approach is called a **complex instruction set computer** (CISC). The computer example in this chapter will use a RISC architecture.

12.2.1 Opcodes and Operands

A computer instruction consists of two fields, an *opcode* and an *operand*. The opcode is a unique binary code given to each instruction in the set. The CPU decodes the opcode in order to know which instruction is being executed and then takes the appropriate steps to complete the instruction. Each opcode is assigned a **mnemonic**, which is a descriptive name for the opcode that can be used when discussing the instruction functionally. An operand is additional information for the instruction that may be required. An instruction may have any number of operands including zero. Figure 12.4 shows an example of how the instruction opcodes and operands are placed into program memory.

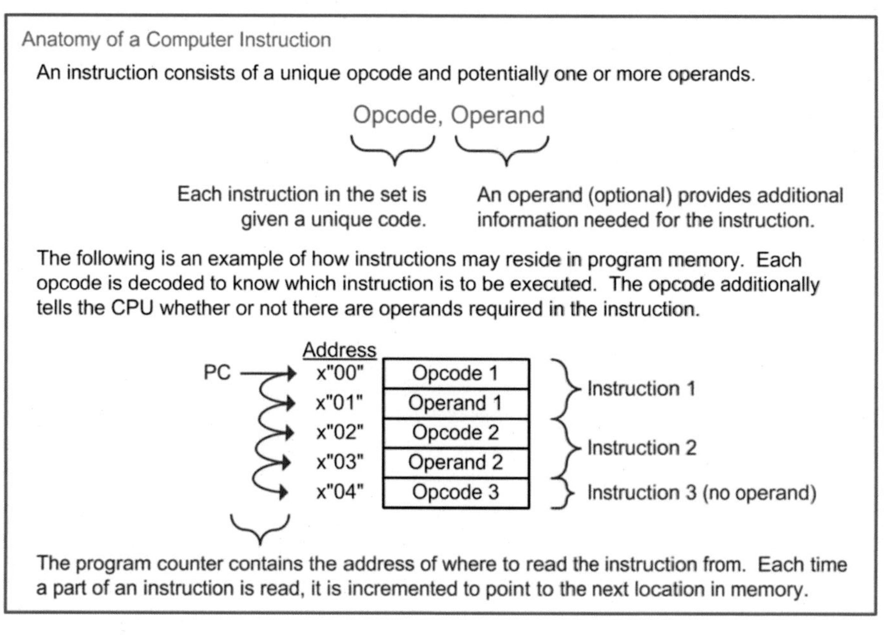

Fig. 12.4
Anatomy of a computer instruction

12.2.2 Addressing Modes

An *addressing mode* describes the way in which the operand of an instruction is used. While modern computer systems may contain numerous addressing modes with varying complexities, we will focus on just a subset of basic addressing modes that are needed to get a simple computer running. These modes are immediate, direct, and inherent.

12.2.2.1 Immediate Addressing (IMM)

Immediate addressing is when the operand of an instruction *is* the information to be used by the instruction. For example, if an instruction existed to put a constant into a register within the CPU using immediate addressing, the operand would *be* the constant. When the CPU reads the operand, it simply inserts the contents into the CPU register, and the instruction is complete.

12.2.2.2 Direct Addressing (DIR)

Direct addressing is when the operand of an instruction contains the *address* of where the information to be used is located. For example, if an instruction existed to put a constant into a register within the CPU using direct addressing, the operand would contain the address of *where* the constant was located in memory. When the CPU reads the operand, it puts this value out on the address bus and performs an additional read to retrieve the contents located at that address. The value read is then put into the CPU register, and the instruction is complete.

12.2.2.3 Inherent Addressing (INH)

Inherent addressing refers to an instruction that does not require an operand because the opcode itself contains all of the necessary information for the instruction to complete. This type of addressing is used on instructions that perform manipulations on data held in CPU registers without the need to access the memory system. For example, if an instruction existed to increment the contents of a register (A), then once the opcode is read by the CPU, it knows everything it needs to know in order to accomplish the task. The CPU simply asserts a series of control signals in order to increment the contents of A, and then the instruction is complete. Notice that no operand is needed for this task. Instead, the location of the register to be manipulated (i.e., A) is inherent within the opcode.

12.2.3 Classes of Instructions

There are three general classes of instructions: (1) loads and stores; (2) data manipulations; and (3) branches. To illustrate how these instructions are executed, examples will be given based on the computer architecture shown in Fig. 12.3.

12.2.3.1 Loads and Stores

This class of instructions accomplishes moving information between the CPU and memory. A **load** is an instruction that moves information from memory *into* a CPU register. When a load instruction uses immediate addressing, the operand of the instruction *is* the data to be loaded into the CPU register. As an example, let's look at an instruction to load the general-purpose register A using immediate addressing. Let's say that the opcode of the instruction is x"86", has a mnemonic LDA_IMM, and is inserted into program memory starting at x"00". Example 12.2 shows the steps involved in executing the LDA_IMM instruction.

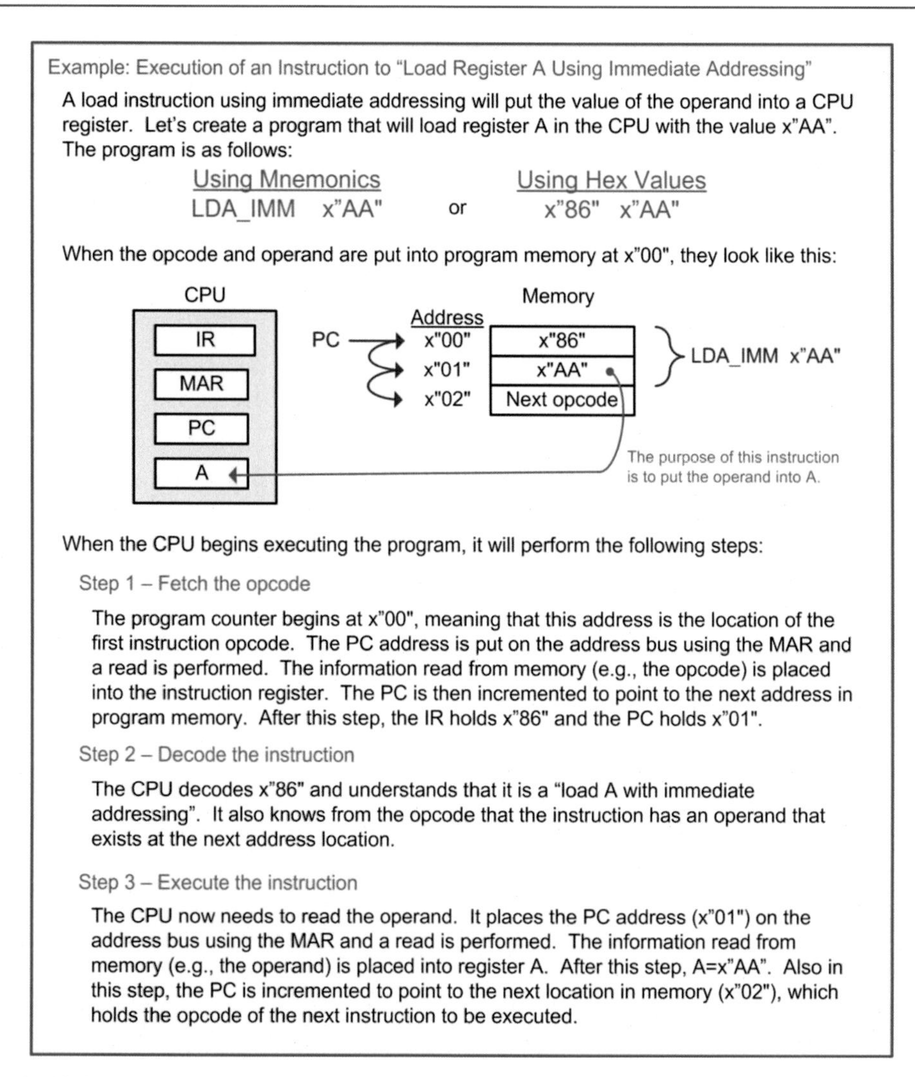

Example: Execution of an Instruction to "Load Register A Using Immediate Addressing"

A load instruction using immediate addressing will put the value of the operand into a CPU register. Let's create a program that will load register A in the CPU with the value x"AA". The program is as follows:

Using Mnemonics		Using Hex Values
LDA_IMM x"AA"	or	x"86" x"AA"

When the opcode and operand are put into program memory at x"00", they look like this:

When the CPU begins executing the program, it will perform the following steps:

Step 1 – Fetch the opcode

The program counter begins at x"00", meaning that this address is the location of the first instruction opcode. The PC address is put on the address bus using the MAR and a read is performed. The information read from memory (e.g., the opcode) is placed into the instruction register. The PC is then incremented to point to the next address in program memory. After this step, the IR holds x"86" and the PC holds x"01".

Step 2 – Decode the instruction

The CPU decodes x"86" and understands that it is a "load A with immediate addressing". It also knows from the opcode that the instruction has an operand that exists at the next address location.

Step 3 – Execute the instruction

The CPU now needs to read the operand. It places the PC address (x"01") on the address bus using the MAR and a read is performed. The information read from memory (e.g., the operand) is placed into register A. After this step, A=x"AA". Also in this step, the PC is incremented to point to the next location in memory (x"02"), which holds the opcode of the next instruction to be executed.

Example 12.2
Execution of an instruction to "Load Register A Using Immediate Addressing"

Now let's look at a load instruction using direct addressing. In direct addressing, the operand of the instruction is the *address* of where the data to be loaded resides. As an example, let's look at an instruction to load the general-purpose register A. Let's say that the opcode of the instruction is x"87", has a mnemonic LDA_DIR, and is inserted into program memory starting at x"08". The value to be loaded into A resides at address x"80", which has already been initialized with x"AA" before this instruction. Example 12.3 shows the steps involved in executing the LDA_DIR instruction.

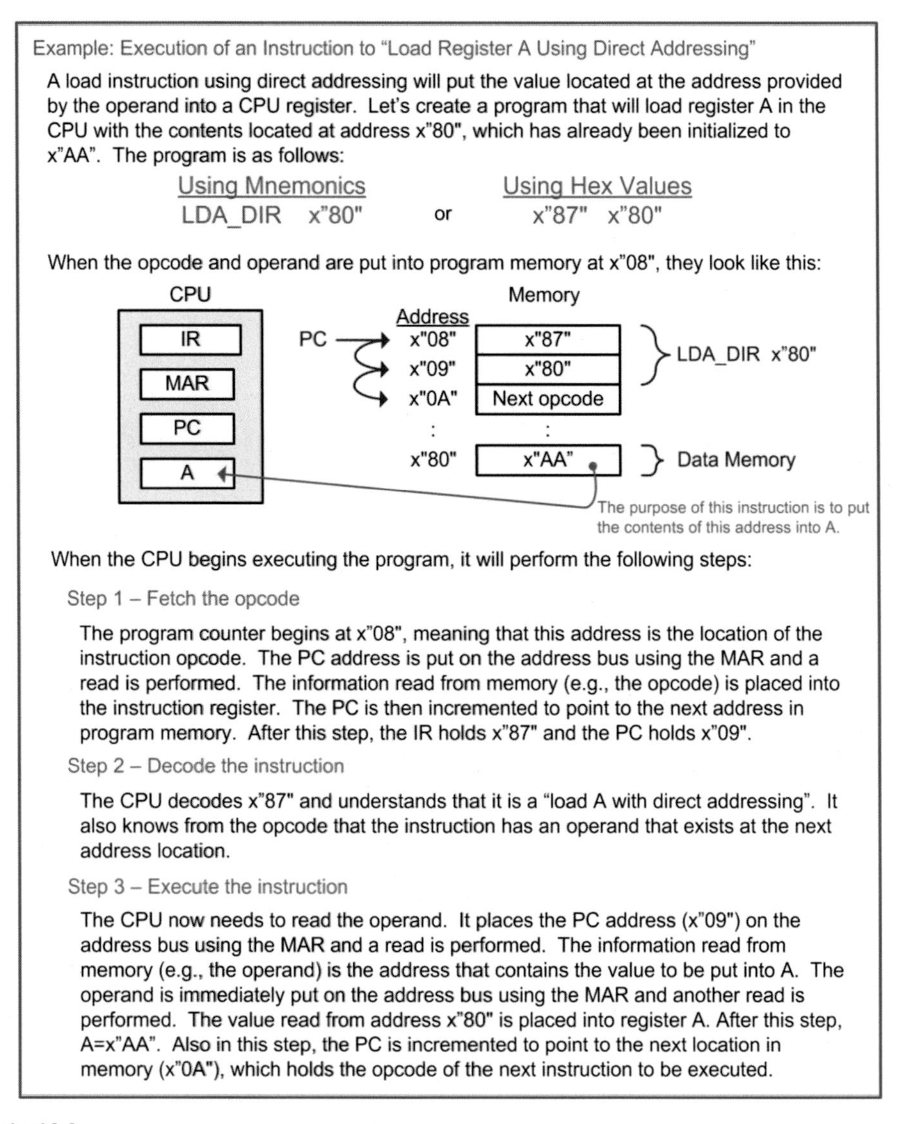

Example: Execution of an Instruction to "Load Register A Using Direct Addressing"

A load instruction using direct addressing will put the value located at the address provided by the operand into a CPU register. Let's create a program that will load register A in the CPU with the contents located at address x"80", which has already been initialized to x"AA". The program is as follows:

Using Mnemonics		Using Hex Values
LDA_DIR x"80"	or	x"87" x"80"

When the opcode and operand are put into program memory at x"08", they look like this:

When the CPU begins executing the program, it will perform the following steps:

Step 1 – Fetch the opcode

The program counter begins at x"08", meaning that this address is the location of the instruction opcode. The PC address is put on the address bus using the MAR and a read is performed. The information read from memory (e.g., the opcode) is placed into the instruction register. The PC is then incremented to point to the next address in program memory. After this step, the IR holds x"87" and the PC holds x"09".

Step 2 – Decode the instruction

The CPU decodes x"87" and understands that it is a "load A with direct addressing". It also knows from the opcode that the instruction has an operand that exists at the next address location.

Step 3 – Execute the instruction

The CPU now needs to read the operand. It places the PC address (x"09") on the address bus using the MAR and a read is performed. The information read from memory (e.g., the operand) is the address that contains the value to be put into A. The operand is immediately put on the address bus using the MAR and another read is performed. The value read from address x"80" is placed into register A. After this step, A=x"AA". Also in this step, the PC is incremented to point to the next location in memory (x"0A"), which holds the opcode of the next instruction to be executed.

Example 12.3
Execution of an instruction to "Load Register A Using Direct Addressing"

A **store** is an instruction that moves information from a CPU register *into* memory. The operand of a store instruction indicates the address of where the contents of the CPU register will be written. As an example, let's look at an instruction to store the general-purpose register A into memory address x"E0". Let's say that the opcode of the instruction is x"96", has a mnemonic STA_DIR, and is inserted into program memory starting at x"04". The initial value of A is x"CC" before the instruction is executed. Example 12.4 shows the steps involved in executing the STA_DIR instruction.

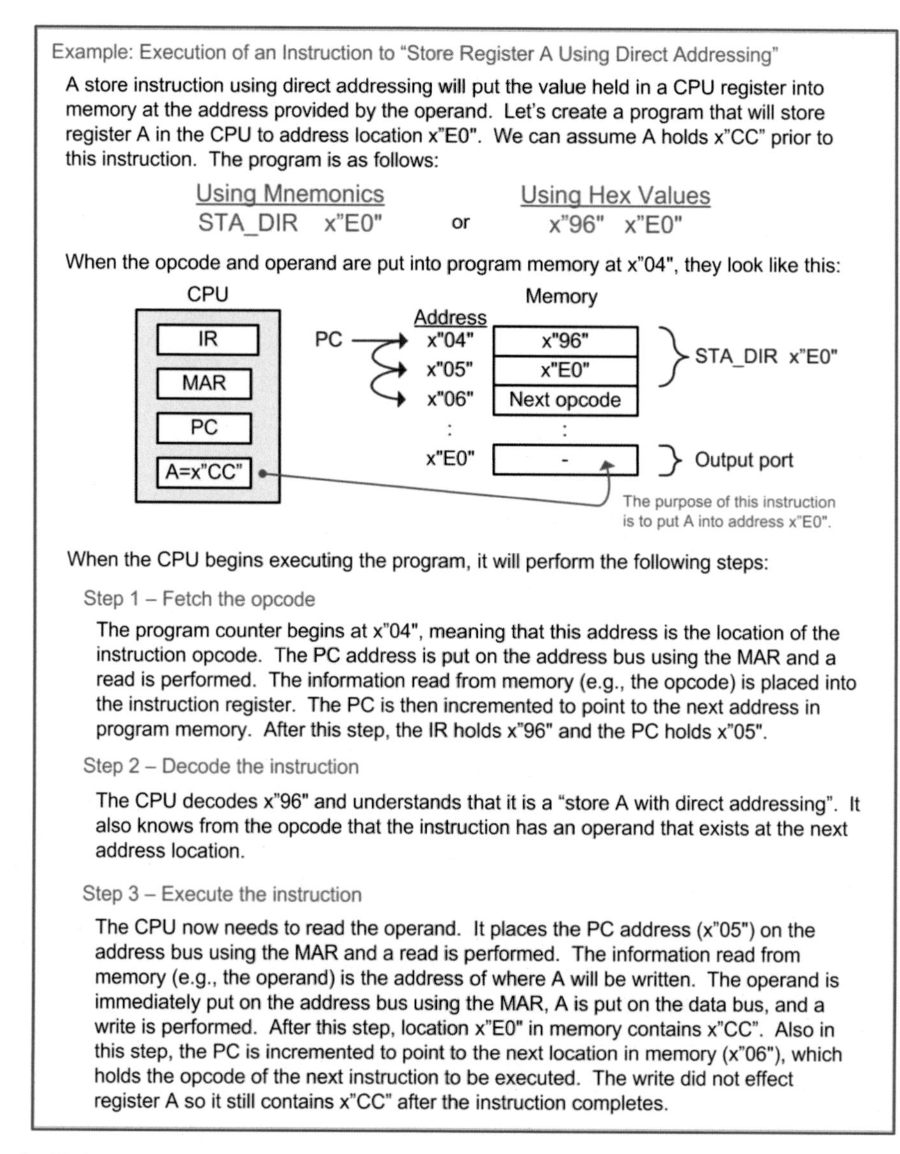

Example: Execution of an Instruction to "Store Register A Using Direct Addressing"

A store instruction using direct addressing will put the value held in a CPU register into memory at the address provided by the operand. Let's create a program that will store register A in the CPU to address location x"E0". We can assume A holds x"CC" prior to this instruction. The program is as follows:

Using Mnemonics		Using Hex Values
STA_DIR x"E0"	or	x"96" x"E0"

When the opcode and operand are put into program memory at x"04", they look like this:

When the CPU begins executing the program, it will perform the following steps:

Step 1 – Fetch the opcode

The program counter begins at x"04", meaning that this address is the location of the instruction opcode. The PC address is put on the address bus using the MAR and a read is performed. The information read from memory (e.g., the opcode) is placed into the instruction register. The PC is then incremented to point to the next address in program memory. After this step, the IR holds x"96" and the PC holds x"05".

Step 2 – Decode the instruction

The CPU decodes x"96" and understands that it is a "store A with direct addressing". It also knows from the opcode that the instruction has an operand that exists at the next address location.

Step 3 – Execute the instruction

The CPU now needs to read the operand. It places the PC address (x"05") on the address bus using the MAR and a read is performed. The information read from memory (e.g., the operand) is the address of where A will be written. The operand is immediately put on the address bus using the MAR, A is put on the data bus, and a write is performed. After this step, location x"E0" in memory contains x"CC". Also in this step, the PC is incremented to point to the next location in memory (x"06"), which holds the opcode of the next instruction to be executed. The write did not effect register A so it still contains x"CC" after the instruction completes.

Example 12.4
Execution of an instruction to "Store Register A Using Direct Addressing"

12.2.3.2 Data Manipulations

This class of instructions refers to ALU operations. These operations act on data that resides in the CPU registers. These instructions include arithmetic, logic operators, shifts and rotates, and tests and compares. Data manipulation instructions typically use inherent addressing because the operations are conducted on the contents of CPU registers and don't require additional memory access. As an example, let's look at an instruction to perform addition on registers A and B. The sum will be placed back in A. Let's say that the opcode of the instruction is x"42", has a mnemonic ADD_AB, and is inserted into program memory starting at x"04". Example 12.5 shows the steps involved in executing the ADD_AB instruction.

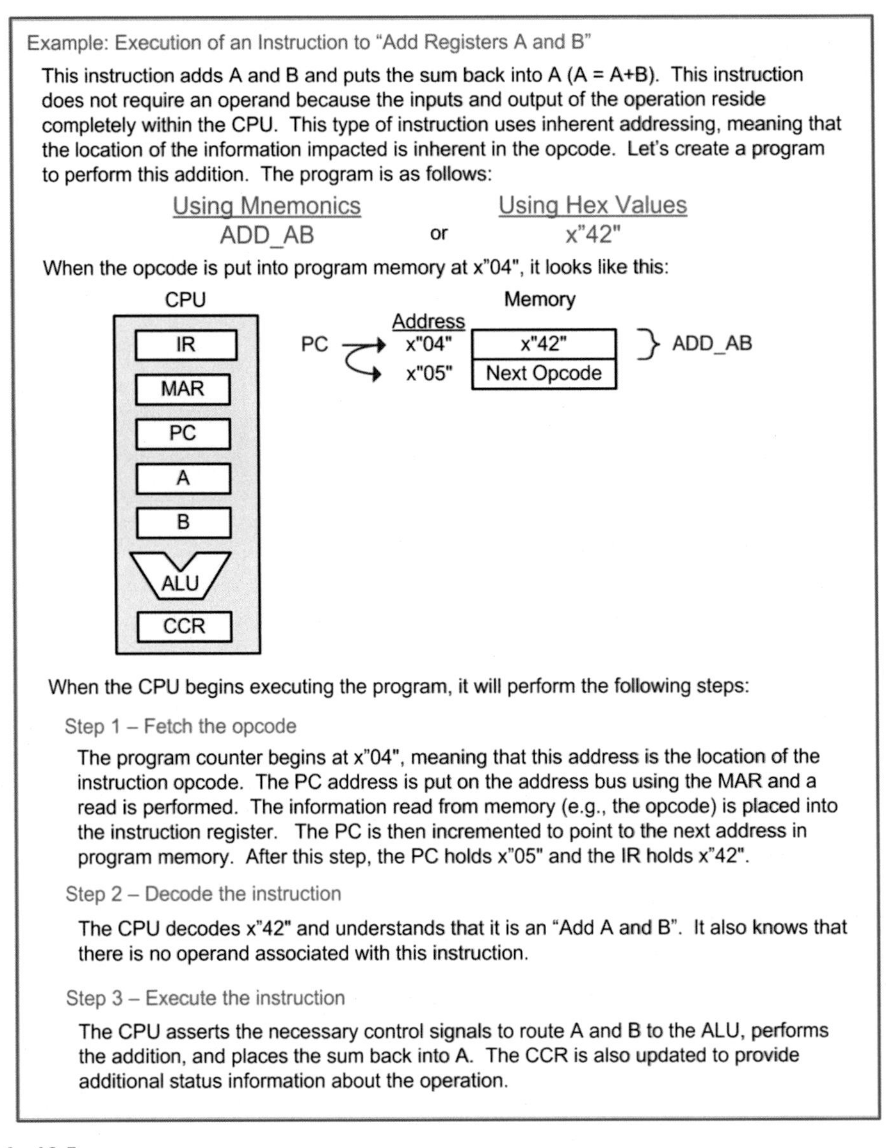

Example: Execution of an Instruction to "Add Registers A and B"

This instruction adds A and B and puts the sum back into A (A = A+B). This instruction does not require an operand because the inputs and output of the operation reside completely within the CPU. This type of instruction uses inherent addressing, meaning that the location of the information impacted is inherent in the opcode. Let's create a program to perform this addition. The program is as follows:

Using Mnemonics		Using Hex Values
ADD_AB	or	x"42"

When the opcode is put into program memory at x"04", it looks like this:

When the CPU begins executing the program, it will perform the following steps:

Step 1 – Fetch the opcode

The program counter begins at x"04", meaning that this address is the location of the instruction opcode. The PC address is put on the address bus using the MAR and a read is performed. The information read from memory (e.g., the opcode) is placed into the instruction register. The PC is then incremented to point to the next address in program memory. After this step, the PC holds x"05" and the IR holds x"42".

Step 2 – Decode the instruction

The CPU decodes x"42" and understands that it is an "Add A and B". It also knows that there is no operand associated with this instruction.

Step 3 – Execute the instruction

The CPU asserts the necessary control signals to route A and B to the ALU, performs the addition, and places the sum back into A. The CCR is also updated to provide additional status information about the operation.

Example 12.5
Execution of an instruction to "Add Registers A and B"

12.2.3.3 Branches

In the previous examples the program counter was always incremented to point to the address of the next instruction in program memory. This behavior only supports a linear execution of instructions. To provide the ability to specifically set the value of the program counter, instructions called *branches* are used. There are two types of branches: **unconditional** and **conditional**. In an unconditional branch, the program counter is always loaded with the value provided in the operand. As an example, let's look at an instruction to *branch always* to a specific address. This allows the program to perform loops. Let's say that the opcode of the instruction is x"20", has a mnemonic BRA, and is inserted into program memory starting at x"06". Example 12.6 shows the steps involved in executing the BRA instruction.

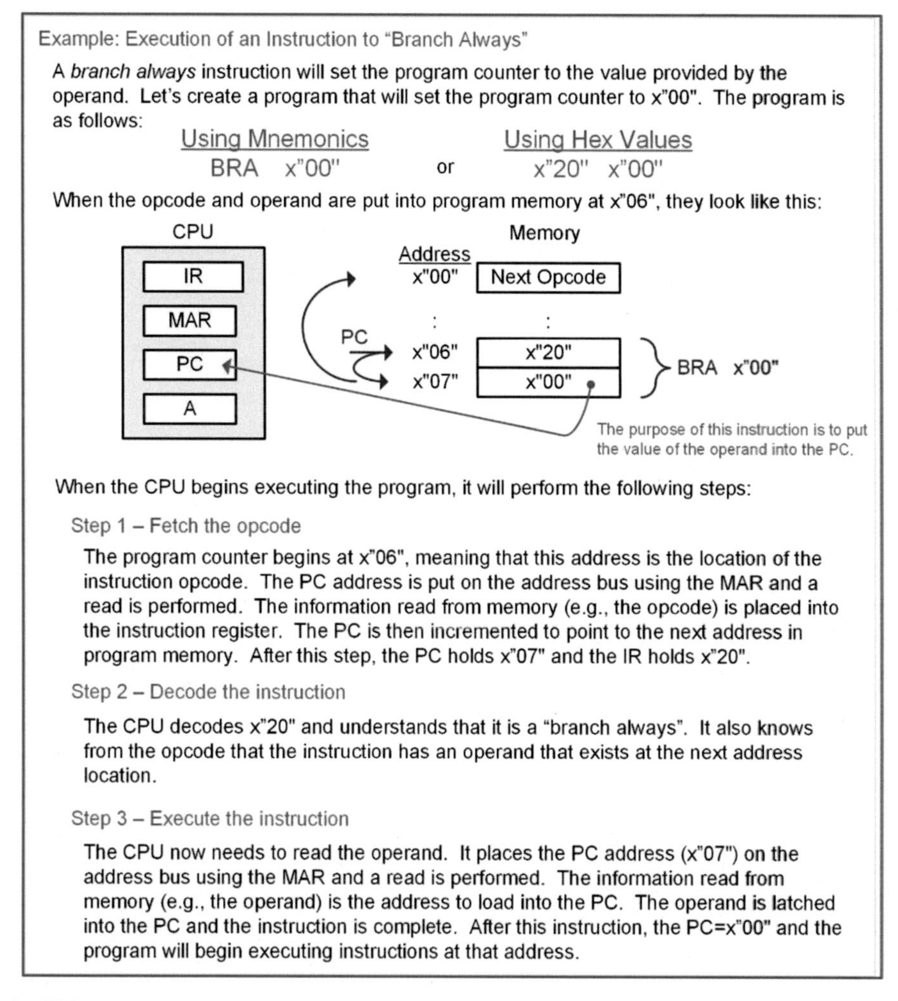
Example: Execution of an Instruction to "Branch Always"

A *branch always* instruction will set the program counter to the value provided by the operand. Let's create a program that will set the program counter to x"00". The program is as follows:

Using Mnemonics		Using Hex Values
BRA x"00"	or	x"20" x"00"

When the opcode and operand are put into program memory at x"06", they look like this:

When the CPU begins executing the program, it will perform the following steps:

Step 1 – Fetch the opcode

The program counter begins at x"06", meaning that this address is the location of the instruction opcode. The PC address is put on the address bus using the MAR and a read is performed. The information read from memory (e.g., the opcode) is placed into the instruction register. The PC is then incremented to point to the next address in program memory. After this step, the PC holds x"07" and the IR holds x"20".

Step 2 – Decode the instruction

The CPU decodes x"20" and understands that it is a "branch always". It also knows from the opcode that the instruction has an operand that exists at the next address location.

Step 3 – Execute the instruction

The CPU now needs to read the operand. It places the PC address (x"07") on the address bus using the MAR and a read is performed. The information read from memory (e.g., the operand) is the address to load into the PC. The operand is latched into the PC and the instruction is complete. After this instruction, the PC=x"00" and the program will begin executing instructions at that address.

Example 12.6
Execution of an instruction to "Branch Always"

In a conditional branch, the program counter is only updated if a particular condition is true. The conditions come from the status flags in the condition code register (NZVC). This allows a program to selectively execute instructions based on the result of a prior operation. Let's look at an example instruction that will branch only if the Z flag is asserted. This instruction is called a *branch if equal to zero*. Let's say that the opcode of the instruction is x"23", has a mnemonic BEQ, and is inserted into program memory starting at x"05". Example 12.7 shows the steps involved in executing the BEQ instruction.

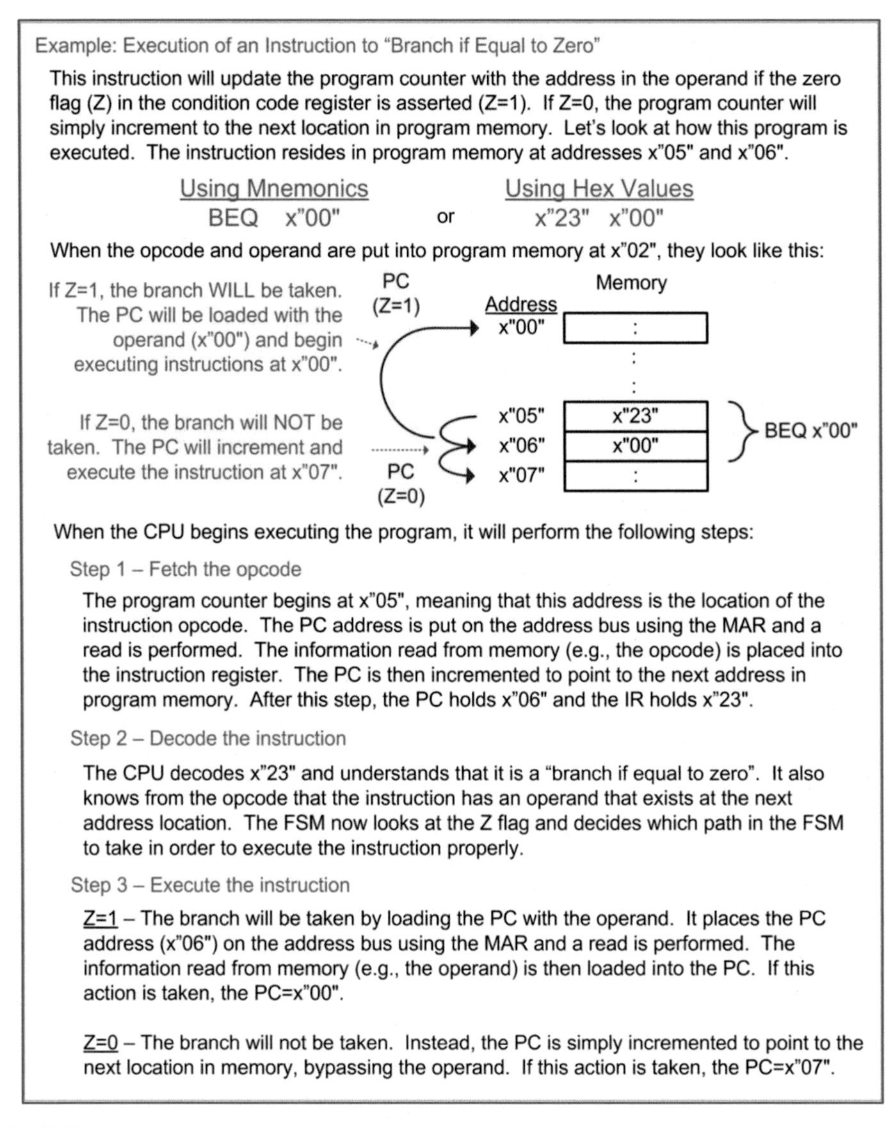

Example 12.7
Execution of an instruction to "Branch if Equal to Zero"

Conditional branches allow computer programs to make *decisions* about which instructions to execute based on the results of previous instructions. This gives computers the ability to react to input signals or act based on the results of arithmetic or logic operations. Computer instruction sets typically contain conditional branches based on the NZVC flags in the condition code registers. The following instructions are a set of possible branches that could be created using the values of the NZVC flags:

- BMI – Branch if minus (N = 1)
- BPL – Branch if plus (N = 0)
- BEQ – Branch if equal to Zero (Z = 1)
- BNE – Branch if not equal to Zero (Z = 0)

- BVS – Branch if two's complement overflow occurred, or V is set (V = 1)
- BVC – Branch if two's complement overflow did not occur, or V is clear (V = 0)
- BCS – Branch if a carry occurred, or C is set (C = 1)
- BCC – Branch if a carry did not occur, or C is clear (C = 0)

Combinations of these flags can be used to create more conditional branches.

- BHI – Branch if higher (C = 1 and Z = 0)
- BLS – Branch if lower or the same (C = 0 and Z = 1)
- BGE – Branch if greater than or equal ((N = 0 and V = 0) or (N = 1 and V = 1)), only valid for signed numbers
- BLT – Branch if less than ((N = 1 and V = 0) or (N = 0 and V = 1)), only valid for signed numbers
- BGT – Branch if greater than ((N = 0 and V = 0 and Z = 0) or (N = 1 and V = 1 and Z = 0)), only valid for signed numbers
- BLE – Branch if less than or equal ((N = 1 and V = 0) or (N = 0 and V = 1) or (Z = 1)), only valid for signed numbers

Concept Check

CC12.2 Software development consists of choosing which instructions, and in what order, will be executed to accomplish a certain task. The group of instructions is called the *program* and is inserted into program memory. Which of the following might a software developer care about?

A) Minimizing the number of instructions that need to be executed to accomplish the task in order to increase the computation rate.

B) Minimizing the number of registers used in the CPU to save power.

C) Minimizing the overall size of the program to reduce the amount of program memory needed.

D) Both A and C.

12.3 Computer Implementation: An 8-Bit Computer Example

12.3.1 Top-Level Block Diagram

Let's now look at the detailed implementation and instruction execution of a computer system in VHDL. In order to illustrate the detailed operation, we will use a simple 8-bit computer system design. Example 12.8 shows the block diagram for the 8-bit computer system. This block diagram also contains the VHDL file and entity names, which will be used when the behavioral model is implemented.

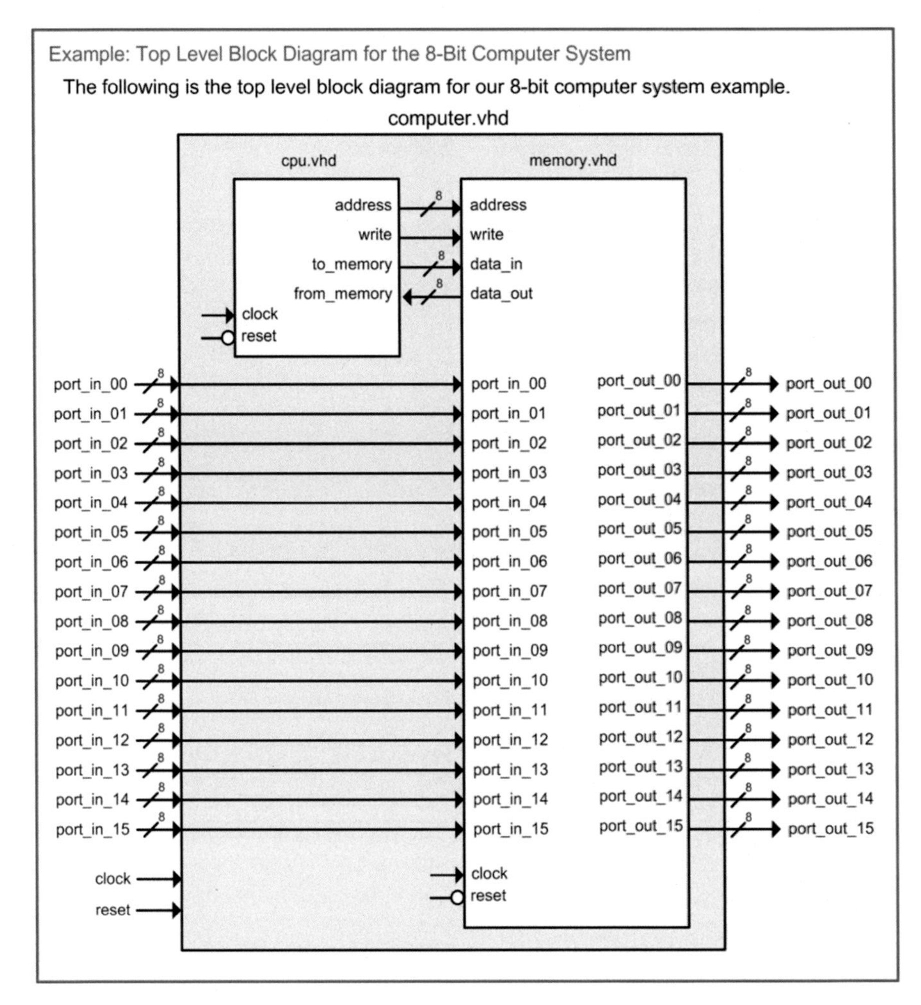

Example 12.8
Top-level block diagram for the 8-bit computer system

We will use the memory map shown in Example 12.1 for our example computer system. This mapping provides 128 bytes of program memory, 96 bytes of data memory, 16x output ports, and 16x input ports. To simplify the operation of this example computer, the address bus is limited to 8 bits. This only provides 256 locations of memory access but allows an entire address to be loaded into the CPU as a single operand of an instruction.

12.3.2 Instruction Set Design

Example 12.9 shows a basic instruction set for our example computer system. This set provides a variety of loads and stores, data manipulations, and branch instructions that will allow the computer to be programmed to perform more complex tasks through software development. These instructions are sufficient to provide a baseline of functionality in order to get the computer system operational. Additional instructions can be added as desired to increase the complexity of the system.

Example: Instruction Set for the 8-Bit Computer System

The following is a base set of instructions that the 8-bit computer system will be able to perform. Each instruction is given a descriptive mnemonic, which allows the system implementation and the programming to be more intuitive. Each instruction is also provided with a unique binary opcode. Some instructions have an operand, which provides additional information necessary for the instruction. If an instruction contains an operand, a description is provided as to how it is used (e.g., as data or as an address).

Mnemonic	Opcode, Operand	Description
"Loads and Stores"		
LDA_IMM	x"86", *<data>*	Load Register A using Immediate Addressing
LDA_DIR	x"87", *<addr>*	Load Register A using Direct Addressing
LDB_IMM	x"88", *<data>*	Load Register B with Immediate Addressing
LDB_DIR	x"89", *<addr>*	Load Register B with Direct Addressing
STA_DIR	x"96", *<addr>*	Store Register A to Memory using Direct Addressing
STB_DIR	x"97", *<addr>*	Store Register B to Memory using Direct Addressing
"Data Manipulations"		
ADD_AB	x"42"	A = A + B (plus)
"Branches"		
BRA	x"20", *<addr>*	Branch Always to Address Provided
BEQ	x"23", *<addr>*	Branch to Address Provided if Z=1

Example 12.9
Instruction set for the 8-bit computer system

12.3.3 Memory System Implementation

Let's now look at the memory system details. The memory system contains program memory, data memory, and input/output ports. Example 12.10 shows the block diagram of the memory system. The program and data memory will be implemented using lower-level components (rom_128x8_sync.vhd and rw_96x8_sync.vhd), while the input and output ports can be modeled using a combination of RTL processes and combinational logic. The program and data memory components contain dedicated circuitry to handle their addressing ranges. Each output port also contains dedicated circuitry to handle its unique address. A multiplexer is used to handle the signal routing back to the CPU based on the address provided.

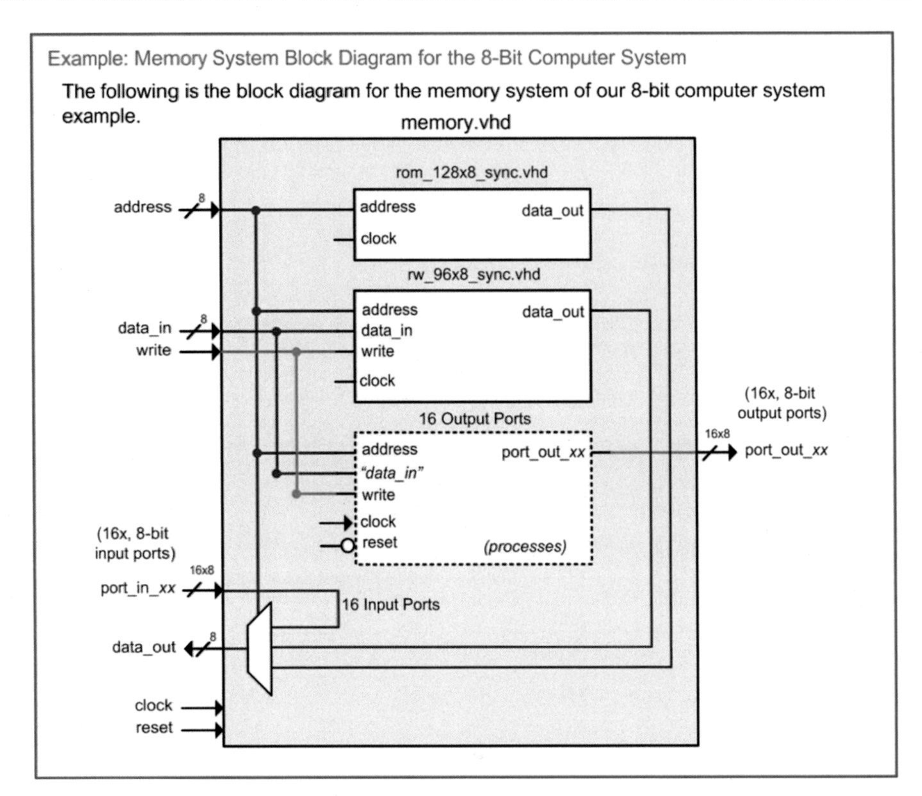

Example 12.10
Memory system block diagram for the 8-bit computer system

12.3.3.1 Program Memory Implementation in VHDL

The program memory can be implemented in VHDL using the modeling techniques presented in Chap. 11. To make the VHDL more readable, the instruction mnemonics can be declared as constants. This allows the mnemonic to be used when populating the program memory array. The following VHDL shows how the mnemonics for our basic instruction set can be defined as constants:

```
constant LDA_IMM  : std_logic_vector (7 downto 0) := x"86";
constant LDA_DIR  : std_logic_vector (7 downto 0) := x"87";
constant LDB_IMM  : std_logic_vector (7 downto 0) := x"88";
constant LDB_DIR  : std_logic_vector (7 downto 0) := x"89";
constant STA_DIR  : std_logic_vector (7 downto 0) := x"96";
constant STB_DIR  : std_logic_vector (7 downto 0) := x"97";
constant ADD_AB   : std_logic_vector (7 downto 0) := x"42";
constant BRA      : std_logic_vector (7 downto 0  := x"20";
constant BEQ      : std_logic_vector (7 downto 0) := x"23";
```

Now the program memory can be declared as an array type with initial values to define the program. The following VHDL shows how to declare the program memory and an example program to perform a load, store, and a branch always. This program will continually write x"AA" to port_out_00.

```
type rom_type is array (0 to 127) of std_logic_vector(7 downto 0);

constant ROM : rom_type := (0      => LDA_IMM,
                            1      => x"AA",
                            2      => STA_DIR,
                            3      => x"E0",
                            4      => BRA,
                            5      => x"00",
                            others => x"00");
```

The address mapping for the program memory is handled in two ways. First, notice that the array type defined above uses indices from 0 to 127. This provides the appropriate addresses for each location in the memory. The second step is to create an internal enable line that will only allow assignments from ROM to data_out when a valid address is entered. Consider the following VHDL to create an internal enable (EN) that will only be asserted when the address falls within the valid program memory range of 0–127:

```
enable : process (address)
  begin
  if ((to_integer(unsigned(address)) >= 0) and
      (to_integer(unsigned(address)) <= 127)) then
     EN <= '1';
  else
     EN <= '0';
  end if;
end process;
```

If this enable signal is not created, the simulation and synthesis will fail because data_out assignments will be attempted for addresses outside of the defined range of the ROM array. This enable line can now be used in the behavioral model for the ROM process as follows:

```
memory : process (clock)
  begin
  if (clock'event and clock='1') then
     if (EN='1') then
        data_out <= ROM(to_integer(unsigned(address)));
     end if;
  end if;
end process;
```

12.3.3.2 Data Memory Implementation in VHDL

The data memory is created using a similar strategy as the program memory. An array signal is declared with an address range corresponding to the memory map for the computer system (i.e., 128–223). An internal enable is again created that will prevent data_out assignments for addresses outside of this valid range. The following is the VHDL to declare the R/W memory array:

```
type rw_type is array (128 to 223) of std_logic_vector(7 downto 0);
signal RW : rw_type;
```

The following is the VHDL to model the local enable and signal assignments for the R/W memory:

```
enable : process (address)
 begin
   if ( (to_integer(unsigned(address)) >= 128) and
        (to_integer(unsigned(address)) <= 223)) then
      EN <= '1';
   else
      EN <= '0';
   end if;
end process;

memory : process (clock)
  begin
   if (clock'event and clock='1') then
        if (EN='1' and write='1') then
           RW(to_integer(unsigned(address))) <= data_in;
        elsif (EN='1' and write='0') then
           data_out <= RW(to_integer(unsigned(address)));
        end if;
   end if;
end process;
```

12.3.3.3 Implementation of Output Ports in VHDL

Each output port in the computer system is assigned a unique address. Each output port also contains storage capability. This allows the CPU to update an output port by writing to its specific address. Once the CPU is done storing to the output port address and moves to the next instruction in the program, the output port holds its information until it is written to again. This behavior can be modeled using an RTL process that uses the address bus and the write signal to create a synchronous enable condition. Each port is modeled with its own process. The following VHDL shows how the output ports at x"E0" and x"E1" are modeled using address specific processes:

```
-- port_out_00 description : ADDRESS x"E0"
U3 : process (clock, reset)
 begin
   if (reset = '0') then
      port_out_00 <= x"00";
   elsif (clock'event and clock='1') then
      if (address = x"E0" and write = '1') then
         port_out_00 <= data_in;
      end if;
   end if;
 end process;

-- port_out_01 description : ADDRESS x"E1"
U4 : process (clock, reset)
 begin
   if (reset = '0') then
      port_out_01 <= x"00";
   elsif (clock'event and clock='1') then
      if (address = x"E1" and write = '1') then
         port_out_01 <= data_in;
      end if;
   end if;
 end process;

                 :
  "the rest of the output port models go here..."
                 :
```

12.3.3.4 Implementation of Input Ports in VHDL

The input ports do not contain storage but do require a mechanism to selectively route their information to the data_out port of the memory system. This is accomplished using the multiplexer shown in Example 12.10. The only functionality that is required for the input ports is connecting their ports to the multiplexer.

12.3.3.5 Memory data_out Bus Implementation in VHDL

Now that all of the memory functionality has been designed, the final step is to implement the multiplexer that handles routing the appropriate information to the CPU on the data_out bus based on the incoming address. The following VHDL provides a model for this behavior. Recall that a multiplexer is combinational logic, so if the behavior is to be modeled using a process, all inputs must be listed in the sensitivity list. These inputs include the outputs from the program and data memory in addition to all of the input ports. The sensitivity list must also include the address bus as it acts as the select input to the multiplexer. Within the process, an if/then statement is used to determine which sub-system drives data_out. Program memory will drive data_out when the incoming address is in the range of 0–127 (x"00" to x"7F"). Data memory will drive data_out when the address is in the range of 128–223 (x"80" to x"DF"). An input port will drive data_out when the address is in the range of 240–255 (x"F0" to x"FF"). Each input port has a unique address, so the specific addresses are listed as *elsif* clauses.

```
MUX1 : process (address, rom_data_out, rw_data_out,
                port_in_00, port_in_01, port_in_02, port_in_03,
                port_in_04, port_in_05, port_in_06, port_in_07,
                port_in_08, port_in_09, port_in_10, port_in_11,
                port_in_12, port_in_13, port_in_14, port_in_15)

  begin
     if    ( (to_integer(unsigned(address)) >= 0) and
             (to_integer(unsigned(address)) <= 127)) then
            data_out <= rom_data_out;

     elsif ( (to_integer(unsigned(address)) >= 128) and
            (to_integer(unsigned(address)) <= 223)) then
            data_out <= rw_data_out;

     elsif (address = x"F0") then data_out <= port_in_00;
     elsif (address = x"F1") then data_out <= port_in_01;
     elsif (address = x"F2") then data_out <= port_in_02;
     elsif (address = x"F3") then data_out <= port_in_03;
     elsif (address = x"F4") then data_out <= port_in_04;
     elsif (address = x"F5") then data_out <= port_in_05;
     elsif (address = x"F6") then data_out <= port_in_06;
     elsif (address = x"F7") then data_out <= port_in_07;
     elsif (address = x"F8") then data_out <= port_in_08;
     elsif (address = x"F9") then data_out <= port_in_09;
     elsif (address = x"FA") then data_out <= port_in_10;
     elsif (address = x"FB") then data_out <= port_in_11;
     elsif (address = x"FC") then data_out <= port_in_12;
     elsif (address = x"FD") then data_out <= port_in_13;
     elsif (address = x"FE") then data_out <= port_in_14;
     elsif (address = x"FF") then data_out <= port_in_15;

     else data_out <= x"00";

     end if;

end process;
```

12.3.4 CPU Implementation

Let's now look at the central processing unit details. The CPU contains two components, the control unit (control_unit.vhd) and the data path (data_path.vhd). The data path contains all of the registers and the ALU. The ALU is implemented as a sub-component within the data path (alu.vhd). The data path also contains a bus system in order to facilitate data movement between the registers and memory. The bus system is implemented with two multiplexers that are controlled by the control unit. The control unit contains the finite-state machine that generates all control signals for the data path as it performs the fetch-decode-execute steps of each instruction. Example 12.11 shows the block diagram of the CPU in our 8-bit microcomputer example.

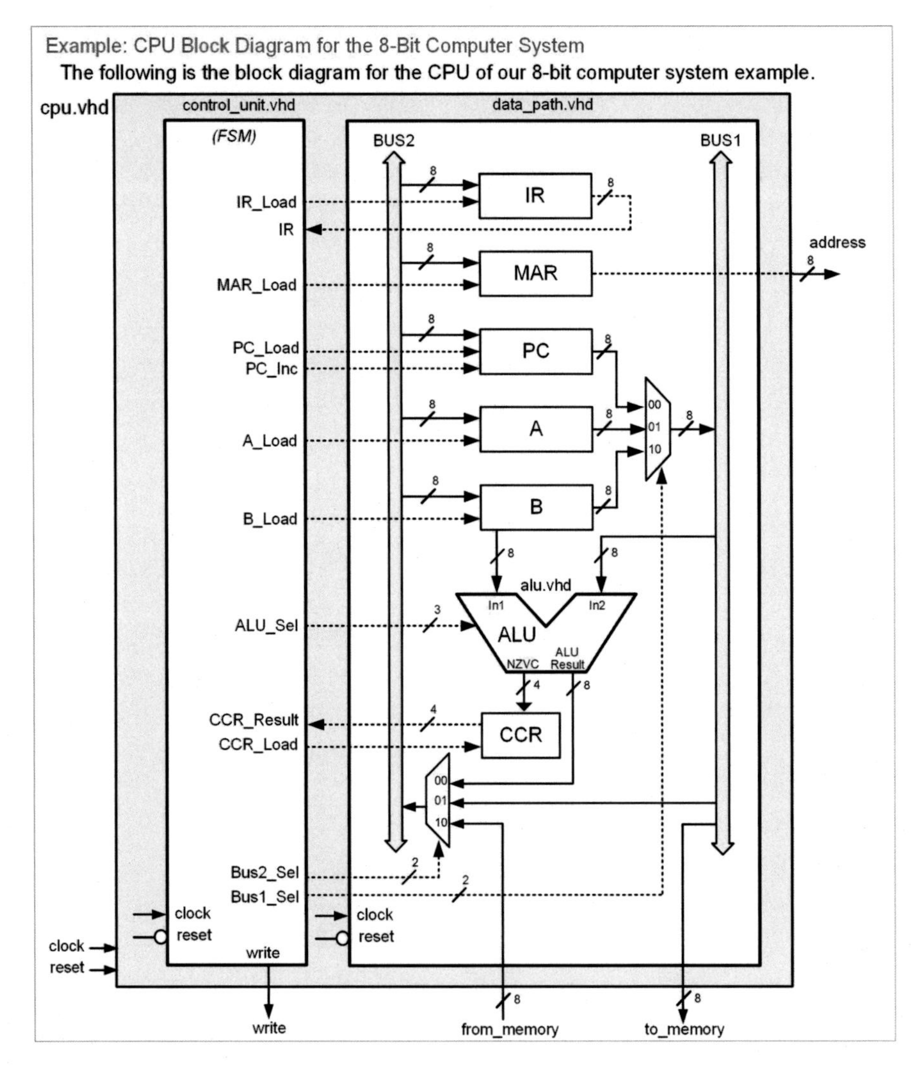

Example 12.11
CPU block diagram for the 8-bit computer system

12.3.4.1 Data Path Implementation in VHDL

Let's first look at the data path bus system that handles internal signal routing. The system consists of two 8-bit busses (Bus1 and Bus2) and two multiplexers. Bus1 is used as the destination of the PC, A, and B register outputs, while Bus2 is used as the input to the IR, MAR, PC, A, and B registers. Bus1 is connected directly to the *to_memory* port of the CPU to allow registers to write data to the memory system. Bus2 can be driven by the *from_memory* port of the CPU to allow the memory system to provide data for the CPU registers. The two multiplexers handle all signal routing and have their select lines (Bus1_Sel and Bus2_Sel) driven by the control unit. The following VHDL shows how the multiplexers are implemented. Again, a multiplexer is combinational logic, so all inputs must be listed in the sensitivity list of its process. Two concurrent signal assignments are also required to connect the MAR to the address port and to connect Bus1 to the to_memory port.

```
MUX_BUS1 : process (Bus1_Sel, PC, A, B)
   begin
     case (Bus1_Sel) is
       when "00"  => Bus1 <= PC;
       when "01"  => Bus1 <= A;
       when "10"  => Bus1 <= B;
       when others => Bus1 <= x"00";
     end case;
end process;

MUX_BUS2 : process (Bus2_Sel, ALU_Result, Bus1, from_memory)
  begin
    case (Bus2_Sel) is
      when "00"  => Bus2 <= ALU_Result;
      when "01"  => Bus2 <= Bus1;
      when "10"  => Bus2 <= from_memory;
      when others => Bus2 <= x"00";
    end case;
end process;

address  <= MAR;
to_memory <= Bus1;
```

Next, let's look at implementing the registers in the data path. Each register is implemented using a dedicated process that is sensitive to clock and reset. This models the behavior of synchronous latches or registers. Each register has a synchronous enable line that dictates when the register is updated. The register output is only updated when the enable line is asserted and a rising edge of the clock is detected. The following VHDL shows how to model the instruction register (IR). Notice that the signal IR is only updated if IR_Load is asserted and there is a rising edge of the clock. In this case, IR is loaded with the value that resides on Bus2.

```
INSTRUCTION_REGISTER : process (Clock, Reset)
  begin
    if (Reset = '0') then
     IR <= x"00";
    elsif (Clock'event and Clock = '1') then
     if (IR_Load = '1') then
        IR <= Bus2;
     end if;
  end if;
 end process;
```

A nearly identical process is used to model the memory address register. A unique signal is declared called *MAR* in order to make the VHDL more readable. MAR is always assigned to address in this system.

```
MEMORY_ADDRESS_REGISTER : process (Clock, Reset)
  begin
    if (Reset = '0') then
      MAR <= x"00";
    elsif (Clock'event and Clock = '1') then
      if (MAR_Load = '1') then
        MAR <= Bus2;
      end if;
    end if;
  end process;
```

Now let's look at the program counter process. This register contains additional functionality beyond simply latching in the value of Bus2. The program counter also has an increment feature. In order to use the "+" operator, we can declare a temporary unsigned vector called PC_uns. The PC process can model the appropriate behavior using PC_uns and then type cast it back to the original PC signal.

```
PROGRAM_COUNTER : process (Clock, Reset)
  begin
    if (Reset = '0') then
      PC_uns <= x"00";
    elsif (Clock'event and Clock = '1') then
      if (PC_Load = '1') then
        PC_uns <= unsigned(Bus2);
      elsif (PC_Inc = '1') then
        PC_uns <= PC_uns + 1;
      end if;
    end if;
  end process;

  PC <= std_logic_vector(PC_uns);
```

The two general-purpose registers A and B are modeled using individual processes as follows:

```
A_REGISTER : process (Clock, Reset)
  begin
    if (Reset = '0') then
      A <= x"00";
    elsif (Clock'event and Clock = '1') then
      if (A_Load = '1') then
        A <= Bus2;
      end if;
    end if;
  end process;

B_REGISTER : process (Clock, Reset)
  begin
    if (Reset = '0') then
      B <= x"00";
    elsif (Clock'event and Clock = '1') then
      if (B_Load = '1') then
        B <= Bus2;
      end if;
    end if;
  end process;
```

The condition code register latches in the status flags from the ALU (NZVC) when the CCR_Load line is asserted. This behavior is modeled using a similar approach as follows:

```
CONDITION_CODE_REGISTER : process (Clock, Reset)
  begin
    if (Reset = '0') then
      CCR_Result <= x"0";
  elsif (Clock'event and Clock = '1') then
    if (CCR_Load = '1') then
```

```
            CCR_Result <= NZVC;
        end if;
      end if;
    end process;
```

12.3.4.2 ALU Implementation in VHDL

The ALU is a set of combinational logic circuitry that performs arithmetic and logic operations. The output of the ALU operation is called *Result.* The ALU also outputs 4 status flags as a 4-bit bus called *NZVC.* The ALU behavior can be modeled using if/then/elsif statements that decide which operation to perform based on the input control signal *ALU_Sel.* The following VHDL shows an example of how to implement the ALU addition functionality. In order to be able to use numerical operators (i.e., +, −, etc.), the numeric_std package is included. Variables can be used within the process to facilitate using the numerical operators. Recall that variables are updated instantaneously so an assignment can be made to the variable and its result is available immediately. Note that in the following VHDL, each operation also updates the NZVC flags. Each of these flags is updated individually. The N flag can be simply driven with position 7 of the ALU result since this bit is the sign bit for signed numbers. The Z flag can be driven using an if/then condition that checks whether the result was x"00". The V flag is updated based on the type of the operation. For the addition operation, the V flag will be asserted if a POS + POS = NEG or a NEG + NEG = POS. These conditions can be checked by looking at the sign bits of the inputs and the sign bit of the result. Finally, the C flag can be directly driven with position 8 of the Sum_uns variable.

```
ALU_PROCESS : process (In1, In2, ALU_Sel)

    variable Sum_uns : unsigned(8 downto 0);

    begin
      if (ALU_Sel = "000") then - ADDITION

          --- Sum Calculation ----------------------------------
            Sum_uns := unsigned('0' & In1) + unsigned('0' & In2);
            Result <= std_logic_vector(Sum_uns(7 downto 0));

          --- Negative Flag (N) -------------------------------
          NZVC(3) <= Sum_uns(7);

          --- Zero Flag (Z) -----------------------------------
          if (Sum_uns(7 downto 0) = x"00") then
            NZVC(2) <= '1';
          else
            NZVC(2) <= '0';
          end if;

          --- Overflow Flag (V) -------------------------------
          if ((In1(7)='0' and In2(7)='0' and Sum_uns(7)='1') or
               (In1(7)='1' and In2(7)='1' and Sum_uns(7)='0')) then
            NZVC(1) <= '1';
          else
            NZVC(1) <= '0';
          end if;

        --- Carry Flag (C) ------------------------------------
          NZVC(0) <= Sum_uns(8);

      elsif (ALU_Sel = ...
                                :    "other ALU functionality goes here"

      end if;
end process;
```

12.3.4.3 Control Unit Implementation in VHDL

Let's now look at how to implement the control unit state machine. We'll first look at the formation of the VHDL to model the FSM and then turn to the detailed state transitions in order to accomplish a variety of the most common instructions. The control unit sends signals to the data path in order to move data in and out of registers and into the ALU to perform data manipulations. The finite-state machine is implemented with the behavioral modeling techniques presented in Chap. 9. The model contains three processes in order to implement the state memory, next state logic, and output logic of the FSM. User-defined types are created for each of the states defined in the state diagram of the FSM. The states associated with fetching (S_FETCH_0, S_FETCH_1, S_FETCH_2) and decoding the opcode (S_DECODE_3) are performed each time an instruction is executed. A unique path is then added after the decode state to perform the steps associated with executing each individual instruction. The FSM can be created one instruction at a time by adding additional state paths after the decode state. The following VHDL code shows how the user-defined state names are created for six basic instructions (LDA_IMM, LDA_DIR, STA_DIR, ADD_AB, BRA, and BEQ).

```
type state_type is
            (S_FETCH_0, S_FETCH_1, S_FETCH_2,
             S_DECODE_3,
             S_LDA_IMM_4, S_LDA_IMM_5, S_LDA_IMM_6,
             S_LDA_DIR_4, S_LDA_DIR_5, S_LDA_DIR_6, S_LDA_DIR_7, S_LDA_DIR_8,
             S_LDB_IMM_4, S_LDB_IMM_5, S_LDB_IMM_6,
             S_LDB_DIR_4, S_LDB_DIR_5, S_LDB_DIR_6, S_LDB_DIR_7, S_LDB_DIR_8,
             S_STA_DIR_4, S_STA_DIR_5, S_STA_DIR_6, S_STA_DIR_7,
             S_STB_DIR_4, S_STB_DIR_5, S_STB_DIR_6, S_STB_DIR_7,
             S_ADD_AB_4,
             S_BRA_4, S_BRA_5, S_BRA_6,
             S_BEQ_4, S_BEQ_5, S_BEQ_6, S_BEQ_7);

signal current_state, next_state : state_type;
```

Within the architecture of the control unit model, the state memory is implemented as a separate process that will update the current state with the next state on each rising edge of the clock. The reset state will be the first fetch state in the FSM (i.e., S_FETCH_0). The following VHDL shows how the state memory in the control unit can be modeled:

```
STATE_MEMORY : process (Clock, Reset)
 begin
   if (Reset = '0') then
      current_state <= S_FETCH_0;
   elsif (clock'event and clock = '1') then
      current_state <= next_state;
   end if;
end process;
```

The next state logic is also implemented as a separate process. The next state logic depends on the current state, instruction register (IR), and the condition code register (CCR_Result). The following VHDL gives a portion of the next state logic process showing how the state transitions can be modeled:

```
NEXT_STATE_LOGIC : process (current_state, IR, CCR_Result)
  begin
      if (current_state = S_FETCH_0) then
       next_state <= S_FETCH_1;
      elsif (current_state = S_FETCH_1) then
       next_state <= S_FETCH_2;
      elsif (current_state = S_FETCH_2) then
       next_state <= S_DECODE_3;
      elsif (current_state = S_DECODE_3) then
                                                      -- select execution path
         if (IR = LDA_IMM) then                       -- Load A Immediate
            next_state <= S_LDA_IMM_4;
         elsif (IR = LDA_DIR) then                    -- Load A Direct
            next_state <= S_LDA_DIR_4;
         elsif (IR = STA_DIR) then                    -- Store A Direct
            next_state <= S_STA_DIR_4;
         elsif (IR = ADD_AB) then                     -- Add A and B
            next_state <= S_ADD_AB_4;
         elsif (IR = BRA) then                        -- Branch Always
            next_state <= S_BRA_4;
         elsif (IR=BEQ and CCR_Result(2)='1') then    -- BEQ and Z=1
            next_state <= S_BEQ_4;
         elsif (IR=BEQ and CCR_Result(2)='0') then    -- BEQ and Z=0
            next_state <= S_BEQ_7;
           else
            next_state <= S_FETCH_0;
         end if;

      elsif...
                          :
         "paths for each instruction go here..."
                          :
         end if;

end process;
```

Finally, the output logic is modeled as a third, separate process. It is useful to explicitly state the outputs of the control unit for each state in the machine to allow easy debugging and avoid synthesizing latches. Our example computer system has Moore type outputs, so the process only depends on the current state. The following VHDL shows a portion of the output logic process:

```
OUTPUT_LOGIC : process (current_state)
  begin
     case(current_state) is
       when S_FETCH_0 =>  -- Put PC onto MAR to read Opcode
           IR_Load <= '0';
           MAR_Load <= '1';
           PC_Load <= '0';
           PC_Inc  <= '0';
           A_Load  <= '0';
           B_Load  <= '0';
           ALU_Sel  <= "000";
           CCR_Load <= '0';
           Bus1_Sel <= "00"; -- "00"=PC, "01"=A,  "10"=B
           Bus2_Sel <= "01"; -- "00"=ALU_Result, "01"=Bus1, "10"=from_memory
           write  <= '0';

       when S_FETCH_1 =>  -- Increment PC
          IR_Load <= '0';
          MAR_Load <= '0';
          PC_Load <= '0';
          PC_Inc  <= '1';
          A_Load  <= '0';
          B_Load  <= '0';
```

```
            ALU_Sel  <= "000";
            CCR_Load <= '0';
            Bus1_Sel <= "00"; -- "00"=PC,  "01"=A,    "10"=B
            Bus2_Sel <= "00"; -- "00"=ALU, "01"=Bus1, "10"=from_memory
            write    <= '0';
                     :
         "output assignments for all other states go here..."
                     :
      end case;
   end process;
```

Detailed Execution of LDA_IMM

Now let's look at the details of the state transitions and output signals in the control unit FSM when executing a few of the most common instructions. Let's begin with the instruction to load register A using immediate addressing (LDA_IMM). Example 12.12 shows the state diagram for this instruction. The first three states (S_FETCH_0, S_FETCH_1, S_FETCH_2) handle fetching the opcode. The purpose of these states is to read the opcode from the address being held by the program counter and put it into the instruction register. Multiple states are needed to handle putting PC into MAR to provide the address of the opcode, waiting for the memory system to provide the opcode, latching the opcode into IR, and incrementing PC to the next location in program memory. Another state is used to decode the opcode (S_DECODE_3) in order to decide which path to take in the state diagram based on the instruction being executed. After the decode state, a series of three more states are needed (S_LDA_IMM_4, S_LDA_IMM_5, S_LDA_IMM_6) to execute the instruction. The purpose of these states is to read the operand from the address being held by the program counter and put it into A. Multiple states are needed to handle putting PC into MAR to provide the address of the operand, waiting for the memory system to provide the operand, latching the operand into A, and incrementing PC to the next location in program memory. When the instruction completes, the value of the operand resides in A, and PC is pointing to the next location in program memory, which is the opcode of the next instruction to be executed.

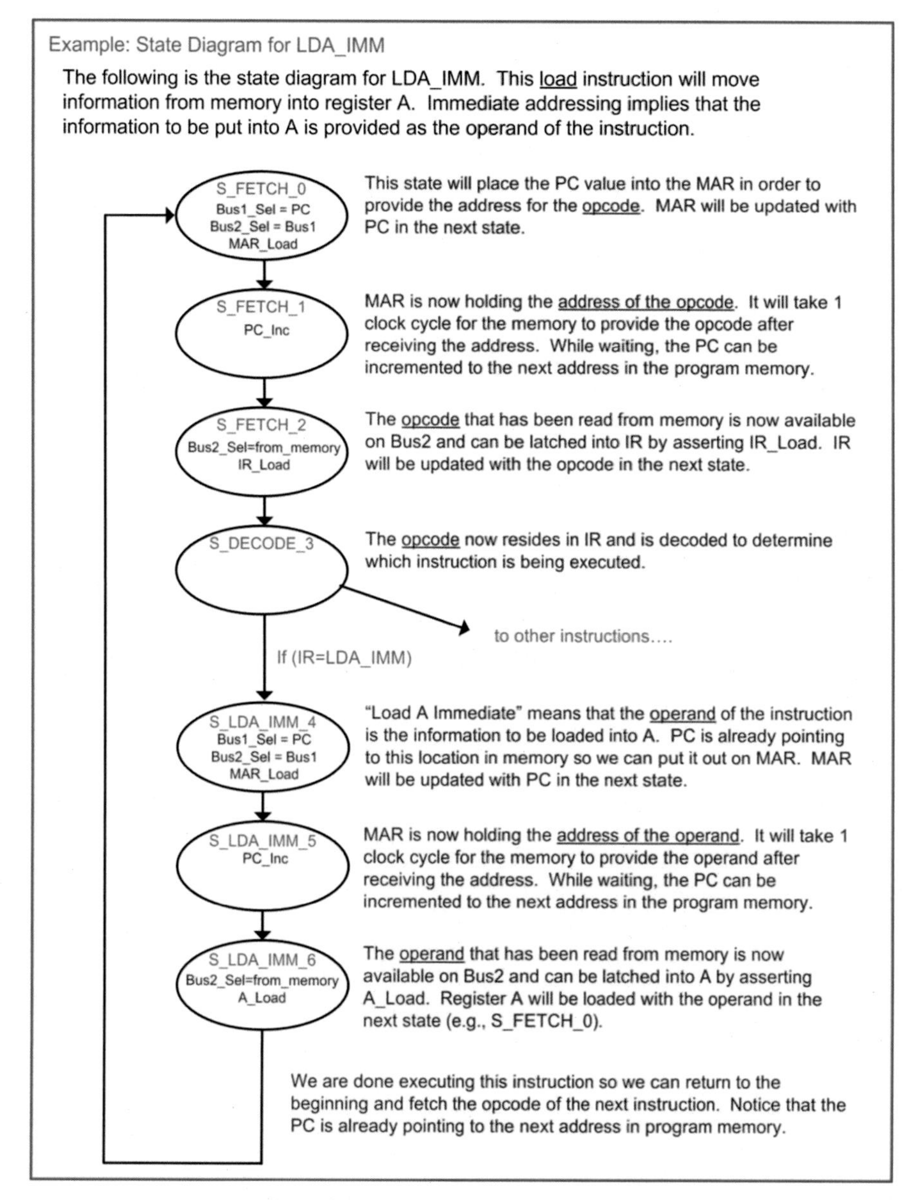

Example 12.12
State diagram for LDA_IMM

Example 12.13 shows the simulation waveform for executing LDA_IMM. In this example, register A is loaded with the operand of the instruction, which holds the value x"AA".

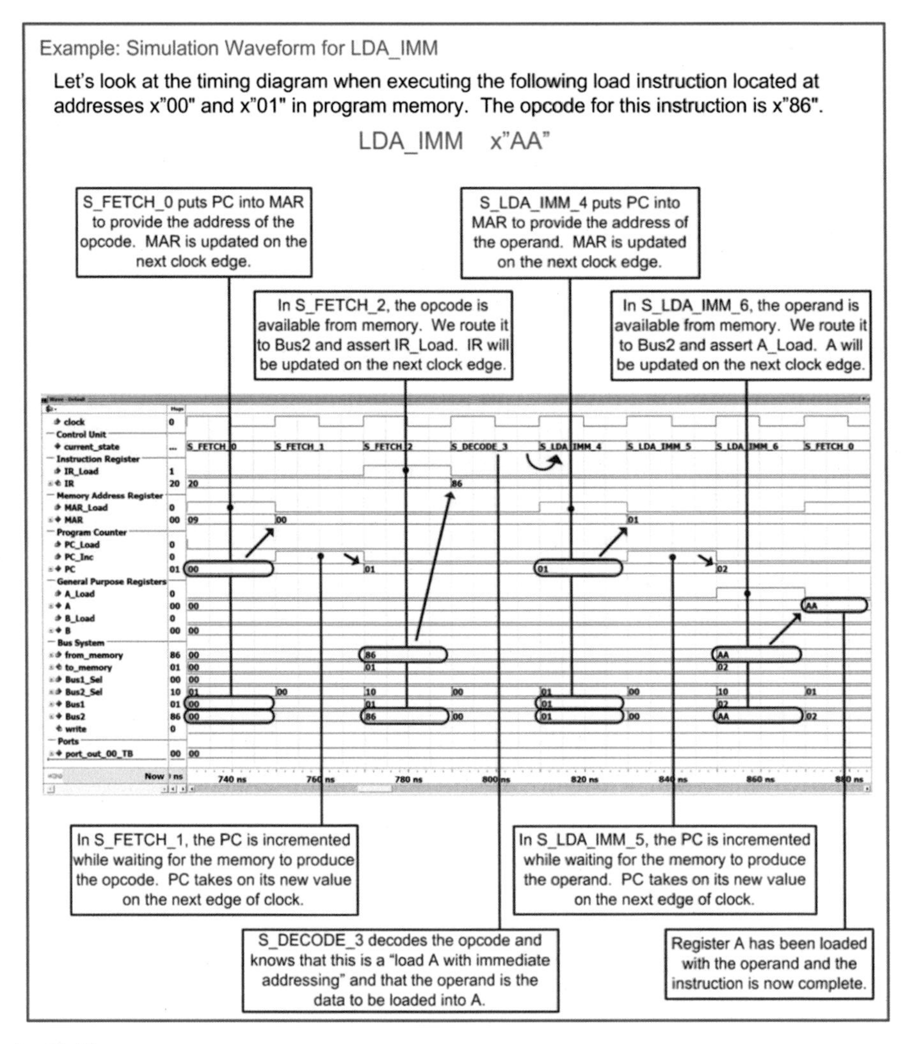

Example 12.13
Simulation waveform for LDA_IMM

Detailed Execution of LDA_DIR

Now let's look at the details of the instruction to load register A using direct addressing (LDA_DIR). Example 12.14 shows the state diagram for this instruction. The first four states to fetch and decode the opcode are the same states as in the previous instruction and are performed each time a new instruction is executed. Once the opcode is decoded, the state machine traverses five new states to execute the instruction (S_LDA_DIR_4, S_LDA_DIR_5, S_LDA_DIR_6, S_LDA_DIR_7, S_LDA_DIR_8). The purpose of these states is to read the operand and then use it as the address of where to read the contents to put into A.

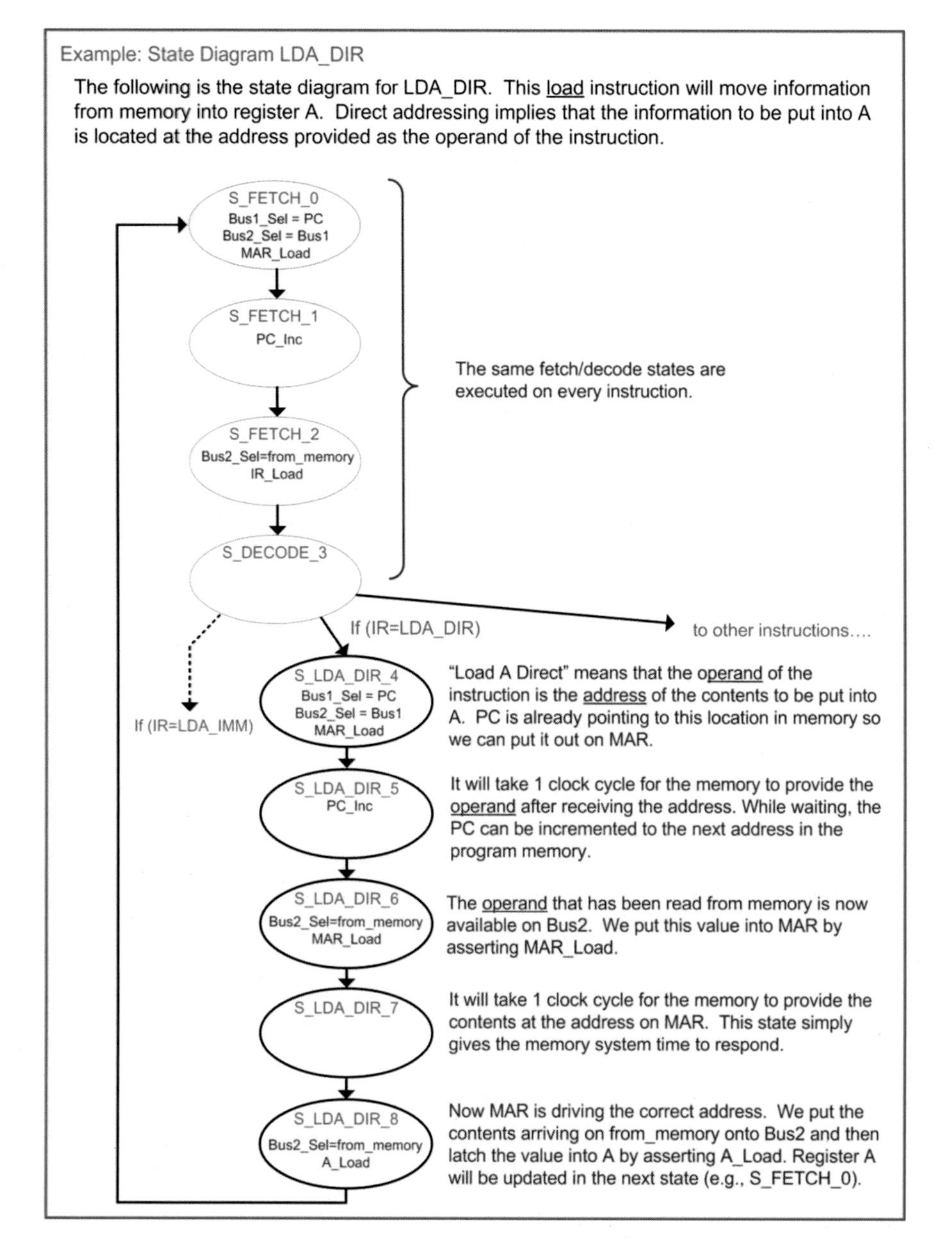

Example 12.14
State diagram for LDA_DIR

Example 12.15 shows the simulation waveform for executing LDA_DIR. In this example, register A is loaded with the contents located at address x"80", which has already been initialized to x"AA".

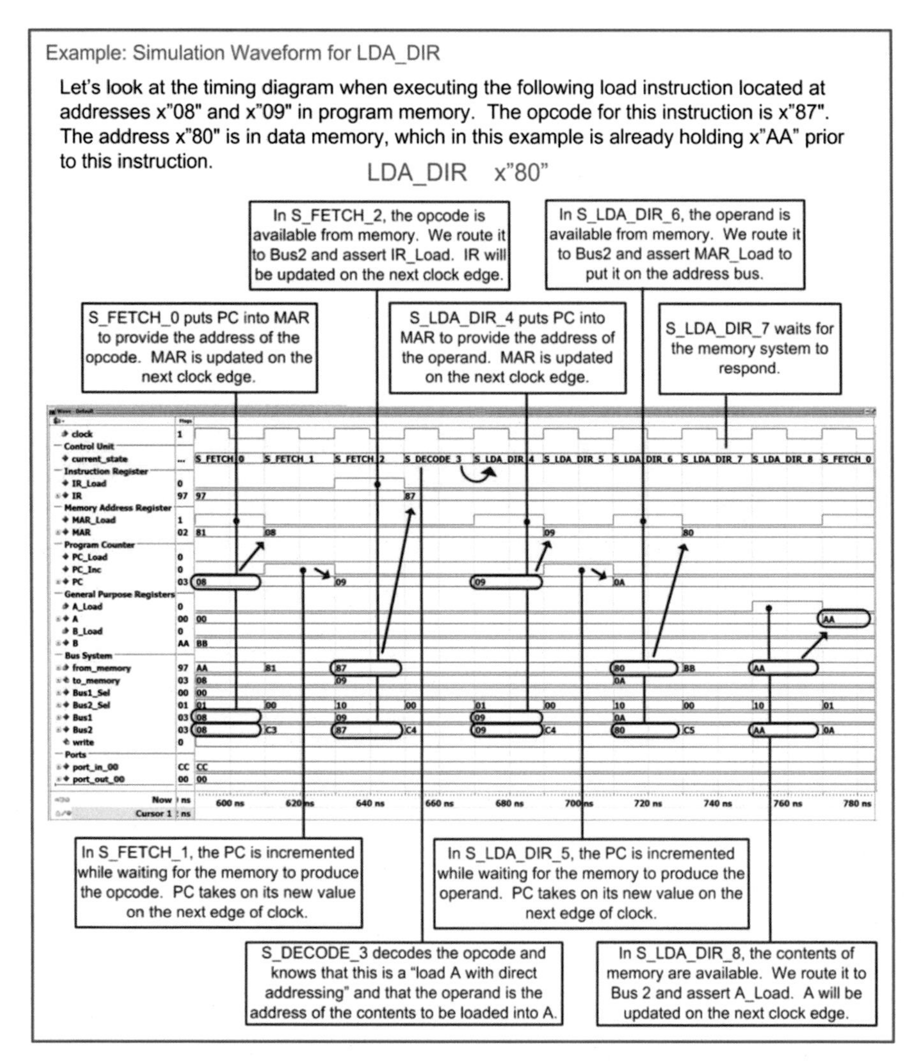

Example 12.15
Simulation waveform for LDA_DIR

Detailed Execution of STA_DIR

Now let's look at the details of the instruction to store register A to memory using direct addressing (STA_DIR). Example 12.16 shows the state diagram for this instruction. The first four states are again the same as prior instructions in order to fetch and decode the opcode. Once the opcode is decoded, the state machine traverses four new states to execute the instruction (S_STA_DIR_4, S_STA_DIR_5, S_STA_DIR_6, S_STA_DIR_7). The purpose of these states is to read the operand and then use it as the address of where to write the contents of A to.

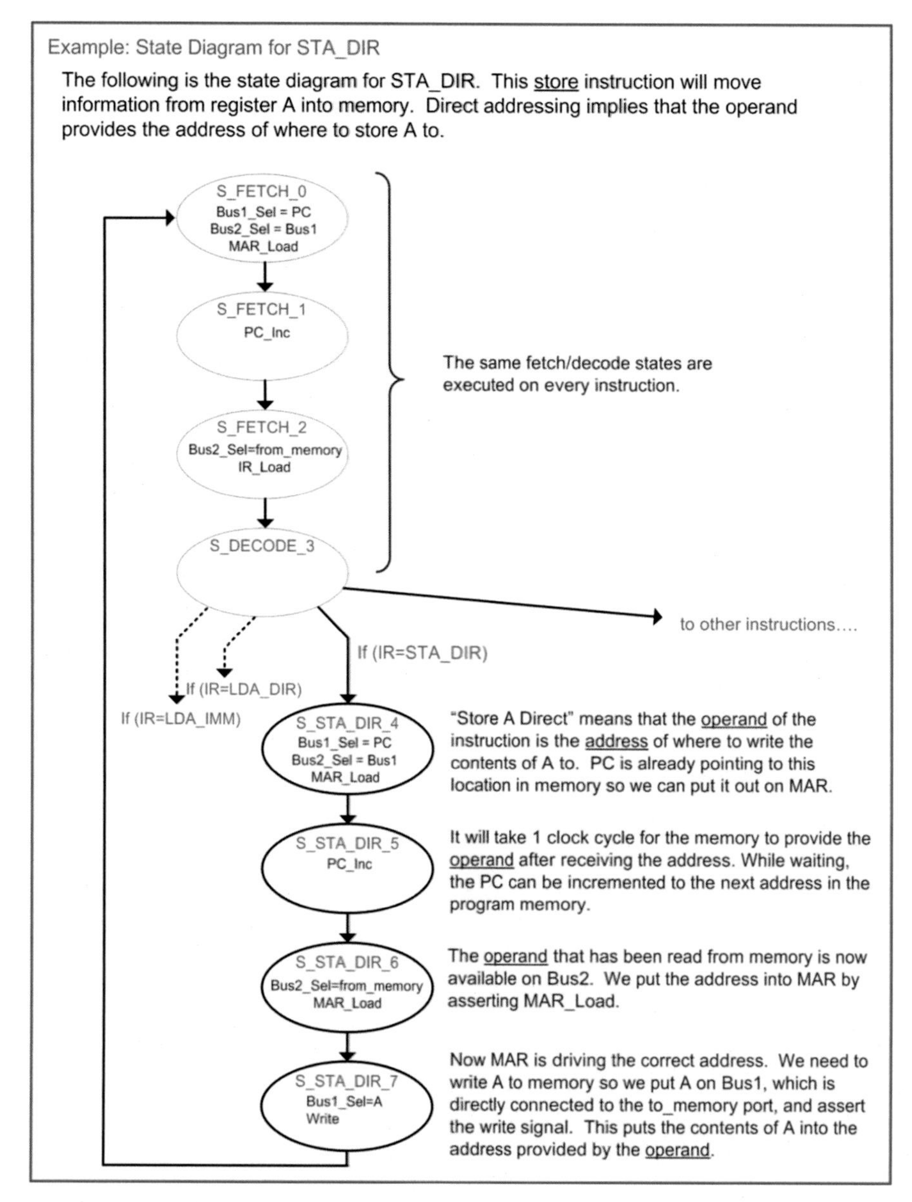

Example 12.16
State diagram for STA_DIR

Example 12.17 shows the simulation waveform for executing STA_DIR. In this example, register A already contains the value x"CC" and will be stored to address x"E0". The address x"E0" is an output port (port_out_00) in our example computer system.

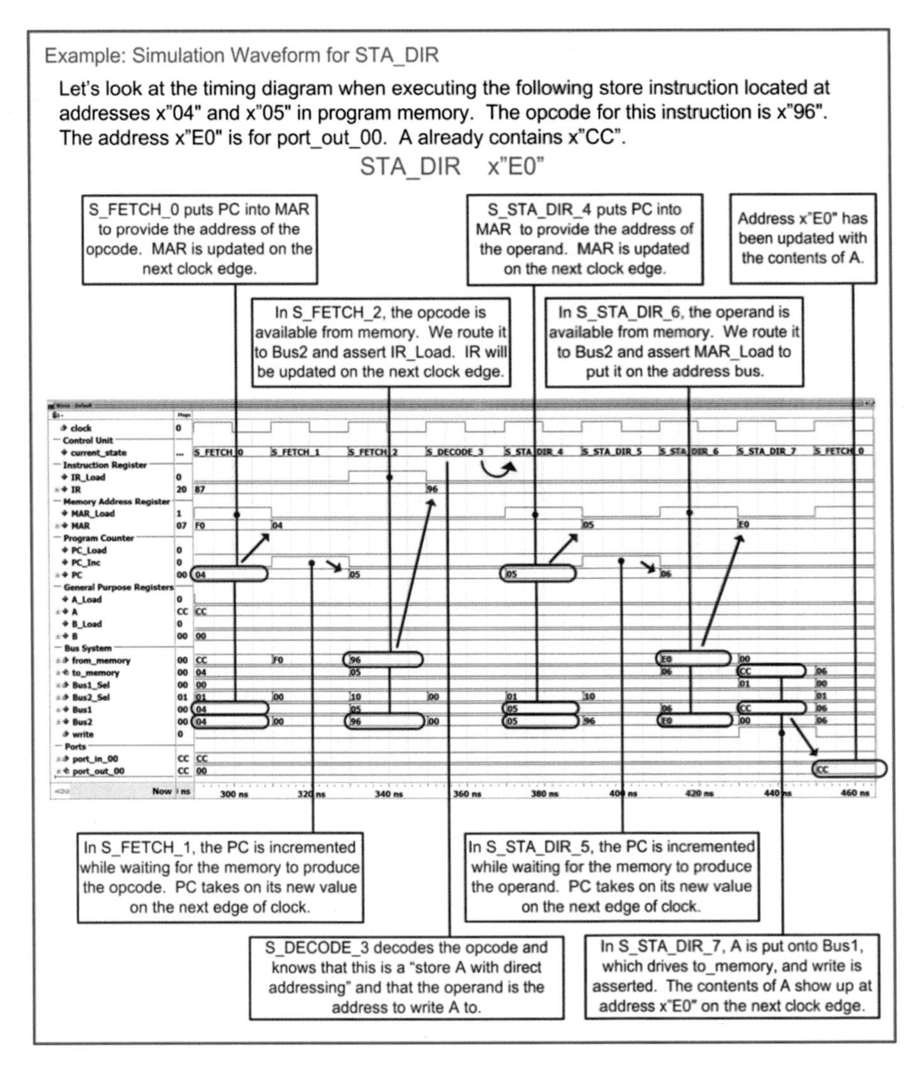

Example 12.17
Simulation waveform for STA_DIR

Detailed Execution of ADD_AB

Now let's look at the details of the instruction to add A to B and store the sum back in A (ADD_AB). Example 12.18 shows the state diagram for this instruction. The first four states are again the same as prior instructions in order to fetch and decode the opcode. Once the opcode is decoded, the state machine only requires one more state to complete the operation (S_ADD_AB_4). The ALU is combinational logic so it will begin to compute the sum immediately as soon as the inputs are updated. The inputs to the ALU are Bus1 and register B. Since B is directly connected to the ALU, all that is required to start the addition is to put A onto Bus1. The output of the ALU is put on Bus2 so that it can be latched into A on the next clock edge. The ALU also outputs the status flags NZVC, which are directly connected to the condition code register. A_Load and CCR_Load are asserted in this state. A and CCR_Result will be updated in the next state (i.e., S_FETCH_0).

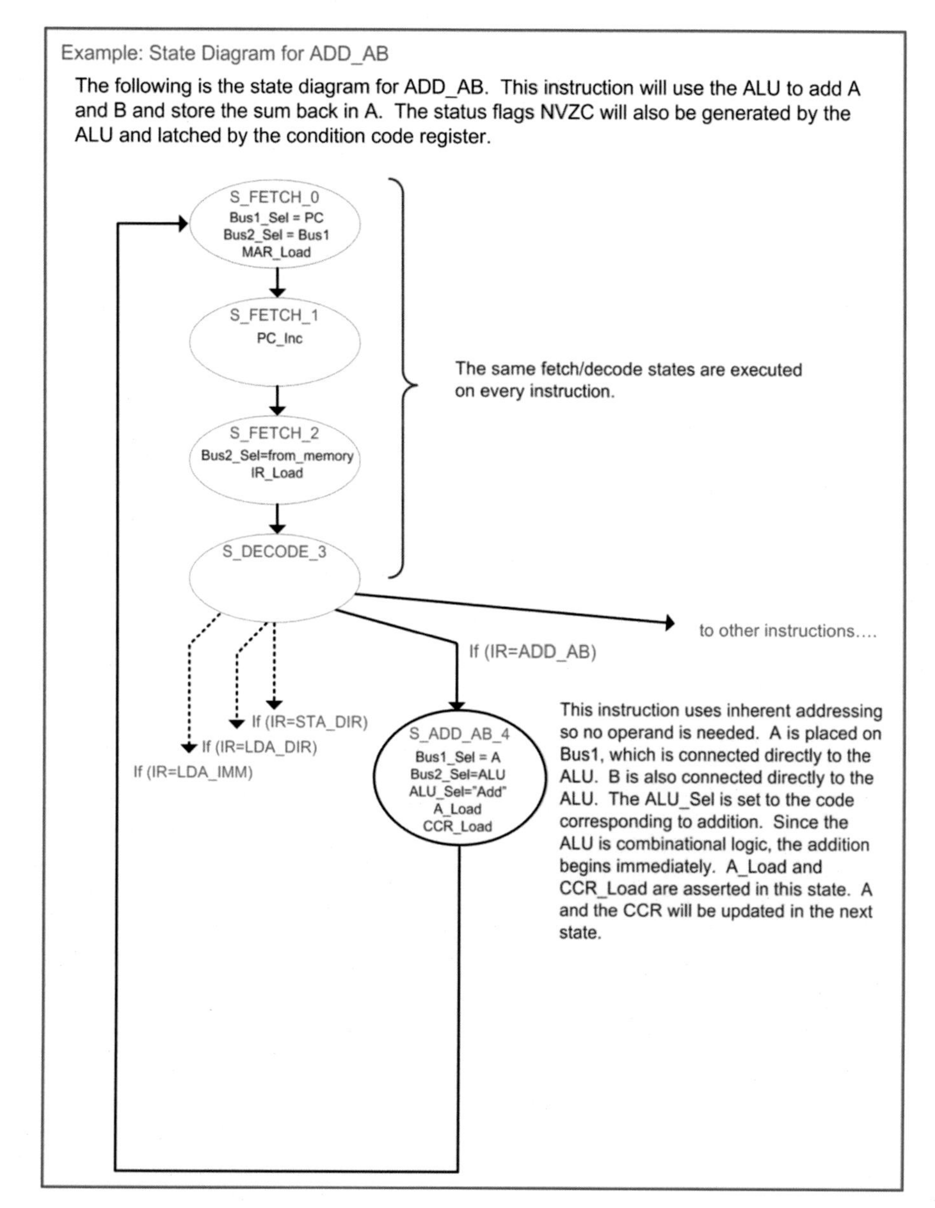

Example 12.18
State diagram for ADD_AB

Example 12.19 shows the simulation waveform for executing ADD_AB. In this example, two load immediate instructions were used to initialize the general-purpose registers to A=x"FF" and B=x"01" prior to the addition. The addition of these values will result in a sum of x"00" and assert the carry (C) and zero (Z) flags in the condition code register.

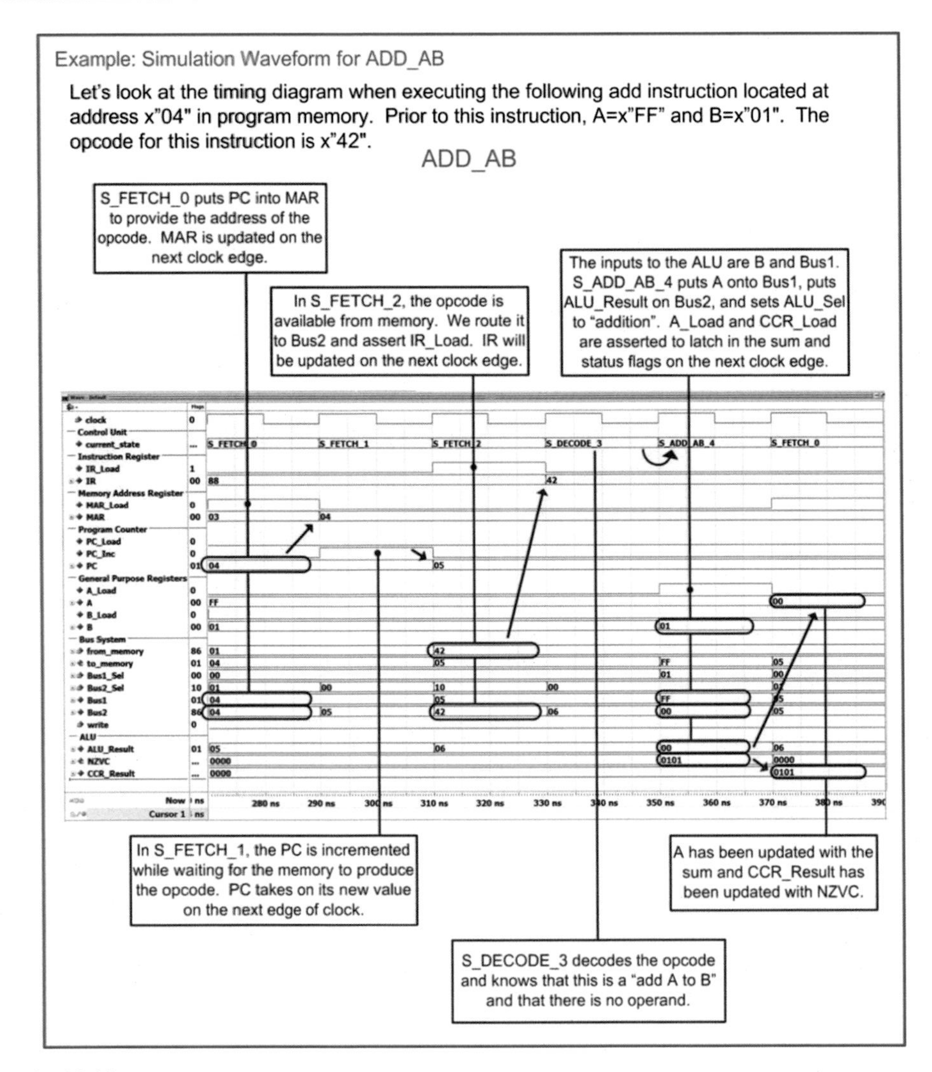

Example 12.19
Simulation waveform for ADD_AB

Detailed Execution of BRA

Now let's look at the details of the instruction to branch always (BRA). Example 12.20 shows the state diagram for this instruction. The first four states are again the same as prior instructions in order to fetch and decode the opcode. Once the opcode is decoded, the state machine traverses four new states to execute the instruction (S_BRA_4, S_BRA_5, S_BRA_6). The purpose of these states is to read the operand and put its value into PC to set the new location in program memory to execute instructions.

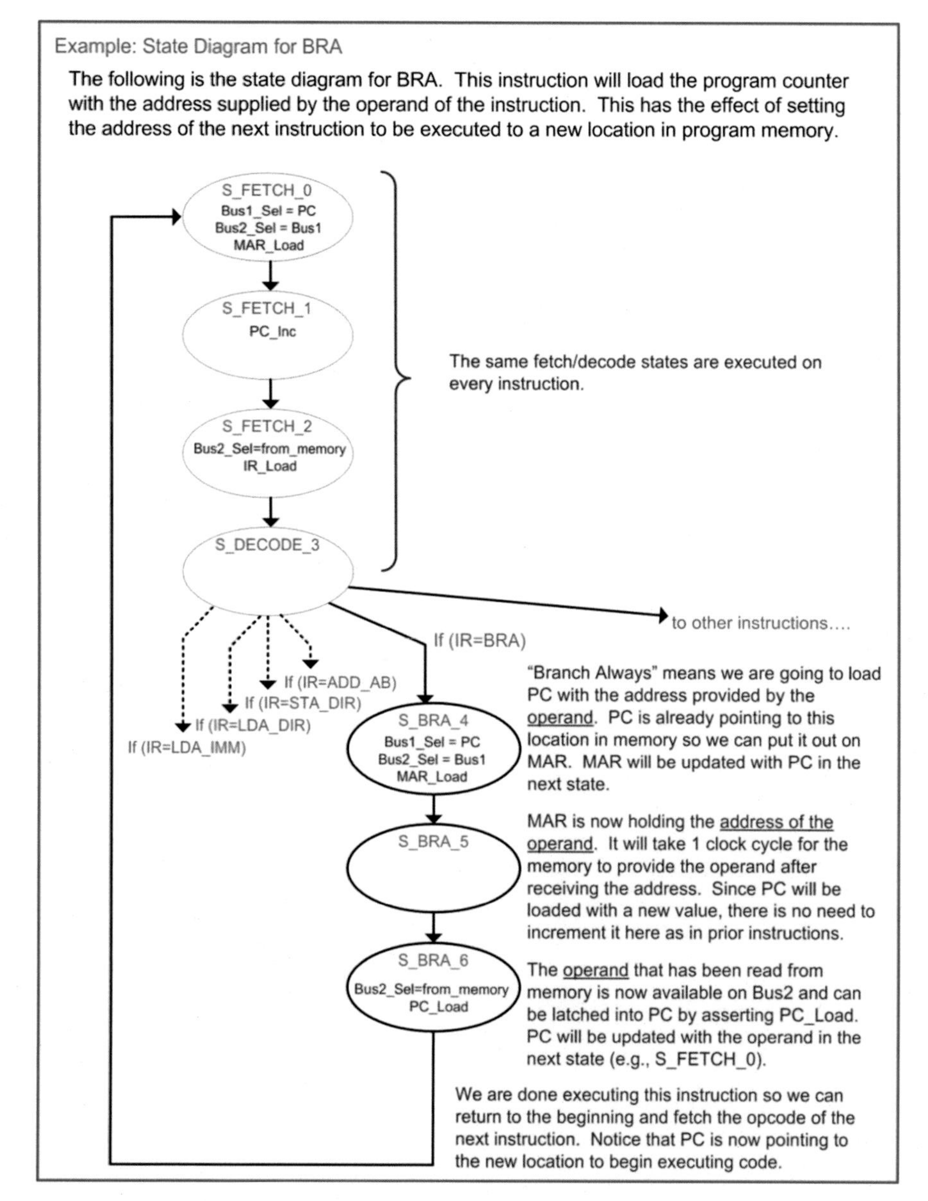

Example 12.20
State diagram for BRA

Example 12.21 shows the simulation waveform for executing BRA. In this example, PC is set back to address x"00".

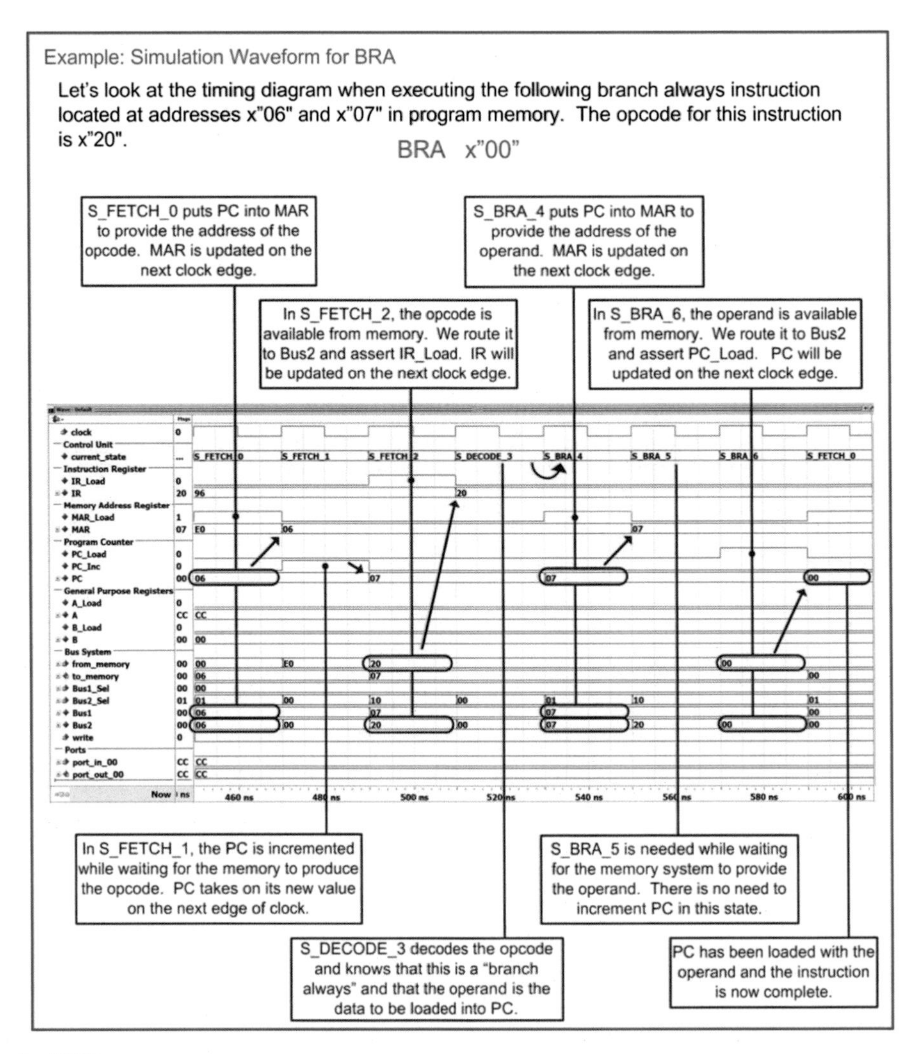

Example 12.21
Simulation waveform for BRA

Detailed Execution of BEQ

Now let's look at the branch if equal to zero (BEQ) instruction. Example 12.22 shows the state diagram for this instruction. Notice that in this conditional branch, the path that is taken through the FSM depends on both IR and CCR. In the case that $Z = 1$, the branch is taken, meaning that the operand is loaded into PC. In the case that $Z = 0$, the branch is not taken, meaning that PC is simply incremented to bypass the operand and point to the beginning of the next instruction in program memory.

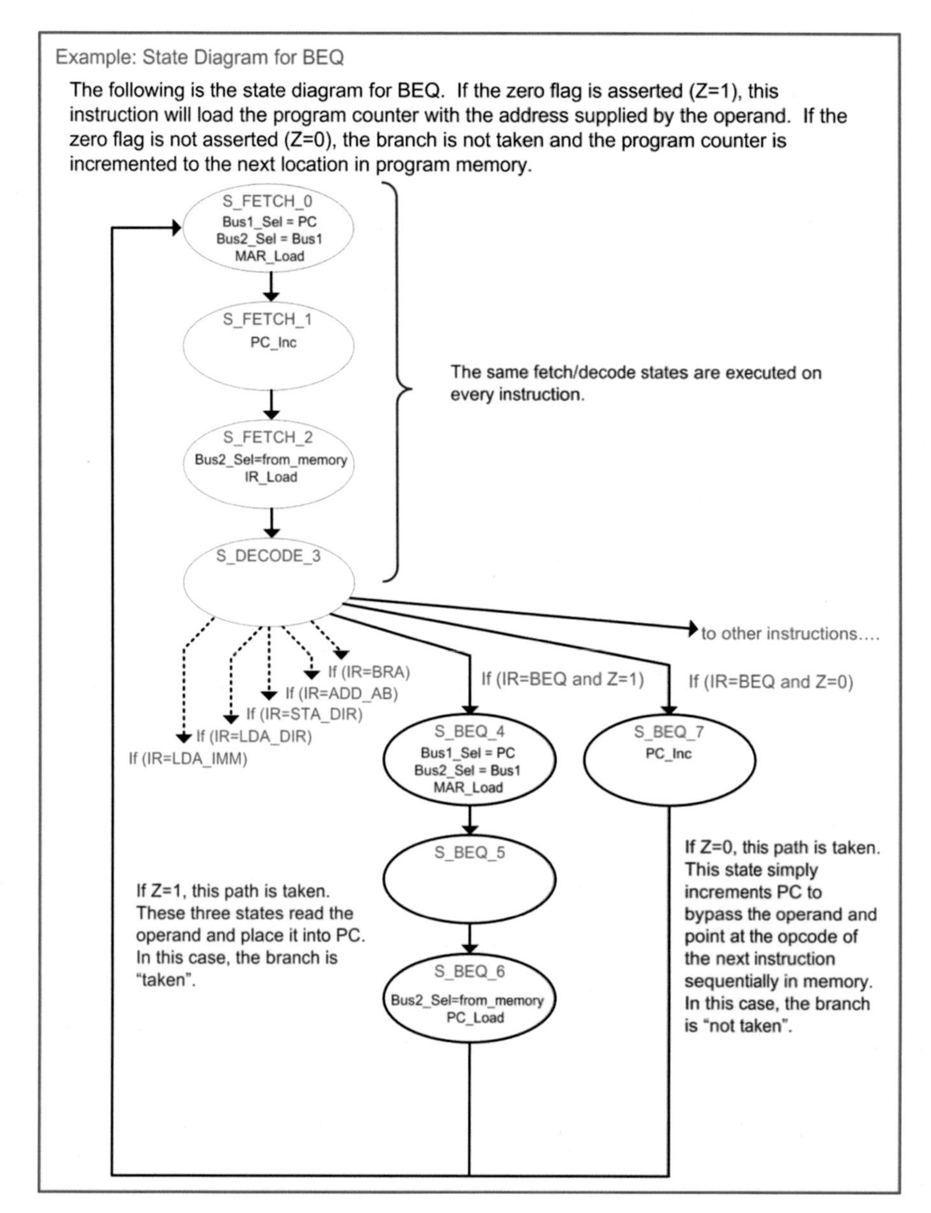

Example 12.22
State diagram for BEQ

Example 12.23 shows the simulation waveform for executing BEQ when the branch *is taken*. Prior to this instruction, an addition was performed on x"FF" and x"01". This resulted in a sum of x"00", which asserted the Z and C flags in the condition code register. Since Z = 1 when BEQ is executed, the branch is taken.

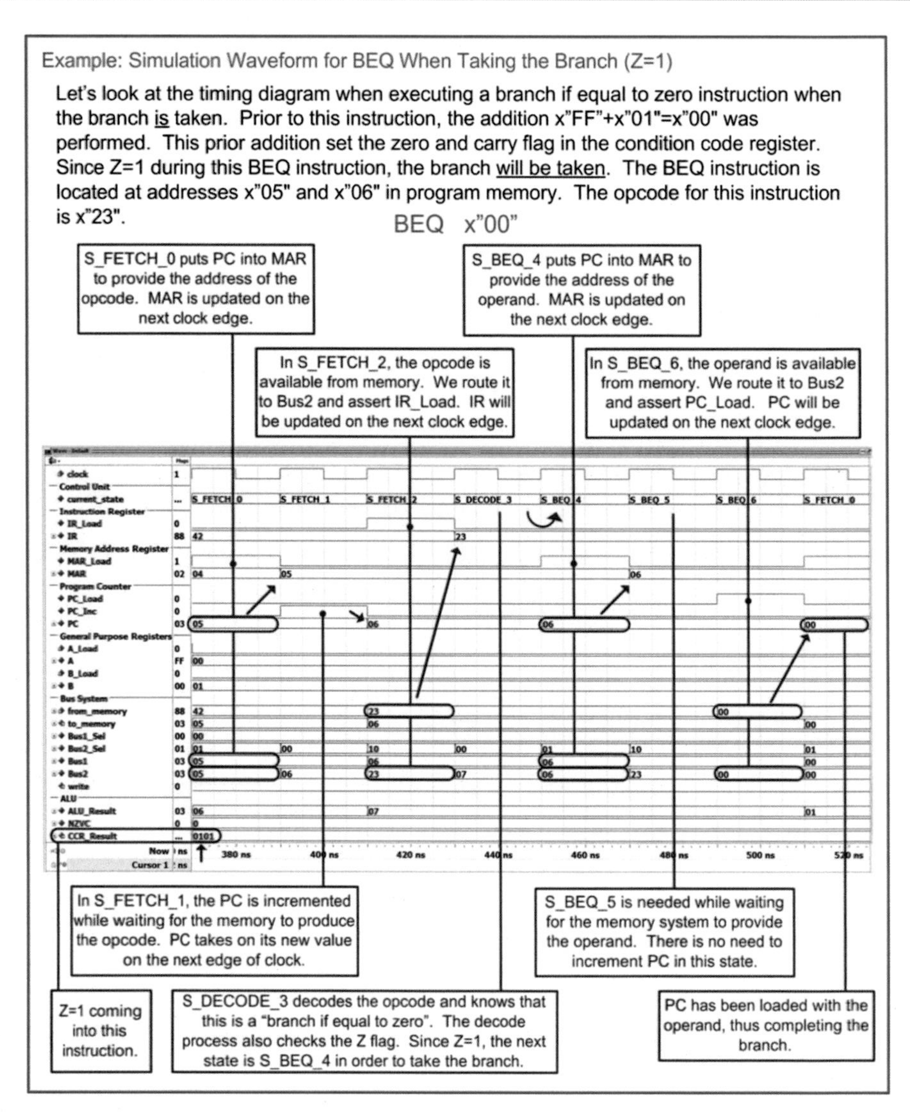

Example 12.23
Simulation waveform for BEQ when taking the branch (Z=1)

Example 12.24 shows the simulation waveform for executing BEQ when the branch *is not taken*. Prior to this instruction, an addition was performed on x"FE" and x"01". This resulted in a sum of x"FF", which did not assert the Z flag. Since Z = 0 when BEQ is executed, the branch is not taken. When not taking the branch, PC must be incremented again in order to bypass the operand and point to the next location in program memory.

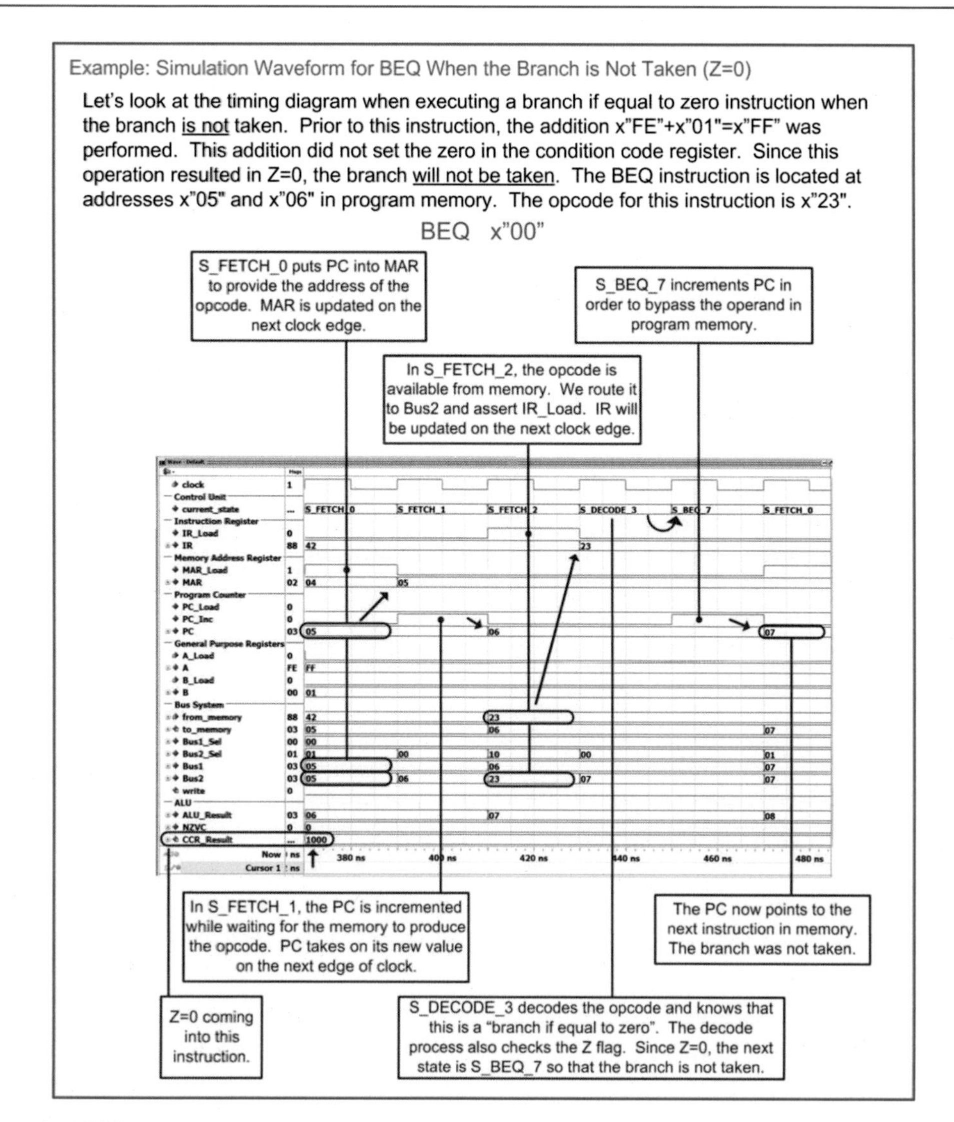

Example 12.24
Simulation waveform for BEQ when the branch is not taken (Z=0)

Concept Check

CC12.3 The 8-bit microcomputer example presented in this section is a very simple architecture used to illustrate the basic concepts of a computer. If we wanted to keep this computer an 8-bit system but increase the depth of the memory, it would require adding more address lines to the address bus. What changes to the computer system would need to be made to accommodate the wider address bus?

A) The width of the program counter would need to be increased to support the wider address bus.

B) The size of the memory address register would need to be increased to support the wider address bus.

C) Instructions that use direct addressing would need additional bytes of operand to pass the wider address into the CPU 8 bits at a time.

D) All of the above.

Summary

- A computer is a collection of hardware components that are constructed to perform a specific set of instructions to process and store data. The main hardware components of a computer are the central processing unit (CPU), program memory, data memory, and input/output ports.
- The CPU consists of registers for fast storage, an arithmetic logic unit (ALU) for data manipulation, and a control state machine that directs all activity to execute an instruction.
- A CPU is typically organized into a *data path* and a *control unit*. The data path contains circuitry used to store and process information. The data path includes registers and the ALU. The control unit is a large state machine that sends control signals to the data path in order to facilitate instruction execution.
- The control unit performs a *fetch-decode-execute* cycle in order to complete instructions.
- The instructions that a computer is designed to execute are called its *instruction set*.
- Instructions are inserted into *program memory* in a sequence that when executed will accomplish a particular task. This sequence of instructions is called a computer *program*.
- An instruction consists of an *opcode* and a potential *operand*. The opcode is the unique binary code that tells the control state machine which instruction is being executed. An operand is additional information that may be needed for the instruction.
- An *addressing mode* refers to the way that the operand is treated. In *immediate* addressing the operand is the actual data to be used. In *direct* addressing the operand is the address of where the data is to be retrieved or stored. In *inherent* addressing all of the information needed to complete the instruction is contained within the opcode, so no operand is needed.
- A computer also contains *data memory* to hold temporary variables during run time.
- A computer also contains input and output ports to interface with the outside world.
- A *memory-mapped* system is one in which the program memory, data memory, and I/O ports are all assigned a unique address. This allows the CPU to simply process information as data and addresses and allows the program to handle where the information is being sent to. A *memory map* is a graphical representation of what address ranges various components are mapped to.
- There are three primary classes of instructions. These are loads and stores, data manipulations, and branches.
- Load instructions move information from memory into a CPU register. A load instruction takes multiple read cycles.
- Store instructions move information from a CPU register into memory. A store instruction takes multiple read cycles and at least one write cycle.
- Data manipulation instructions operate on information being held in CPU registers.

Data manipulation instructions often use inherent addressing.

- Branch instructions alter the flow of instruction execution. *Unconditional branches* always change the location in memory of where the CPU is executing instructions. *Conditional branches* only change the location of instruction execution if a status flag is asserted.
- Status flags are held in the condition code register and are updated by certain instructions. The most commonly used flags are the negative flag (N), zero flag (Z), two's complement overflow flag (V), and carry flag (C).

Exercise Problems

Section 12.1: Computer Hardware

12.1.1 What computer hardware sub-system holds the temporary variables used by the program?

12.1.2 What computer hardware sub-system contains fast storage for holding and/or manipulating data and addresses?

12.1.3 What computer hardware sub-system allows the computer to interface to the outside world?

12.1.4 What computer hardware sub-system contains the state machine that orchestrates the fetch-decode-execute process?

12.1.5 What computer hardware sub-system contains the circuitry that performs mathematical and logic operations?

12.1.6 What computer hardware sub-system holds the instructions being executed?

Section 12.2: Computer Software

12.2.1 In computer software, what are the names of the most basic operations that a computer can perform?

12.2.2 Which element of computer software is the binary code that tells the CPU which instruction is being executed?

12.2.3 Which element of computer software is a collection of instructions that perform a desired task?

12.2.4 Which element of computer software is the supplementary information required by an instruction such as constants or which registers to use?

12.2.5 Which class of instructions handles moving information between memory and CPU registers?

12.2.6 Which class of instructions alters the flow of program execution?

12.2.7 Which class of instructions alters data using either arithmetic or logical operations?

Section 12.3: Computer Implementation: An 8-Bit Computer Example

12.3.1 Design the example 8-bit computer system presented in this chapter in VHDL with the ability to execute the three instructions LDA_IMM, STA_DIR, and BRA. Simulate your computer system using the following program that will continually write the patterns x"AA" and x"BB" to output ports port_out_00 and port_out_01:

```
constant ROM : rom_type := (
    0      => LDA_IMM,
    1      => x"AA",
    2      => STA_DIR,
    3      => x"E0",
    4      => STA_DIR,
    5      => x"E1",
    6      => LDA_IMM,
    7      => x"BB",
    8      => STA_DIR,
    9      => x"E0",
    10     => STA_DIR,
    11     => x"E1",
    12     => BRA,
    13     => x"00",
    others => x"00");
```

12.3.2 Add the functionality to the computer model from 12.3.1 the ability to perform the LDA_DIR instruction. Simulate your computer system using the following program that will continually read from port_in_00 and write its contents to port_out_00:

```
constant ROM : rom_type := (
    0      => LDA_DIR,
    1      => x"F0",
    2      => STA_DIR,
    3      => x"E0",
    4      => BRA,
    5      => x"00",
    others => x"00");
```

12.3.3 Add the functionality to the computer model from 12.3.2 the ability to perform the instructions LDB_IMM, LDB_DIR, and STB_DIR. Modify the example programs given in exercise 12.3.1 and 12.3.2 to use register B in order to simulate your implementation.

12.3.4 Add the functionality to the computer model from 12.3.3 the ability to perform the addition instruction ADD_AB. Test your addition instruction by simulating the following program. The first addition instruction will perform x"FE" + x"01" = x"FF" and assert the negative (N) flag. The second addition instruction will perform x"FF" + x"01" = x"00" and assert the carry (C) and zero (Z) flags. The third addition instruction will perform x"7F" + x"7F" = x"FE" and assert the two's complement overflow (V) and negative (N) flags.

```
constant ROM : rom_type := (
    0     => LDA_IMM,  -- A=x"FE"
    1     => x"FE",
    2     => LDB_IMM,  -- B=x"01"
    3     => x"01",
    4     => ADD_AB,   -- A=A+B
    5     => LDA_IMM,  -- A=x"FF"
    6     => x"FF",
    7     => LDB_IMM,  -- B=x"01"
    8     => x"01",
    9     => ADD_AB,   -- A=A+B
    10    => LDA_IMM,  -- A=x"7F"
    11    => x"7F",
    12    => LDB_IMM,  -- B=x"7F"
    13    => x"7F",
    14    => ADD_AB,   -- A=A+B
    15    => BRA,
    16    => x"00",
    others => x"00");
```

12.3.5 Add the functionality to the computer model from 12.3.4 the ability to perform the *branch if equal to zero* instruction BEQ. Simulate your implementation using the following program. The first addition in this program will perform x"FE" + x"01" = x"FF" (Z=0). The subsequent BEQ instruction should NOT take the branch. The second addition in this program will perform x"FF" + x"01" = x"00" (Z = 1) and SHOULD take the branch. The final instruction in this program is a BRA that is inserted for safety. In the event that the BEQ is not operating properly, the BRA will set the program counter back to x"00" and prevent the program from running away.

```
constant ROM : rom_type := (
    0     => LDA_IMM,
    1     => x"FE",
    2     => LDB_IMM,
    3     => x"01",
    4     => ADD_AB,
    5     => BEQ,
    6     => x"00",  -- should not
                     -- branch

    7     => LDA_IMM,
    8     => x"FF",
    9     => LDB_IMM,
    10    => x"01",
    11    => ADD_AB,
    12    => BEQ,
    13    => x"00",  -- should
                     -- branch

    14    => BRA,
    15    => x"00",

    others => x"00");
```

Chapter 13: Floating-Point Systems

This chapter introduces the concept of floating-point numbers and how to build systems that use floating-point representations to perform mathematical operations. Floating-point numbers represent *real* numbers (i.e., ones that have a fractional component) using an encoding technique that first converts the original number into scientific notation (SN) and then encodes the three distinct fields of the notation with binary codes (e.g., the sign, the mantissa, and the exponent). The standard that guides the encoding/decoding of floating-point numbers is IEEE 754. This standard is used in all modern computers and is also provided as a set of packages for VHDL. This chapter begins with a detailed explanation of the IEEE 754 encoding approach including the anatomy, the algorithms for converting between decimal and floating-point formats, range, and precision. The chapter then discusses arithmetic using floating-point numbers and then moves into VHDL modeling for floating-point systems using the IEEE 754 standard. The goal of this chapter is to provide an understanding of floating-point numbers and the basic principles of how to begin building digital systems that use floating-point numbers in VHDL.

Learning Outcomes—After completing this chapter, you will be able to:

13.1 Describe the IEEE 754 standard including the encoding algorithm, data types, range, and precision of floating-point representation.
13.2 Perform conversions between decimal and IEEE 754 formats by hand.
13.3 Perform basic arithmetic with floating-point numbers by hand.
13.4 Design systems that use floating-point numbers for mathematical operations in VHDL.

13.1 Overview of Floating-Point Numbers

13.1.1 Limitations of Fixed-Point Numbers

The first step in understanding the need for floating-point numbers is to look at the limitations of *fixed-point* representation of real numbers. In a fixed-point encoding approach, a predetermined number of bits is used to represent a real number, and the radix point is *fixed* within the number. The radix point can be placed anywhere within the binary number, but once this location is decided, it cannot move. If the radix point is placed directly in the middle of the bit field, then the upper half of the bits represents the whole part of the number, and the lower half of the bits represents the fractional part. Fixed-point encoding is straightforward and has the advantage that it can use arithmetic circuits built for integer numbers. Figure 13.1 provides an overview of fixed-point encoding in binary.

B. J. LaMeres, *Quick Start Guide to VHDL*, https://doi.org/10.1007/978-3-031-42543-1_13

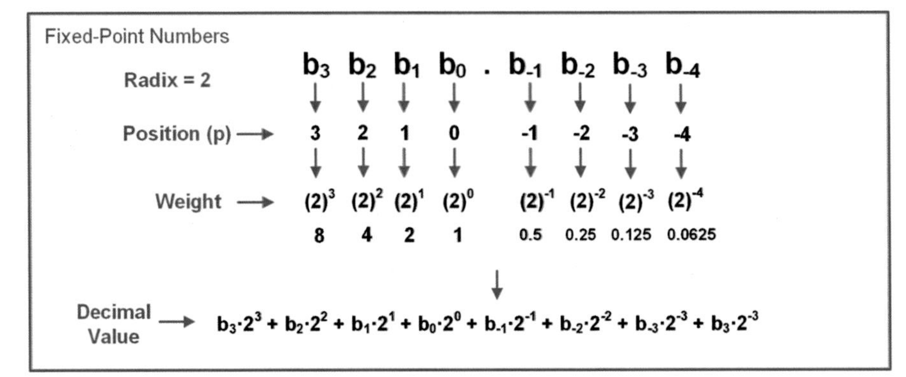

Fig. 13.1
Overview of fixed-point representation for real numbers

The main disadvantage of fixed-point encoding is that once the location of the radix point is determined, it can never be moved. This can lead to suboptimal encoding when a system is built that operates on smaller numbers (i.e., fractional numbers that are close to zero) where it would be better to have the radix point located further to the left in the binary representation. Locating the radix point further to the left would allow more bits in the field to be used for fractional accuracy. Conversely, this also leads to suboptimal encoding when a system is built to operate on larger numbers (i.e., those that don't need a great deal of fractional accuracy) where it would be better to have the radix point located further to the right in the binary representation. Fixed-point encoding leads to a generally suboptimal use of the bits within the field, in addition to a limited range of values it can encode. As arithmetic is performed on fixed-point numbers, accuracy tends to be diminished as fractional values are truncated.

13.1.2 The Anatomy of a Floating-Point Number

An alternative to fixed-point encoding is a floating-point approach in which the radix point can be dynamically moved within the bit field based on a location value encoded into the number itself. This is the motivation for floating-point representation. The term *float* refers to the radix point *floating* within the number based on a value encoded in the word. The general concept of floating-point representation is to first convert the number into binary SN and then encode the three parts of the number separately. The first field encoded is the *sign* of the number where 0 represents a positive number and 1 represents a negative number. The second field encoded is the *mantissa*, which is part of the SN number that is separate from the exponent multiplier portion. And finally, the *exponent* of the multiplier is encoded. The three distinct fields of the SN number are then put into a single binary word with the sign bit as the MSB followed by the exponent field and then the mantissa. Figure 13.2 shows the general overview of floating-point representation. Note that this encoding approach yields a *signed magnitude* number rather than a two's complement encoding approach.

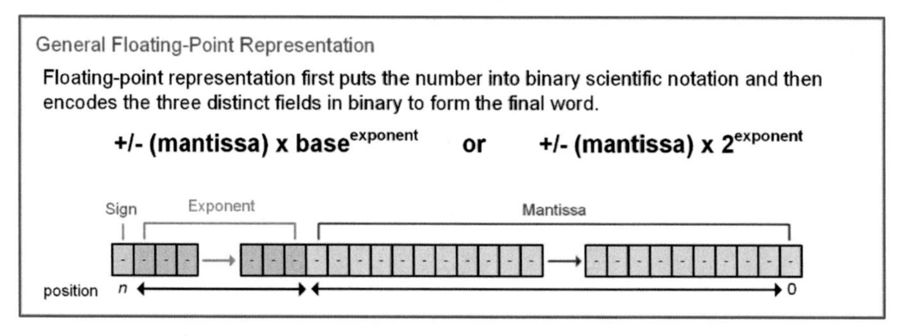

Fig. 13.2
General format of a floating-point number

13.1.3 The IEEE 754 Standard

When the adoption of integrated circuit-based microprocessors began to accelerate in the 1970s, work began on how to effectively implement floating-point numbers and arithmetic circuitry to accompany the new CPUs. As is always the case, competing ideas for how to implement floating-point numbers emerged, and incompatibility between different manufactures presented a barrier for wide-scale interoperability. In order to create a consistent approach for encoding floating-point numbers, an IEEE standard was created. The *IEEE 754 Standard for Floating-Point Arithmetic* was first released in 1985 and defined a variety of formats and special types for floating-point numbers. The standard was updated in 2008 and again in 2019 with minor revisions. The IEEE 754 standard is the most widely used approach for encoding floating-point numbers. While the current IEEE 754 standard has grown to include many different types, special functions, and operations, the most commonly used set of features are as follows:

- Encoding for a *single-precision* (32-bits) and *double-precision* (64-bits) floating-point numbers
- Mandatory mathematical operations support (add, subtract, multiply, divide, square root, fused-multiply-add)
- Conversions to/from IEEE 754 encoding to other formats (integer, character, hex character)
- Representation of non-numbers (e.g., positive and negative infinity, positive and negative zero, and not-a-number to indicate exceptions)
- Rounding procedures (round to the nearest value, round toward infinity, round toward negative infinity, round toward zero)

Not every system that uses IEEE 754 implements the full set of features defined in the standard. The number of features implemented is related to the available computing hardware resources. Today, floating-point implementations range from small microcontrollers that only support limited floating-point usage all the way to high-end servers that implement the complete standard. The remaining sections in this chapter will present the most commonly used features from the IEEE standard.

13.1.4 Single-Precision Floating-Point Representation (32-Bit)

The IEEE 754 standard defines an encoding scheme that uses 32 bits and is referred to as a *single-precision* number. This is formally named **binary32** in IEEE 754, but more commonly referred to as an FP32 number, or simply a *float*. In a single-precision number, 1 bit is used for the sign bit (bit position 31), 8 bits are allocated for the exponent (bits 30 → 23), and 23 bits are allocated for the mantissa (22 → 0). Figure 13.3 shows the anatomy of an IEEE single-precision number.

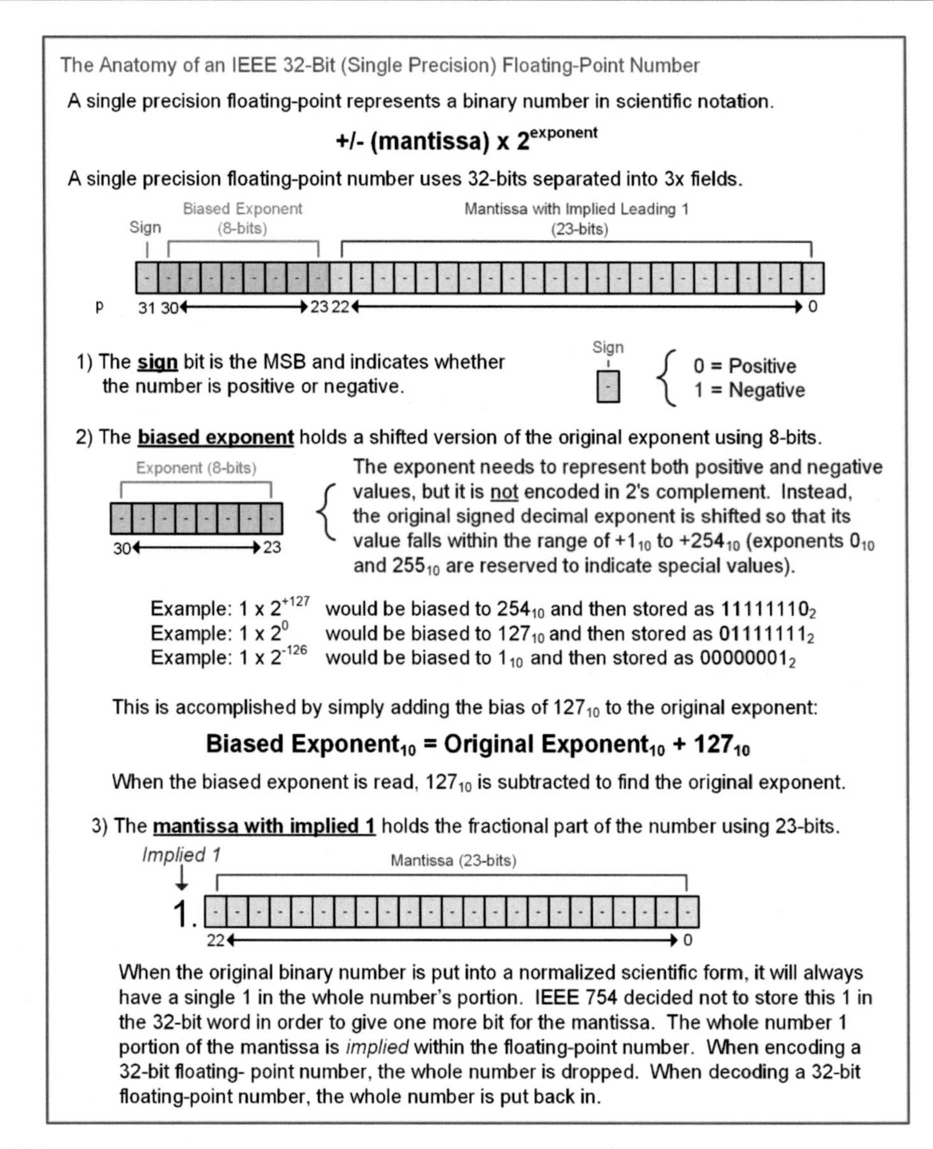

Fig. 13.3
The anatomy of an IEEE 32-bit (single-precision) floating-point number

The sign bit is encoded using the scheme of 0 = positive and 1 = negative. An important concept of single-precision numbers is that the sign bit is not treated the same as in two's complement encoded numbers where it is used as part of the encoding scheme. In single-precision numbers, the sign bit is a stand-alone indicator of whether the number is positive or negative. This is technically a *signed magnitude* representation approach.

The 23 mantissa bits represent the significant bits of the number separate from the exponent multiplier. It is common to see the mantissa referred to as the *significand* for this reason. IEEE 754 uses the concept of an *implied 1*, which takes advantage of the fact that in binary SN, a number can always be put into a form where the whole number to the left of the radix point will always be 1. Since this is always true, IEEE 754 does not include the whole number portion of the mantissa in the 23 bits of the field. This gives one additional bit of precision in the final encoded number. When encoding the number, the implied 1 is not included in the binary32 word. When decoding the number, the implied 1 is added back to the mantissa to form the original value. When a number is put into binary SN with a single 1 in the whole number's position, it is said to be *normalized*. Since the mantissa in the IEEE 754 field represents values to the right of the radix point (without considering the exponent), it will often be called the *fractional* part of the number.

The 8 exponent bits represent the decimal value of the exponent in the binary SN. IEEE 754 decided not to use two's complement encoding in the exponent bits to make some mathematical operations easier and instead only encode unsigned numbers. To encode both positive and negative exponents, the original exponent (which can be positive or negative) is *biased* by adding 127_{10} to it. This essentially *shifts* the original exponent from values that can range from -126_{10} to $+127_{10}$ to an unsigned range of 1_{10} to 254_{10}. This results in a *biased exponent* that is encoded in binary and then stored in the final field. When decoding the exponent, the bias of 127_{10} is simply subtracted from the biased exponent to find the original exponent. The IEEE 754 standard reserves the 0_{10} and 255_{10} exponent codes for special functions such as +/− infinity and +/− zero.

The *range* of a single-precision IEEE number is related to the number of bits in its exponent. The range of floating-point numbers is very large and gives it the ability to represent scientific quantities that fixed-point number can't achieve with the same number of bits. Figure 13.4 shows the largest and smallest numbers possible in normal IEEE 754 single-precision encoding given the constraint that the exponent cannot use the reserved codes 0_{10} and 255_{10}.

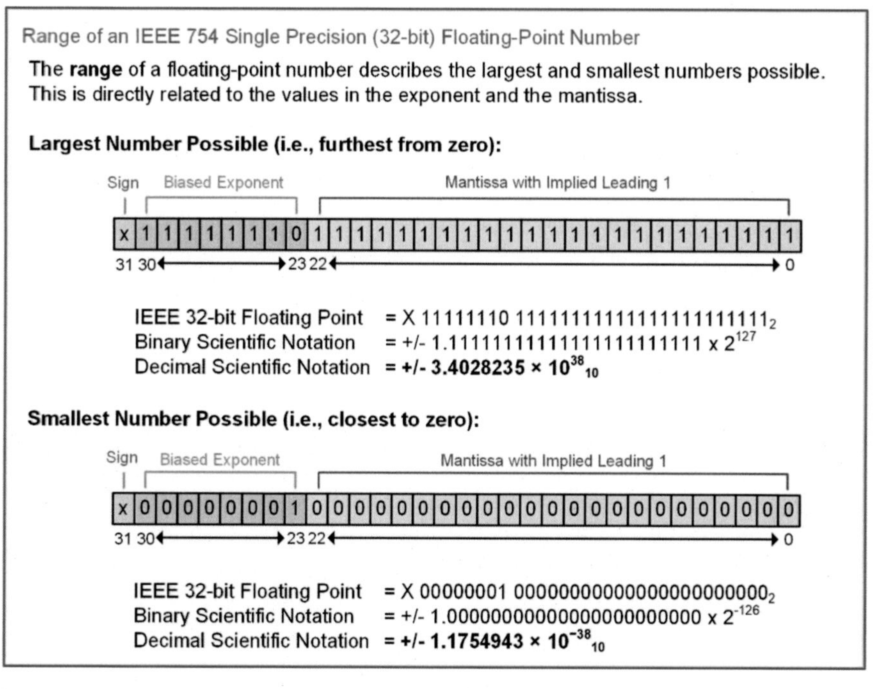

Fig. 13.4
Range of an IEEE 754 single-precision (32-bit) floating-point number

The *precision*, or accuracy, of a single-precision IEEE number is related to the number of bits in its mantissa. Precision is the biggest weakness of floating-point numbers and is most noticeable in 32-bit numbers. The precision of a binary32 value is dictated by the number of mantissa bits available to encode significant figures. The 24 mantissa bits of a binary32 value corresponds to ~7 significant figures in decimal. The loss of precision can become noticeable when performing continual mathematical operations where the error compounds due to LSBs of the result being truncated to fit into the 23-bits available in the mantissa. Floating-point numbers are not guaranteed to represent every possible number exactly. Only numbers that can be represented with the number of bits in the mantissa and exponent are guaranteed to be exact. All other numbers are representations that get as close as possible to the actual value. IEEE 754 defines rounding types that can be used for inexact values. Figure 13.5 shows the precision of an IEEE 754 single-precision floating-point number.

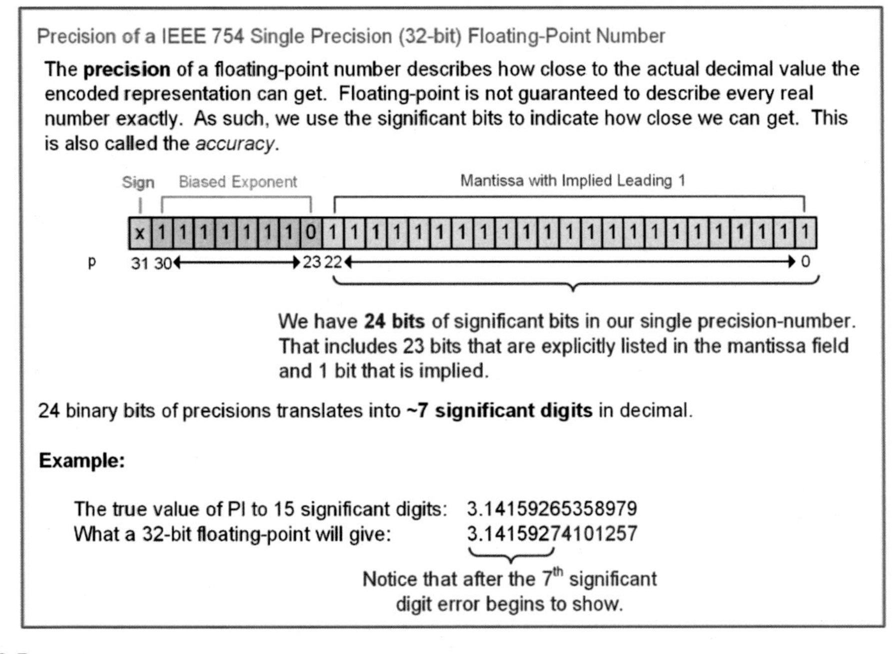

Fig. 13.5
Range of an IEEE 754 single-precision (32-bit) floating-point number

IEEE 754 also supports a special class of numbers called *subnormal*, or denormalized numbers, that allows a representation of values smaller than what can be represented using normal encoding. In the subnormal special case, the implied 1 of the mantissa is not used in the calculation of the value and is instead replaced with a 0 in the whole number position. A subnormal value is indicated by setting the exponent equal to 0. Subnormal numbers can have a positive or negative sign. Subnormal numbers can take on mantissa values from 00...01 to 11...11. The mantissa value of 00...00 with an exponent of 0 is reserved for a different special value that will be covered later. Supporting subnormal values is optional and often associated with how much hardware can be allocated for the floating-point system.

If a calculation results in a value that is larger than an IEEE 754 normalized code can represent, it is called *overflow*. If a calculation results in a value that is smaller than an IEEE 754 normalized code can represent, it is called *underflow*. With these terms, we now have all the definitions needed for the complete number line of a binary32 floating-point representation, which is shown in Fig. 13.6.

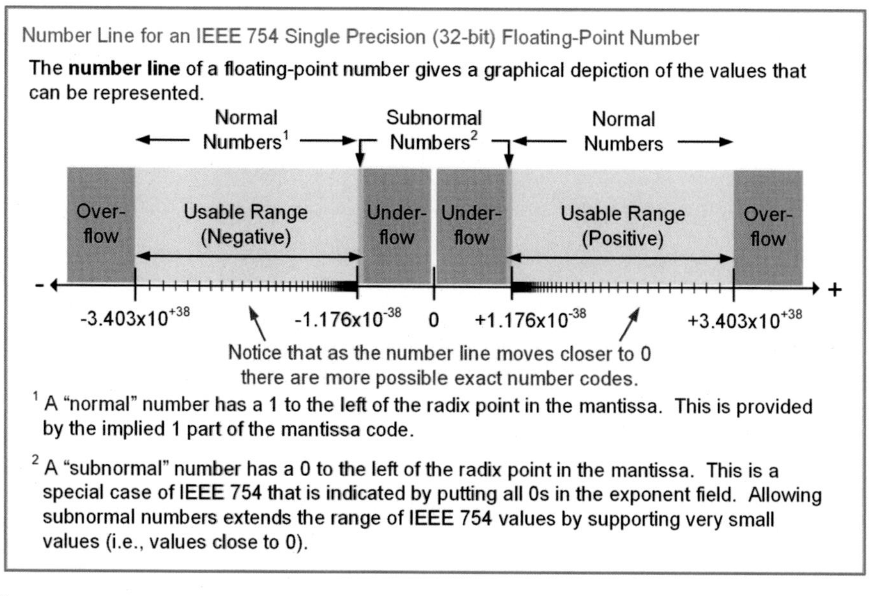

Fig. 13.6
Number line for an IEEE 754 single-precision (32-bit) floating-point number

13.1.5 Double-Precision Floating-Point Representation (64-Bit)

The IEEE 754 standard also defines an encoding scheme for a 64-bit floating-point format that is referred to as a *double-precision* number. This is formally named **binary64** in IEEE 754, but more commonly referred to as an FP64 number, or simply a *double*. In a double-precision number, 1 bit is used for the sign bit (bit position 63), 11 bits are allocated for the exponent (bits 62 → 52), and 52 bits are allocated for the mantissa (51 → 0). Figure 13.7 shows the anatomy of a 64-bit double-precision floating-point number in IEEE 754.

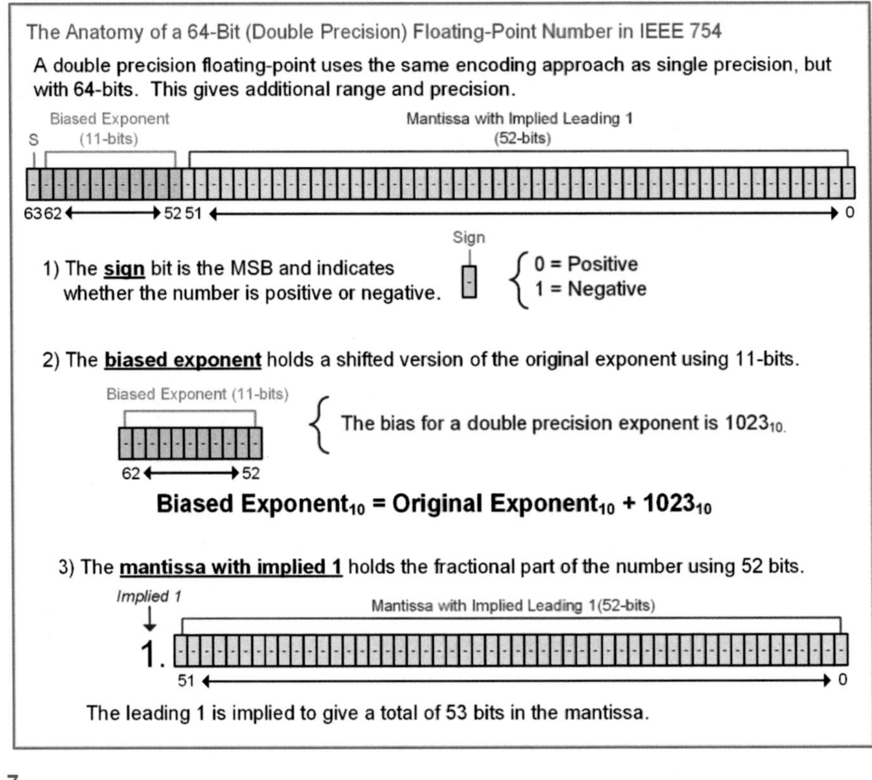

Fig. 13.7
The anatomy of an IEEE 64-bit (double-precision) floating-point number

The double-precision format gives an extremely wide range of numbers due to increasing the number of bits used in the biased exponent. Figure 13.8 shows the range of an IEEE 754 double-precision number.

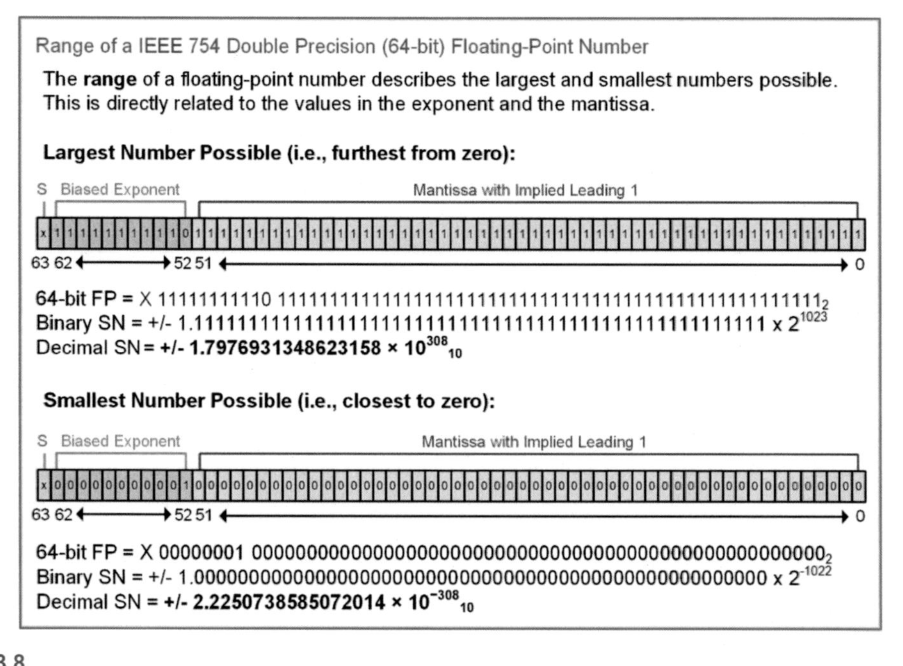

Fig. 13.8

Range of an IEEE 754 double-precision (64-bit) floating-point number

The double-precision encoding technique also addresses the precision issue that is associated with the single-precision format. By increasing the mantissa field to 52 bits, an accuracy of ~16 decimal digits is achieved. Figure 13.9 shows the precision of an IEEE 754 double-precision number.

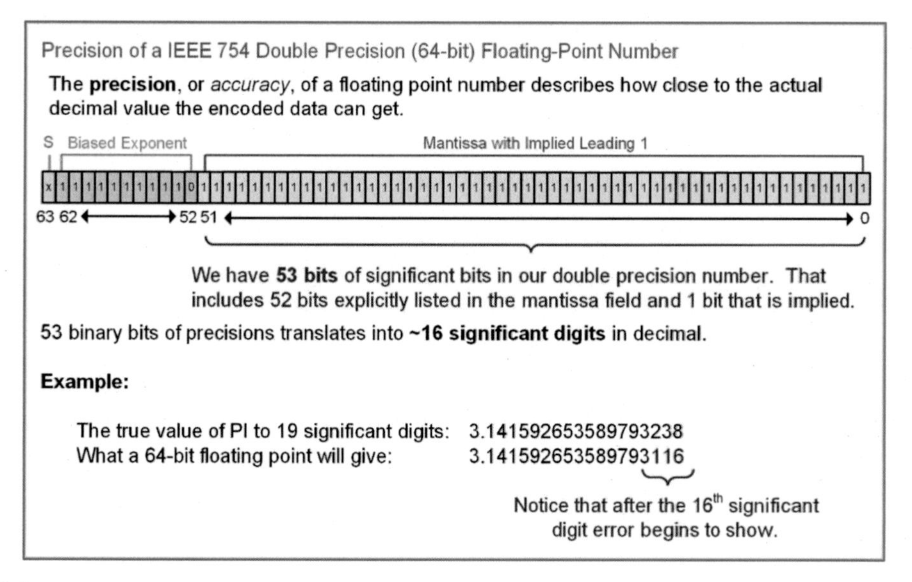

Fig. 13.9
Precision of an IEEE 754 double-precision (64-bit) floating-point number

Figure 13.10 gives the number line of a binary64 floating-point representation.

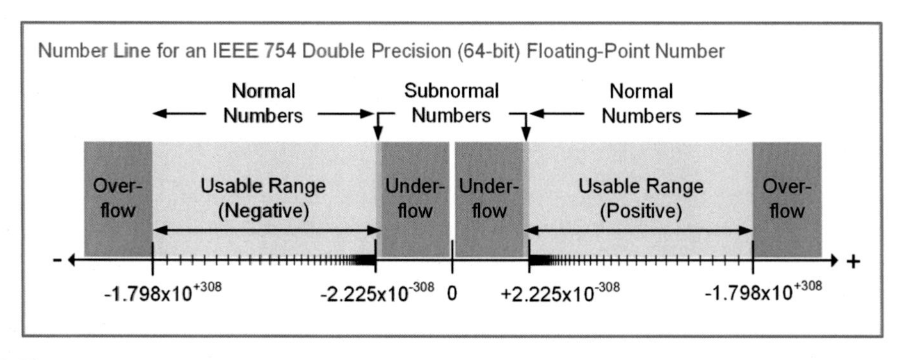

Fig. 13.10
Number line for an IEEE 754 double-precision (64-bit) floating-point number

13.1.6 IEEE 754 Special Values

IEEE 754 defines a set of other special values that are needed for implementation of a useful arithmetic system. IEEE 754 provides unique codes for **+0** and **−0**. To indicate 0, both the exponent and mantissa fields hold all 0s. The sign bit is used normally to indicate whether zero is positive or negative. The code for +0 is considered *exact zero* and used as the result of operations that result in a true zero value (i.e., n-n). While +0 and −0 have unique codes, they are evaluated as equal during compares.

IEEE 754 provides two unique codes for **+infinity** and **−infinity**. Infinity is indicated by setting the exponent to all 1s and the mantissa to all 0s. The sign bit is used normally to indicate whether infinity is positive or negative.

IEEE 754 provides an exception value called ***Not a Number*** (NaN) to indicate an invalid result. A NaN is indicated by an exponent of all 1s and a non-zero mantissa. There are two supported NaN types, a *quiet NaN* (qNaN) and a *signaling NaN* (sNaN). If the first bit of the mantissa is a 1, it indicates a qNaN. A qNaN propagates through an operation yielding a result of qNaN without signaling an exception. If the first bit of the mantissa is a 0, it indicates a sNaN. A sNaN signals an exception immediately upon use. The remaining bits in the mantissa for a sNaN are called the *payload* and hold user-defined exception diagnostic information.

Figure 13.11 gives a list of the IEEE 754 special values and their associated codes.

IEEE 754 Special Values

In IEEE 754, the exponent values of 00...00 (i.e., all 0s) and 11...11 (i.e., all 1s) are reserved to indicate *special values*. When these exponent values are used, the sign bit and mantissa provide additional information about the special value being represented.

Special Value	Sign	Exponent	Mantissa	Notes
+0	0	00...00	00...00	Also called "exact zero"
-0	1	00...00	00...00	0+ and 0- are distinct values, but are evaluated as equal during a compare.
Subnormal (Positive)	0	00...00	00...01 to 11.11	Subnormal numbers use a 0 in place of the implied 1's position in the mantissa. This provides extremely small values to be encoded surrounding 0. These are also referred to as *denormalized* numbers.
Subnormal (Negative)	1	00...00	00...01 to 11.11	
+Infinity	0	11...11	00...00	-
-Infinity	1	11...11	00...00	-
Not a Number (NaN)	x	11...11	00...01 to 01..11	If the non-zero mantissa begins with a 0, it indicates a *signaling NaN* (sNaN). The remaining mantissa bits are referred to as the *payload* and contain user-defined information about the exception.
			10...0 to 11..11	If the non-zero mantissa begins with a 1, it indicates a *quiet NaN* (qNaN). This is the default exception action.

Fig. 13.11
IEEE 754 special values

IEEE 754 also defines the results of operations on special values. If special values are supported, the codes of the inputs are checked prior to performing the operation using a dedicated decoder. If a special value is detected, it will then use the defined results in Fig. 13.12 for the result instead of using the software or circuitry designed to handle regular normal number inputs.

Results of Operations Using IEEE 754 Special Values

IEEE 754 defines the results of the following operations that use special values:

Operation	Result
n ÷ (+Infinity)[1]	+0
n ÷ (-Infinity)[1]	-0
n ÷ (+0)[1]	+Infinity
n ÷ (-0)[1]	-Infinity
(+/-Infinity) x (+/-Infinity)[2]	+/-Infinity
(+Infinity) + (+Infinity)	+Infinity
(-Infinity) + (-Infinity)	-Infinity
(+Infinity) - (+Infinity)	NaN
(-Infinity) + (+Infinity)	NaN
+/-0 x (+/-Infinity)	NaN
(+/-0) ÷ (+/-0)	NaN
(+/-Infinity) ÷ (+/-Infinity)	NaN
NaN[3]	NaN

[1] When the sign of n changes, it will change the sign of the result.

[2] The sign of the result follows the standard rules of multiplication (i.e., POSxPOS=POS, POSxNEG=NEG, NEGxPOS=NEG, NEGxNEG=POS).

[3] The most significant bit of the mantissa indicates whether the NaN is *signaling* (0) or *quiet* (1). A quiet NaN (qNaN) propagates through operations without signaling exceptions and produces a result of qNaN. A signaling NaN (sNaN) signals an exception when used.

Fig. 13.12
Results of operations using IEEE 754 special values

13.1.7 IEEE 754 Rounding Types

The IEEE 754 standard specifies a variety of rounding options for results that don't fall exactly into one of the possible floating-point codes. IEEE 754 also supports the use of *guard bits* to improve accuracy. Guard bits are additional bits added to the mantissa during an operation. After the operation, the result is rounded, and then the guard bits are removed. Inexact numbers that are rounded shall have the same sign as the original unrounded number. NaNs are not rounded.

The first rounding approach is called ***round to nearest – ties to even***. This approach will round a number that falls within two exact representations to the closest numerical value. In the case that a number falls equally within two exact representations, the number is rounded to its even neighbor. The second rounding approach is called ***round to nearest – ties away from zero*** *(a.k.a., ties to away)*. This approach also rounds a number that falls within two exact representations to the closest numerical value, but a number that falls equally within two exact representations will be rounded to the neighbor furthest from zero. An alternate description of this rounding approach is that a number exactly between two values will be rounded to the neighbor with the largest magnitude. The third rounding approach is called ***round toward zero*** and will always round toward the value's neighbor that is lower in magnitude (i.e., closer to zero). The fourth rounding approach is called ***round toward positive*** and will always round toward the more positive value (i.e., the neighbor to its right on the number line). The final rounding approach is called ***round toward negative*** and will always round toward the more negative value (i.e., the neighbor to its left on the number line). Figure 13.13 gives a summary of the IEEE rounding types.

IEEE 754 Rounding Types

IEEE 754 defines five rounding strategies that can be specified.

Mode	Example Values			
	+21.5	**+22.5**	**-21.5**	**-22.5**
Round to nearest, ties to even	+22.0	+22.0	-22.0	-22.0
Round to nearest, ties away from zero	+22.0	+23.0	-22.0	-23.0
Round toward 0	+21.0	+22.0	-21.0	-22.0
Round toward +∞	+22.0	+23.0	-21.0	-22.0
Round toward -∞	+21.0	+22.0	-22.0	-23.0

Fig. 13.13
IEEE 754 rounding types

13.1.8 Other Capabilities of the IEEE 754 Standard

The IEEE 754 standard contains other widths of floating-point numbers that are less commonly used. The **binary16** format uses 16 bits to encode the real number and is commonly referred to as a *half*. The **binary128** format uses 128 bits to encode the real number and is commonly referred to as a *quadruple*. The **binary256** format uses 256 bits to encode the real number and is commonly referred to as an *octuple*.

IEEE 754 supports three decimal encoded floating-point formats called **decimal32**, **decimal64**, and **decimal128**. These encoding approaches are meant to match decimal rounding rules exactly, and they are primarily used in monetary calculations where even a small amount of rounding error can lead to a significant amount of money lost.

There are many nuances in the IEEE 754 standard when it comes to implementation. As such, a designer should consult the latest standard for the exact details of the encoding and decide which features are needed in their system. A designer also will need to decide whether the IEEE 754 features should be implemented in hardware, software, or a combination of both.

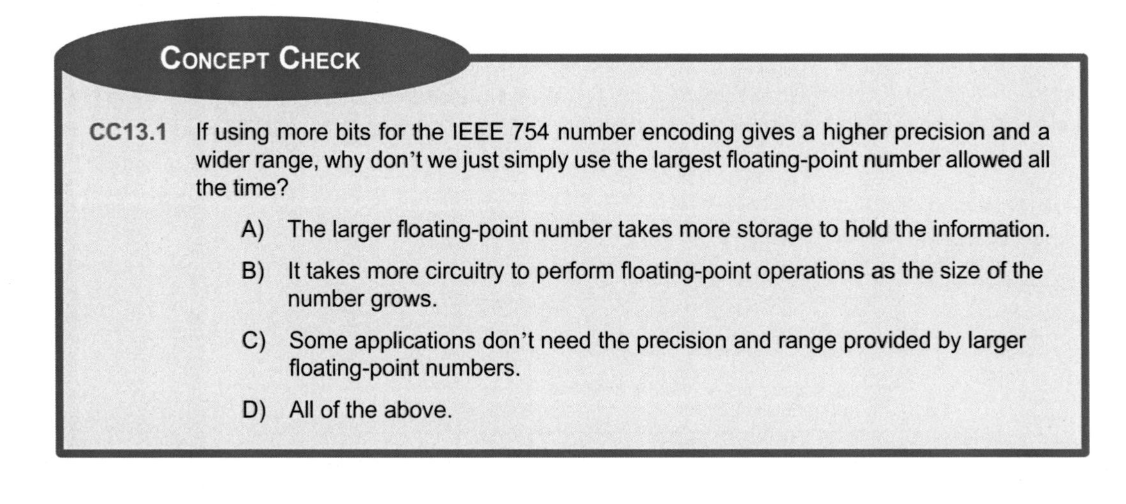

CONCEPT CHECK

CC13.1 If using more bits for the IEEE 754 number encoding gives a higher precision and a wider range, why don't we just simply use the largest floating-point number allowed all the time?

A) The larger floating-point number takes more storage to hold the information.

B) It takes more circuitry to perform floating-point operations as the size of the number grows.

C) Some applications don't need the precision and range provided by larger floating-point numbers.

D) All of the above.

13.2 IEEE 754 Base Conversions

13.2.1 Converting from Decimal into IEEE 754 Single-Precision Numbers

Converting from a decimal number into a single-precision floating-point number by hand consists of these steps:

1) Convert the decimal number into a fixed-point binary representation.
2) Convert the fixed-point number into normalized binary scientific notation.
3) From the binary SN, determine the sign bit.
4) From the binary SN, determine the biased exponent.
5) From the binary SN, determine the mantissa with implied leading 1.
6) Combine the three fields from steps 3–5 into the final 32-bit binary number.

Example 13.1 shows the process of converting a $+22.125_{10}$ into a single-precision floating-point representation.

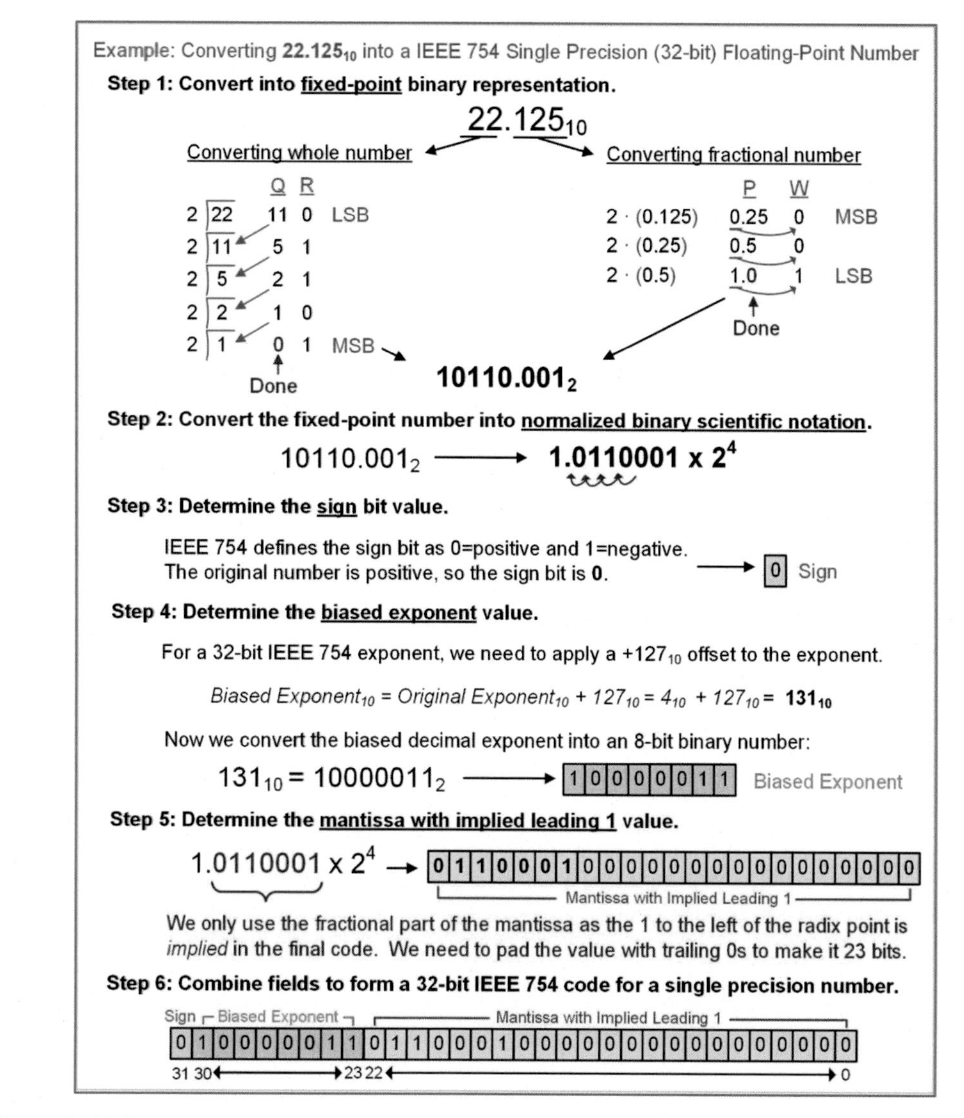

Example 13.1
Converting 22.125_{10} into IEEE 754 single-precision (32-bit) floating-point number

Example 13.2 shows another example of converting a decimal number (-45.4_{10}) into a single-precision floating-point representation, this time illustrating how to handle a fractional component that repeats, in addition to a number with a negative sign.

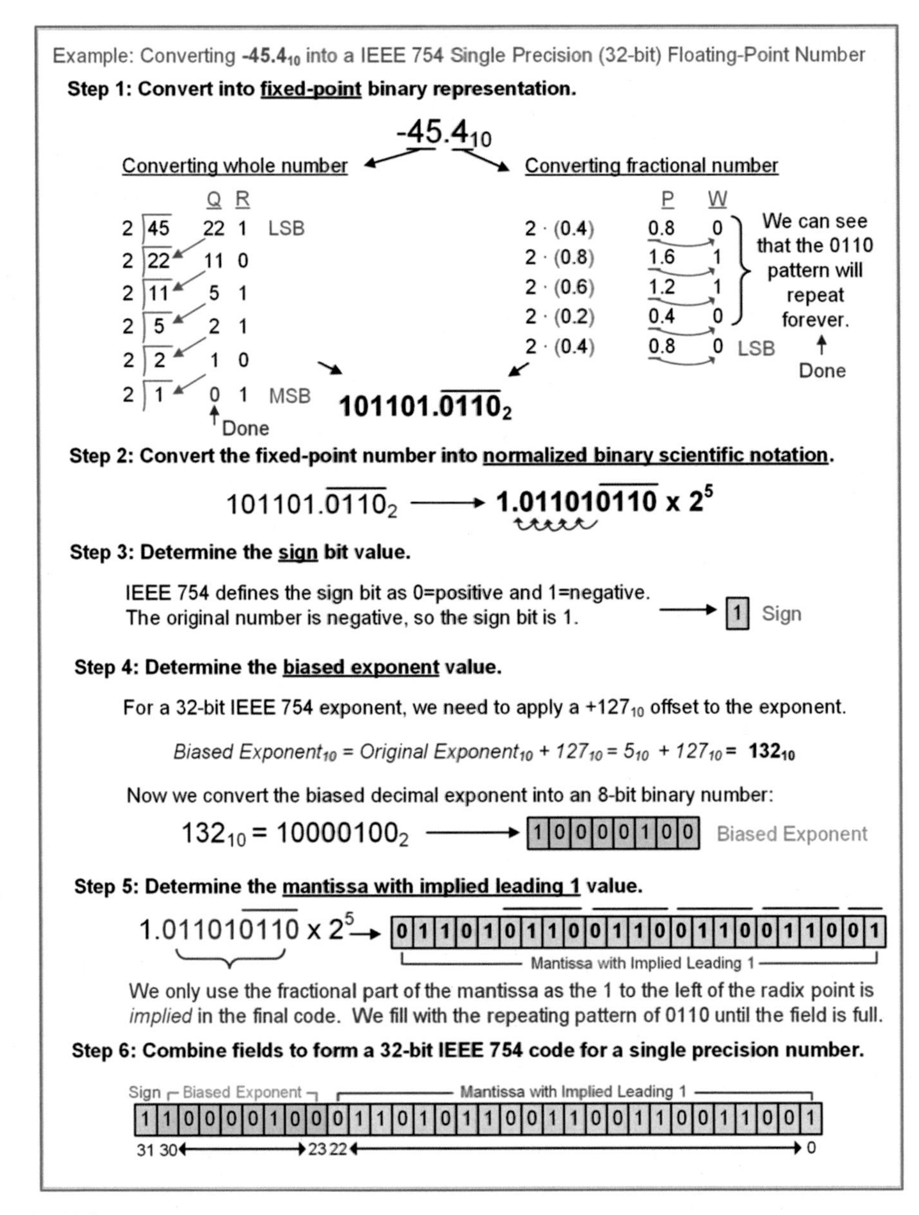

Example 13.2
Converting -45.4_{10} into IEEE 754 single-precision (32-bit) floating-point number

Example 13.3 shows another example of converting a decimal number (-0.05_{10}) into a single-precision floating-point representation, this time on a small decimal number with a repeating fractional component.

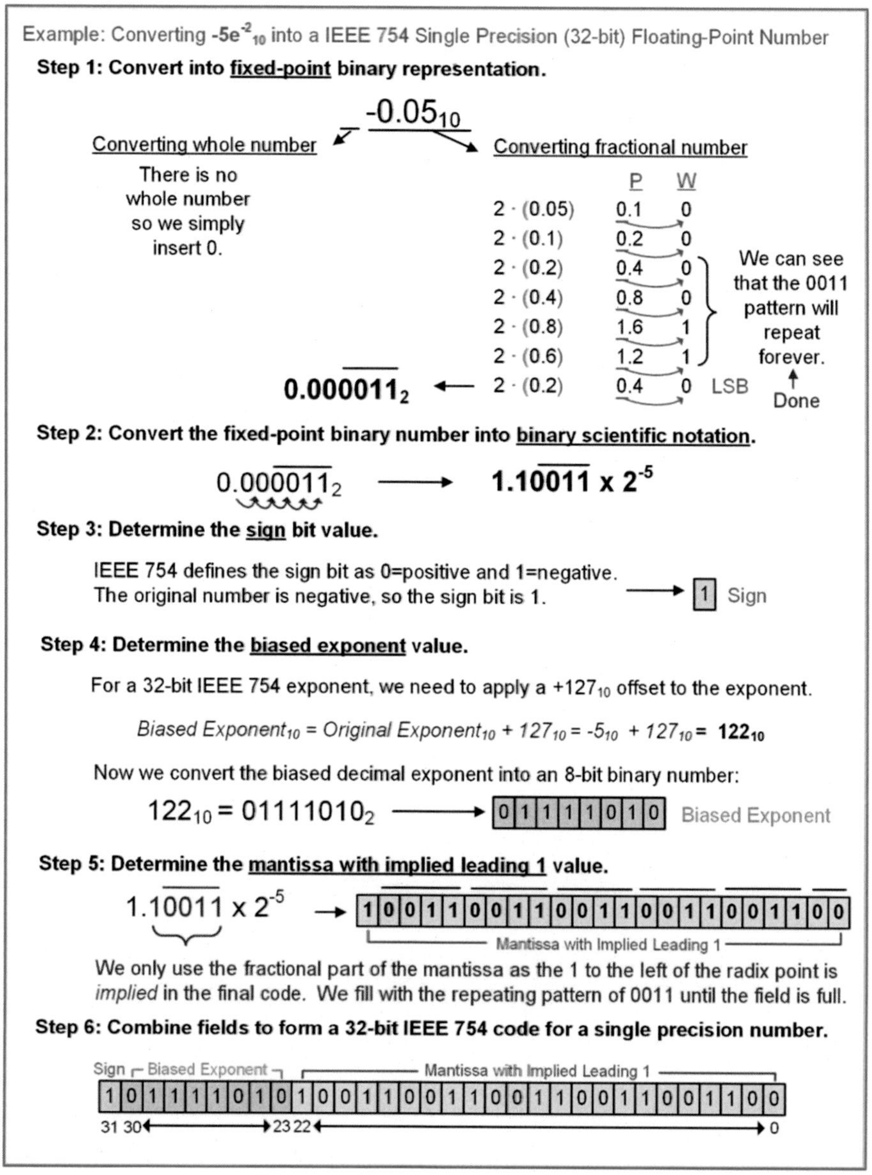

Example: Converting $-5e^{-2}{}_{10}$ into a IEEE 754 Single Precision (32-bit) Floating-Point Number

Step 1: Convert into fixed-point binary representation.

-0.05_{10}

Converting whole number

There is no whole number so we simply insert 0.

Converting fractional number

	P	W
2 · (0.05)	0.1	0
2 · (0.1)	0.2	0
2 · (0.2)	0.4	0
2 · (0.4)	0.8	0
2 · (0.8)	1.6	1
2 · (0.6)	1.2	1
2 · (0.2)	0.4	0 LSB

We can see that the 0011 pattern will repeat forever.

Done

$\mathbf{0.0\overline{0011}_2}$

Step 2: Convert the fixed-point binary number into binary scientific notation.

$0.0\overline{0011}_2 \longrightarrow \mathbf{1.1\overline{0011} \times 2^{-5}}$

Step 3: Determine the sign bit value.

IEEE 754 defines the sign bit as 0=positive and 1=negative.
The original number is negative, so the sign bit is 1. → [1] Sign

Step 4: Determine the biased exponent value.

For a 32-bit IEEE 754 exponent, we need to apply a $+127_{10}$ offset to the exponent.

Biased Exponent$_{10}$ = *Original Exponent*$_{10}$ + 127_{10} = -5_{10} + 127_{10} = $\mathbf{122_{10}}$

Now we convert the biased decimal exponent into an 8-bit binary number:

$122_{10} = 01111010_2$ → [0|1|1|1|1|0|1|0] Biased Exponent

Step 5: Determine the mantissa with implied leading 1 value.

$1.1\overline{0011} \times 2^{-5}$ → [1|0|0|1|1|0|0|1|1|0|0|1|1|0|0|1|1|0|0|1|1|0|0]

Mantissa with Implied Leading 1

We only use the fractional part of the mantissa as the 1 to the left of the radix point is *implied* in the final code. We fill with the repeating pattern of 0011 until the field is full.

Step 6: Combine fields to form a 32-bit IEEE 754 code for a single precision number.

Sign | Biased Exponent | Mantissa with Implied Leading 1

[1|0|1|1|1|1|0|1|0|1|0|0|1|1|0|0|1|1|0|0|1|1|0|0|1|1|0|0|1|1|0|0]

31 30 ↔ 23 22 ↔ 0

Example 13.3
Converting -0.05_{10} into IEEE 754 single-precision (32-bit) floating-point number

13.2.2 Converting from IEEE 754 Single-Precision Numbers into Decimal

Converting from an IEEE 754 single-precision number into decimal by hand is the reverse of the prior section:

1) Reassemble the original mantissa by adding back in the implied 1.
2) Determine the decimal value of the original exponent from the biased exponent.
3) Determine whether the number is positive or negative from sign bit.
4) Assemble the extracted information from steps 1–3 into binary scientific notation.
5) Shift the radix point in the binary SN per the exponent value to get back into fixed-point binary.
6) Convert the fixed-point binary number to decimal.

Example 13.4 shows an example of converting from an IEEE 754 single-precision floating-point number (-34.75_{10}) into decimal.

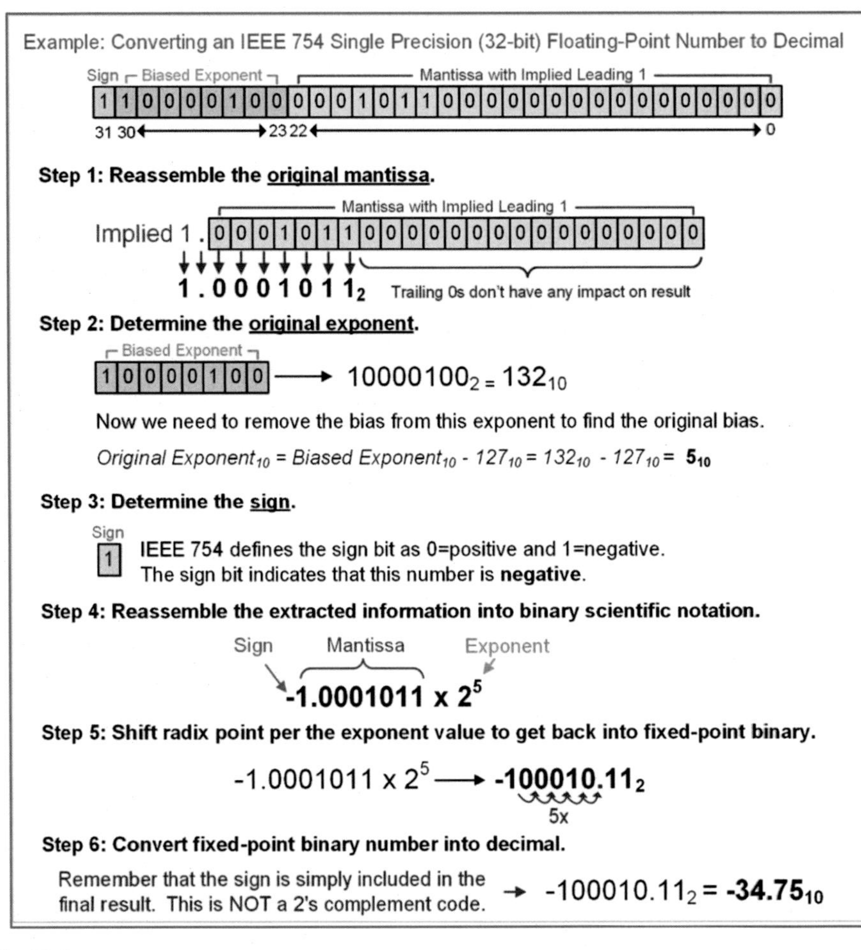

Example 13.4
Converting from an IEEE 754 single-precision number into decimal

Concept Check

CC13.2 Why is having the number in normalized binary scientific notation important for IEEE 754 encoding?

A) It avoids using engineering notation (i.e., when the exponents are multiples of 3), which has always bothered the scientists.

B) It allows the entire word to be used for the mantissa to achieve the greatest precision possible.

C) Being normalized guarantees that the bit to the left of the radix point will be a one and enables the use of the "implied 1" concept for the mantissa. This gives one extra bit for the mantissa, which leads to a higher precision.

D) It minimizes the number of bits used for the exponent so that the exponent can be ignored.

13.3 Floating-Point Arithmetic

13.3.1 Addition and Subtraction of IEEE 754 Numbers

The process of adding and subtracting IEEE 754 numbers follows a similar algorithm as when using base 10 SN. The first step is to ensure that both inputs have the same exponent by manipulating the exponent of one of the inputs until it matches the other. This has the result of moving the decimal point in the mantissa. The addition/subtraction is then performed on the mantissas of the inputs. The common exponent is then applied to the sum/difference. The final step is to *normalize* the result, which means ensuring that the result only has one non-zero digit to the left of the decimal point in the final SN. This is again accomplished by altering the exponent until the decimal point is in the desired location. These steps are summarized below:

1) Make exponents of the input arguments identical.
2) Perform addition/subtraction on the mantissas.
3) Apply the common exponent from step 1 to the result.
4) Normalize the result (if necessary).

Example 13.5 shows an example of performing addition/subtraction on numbers in base 10 SN to illustrate these steps.

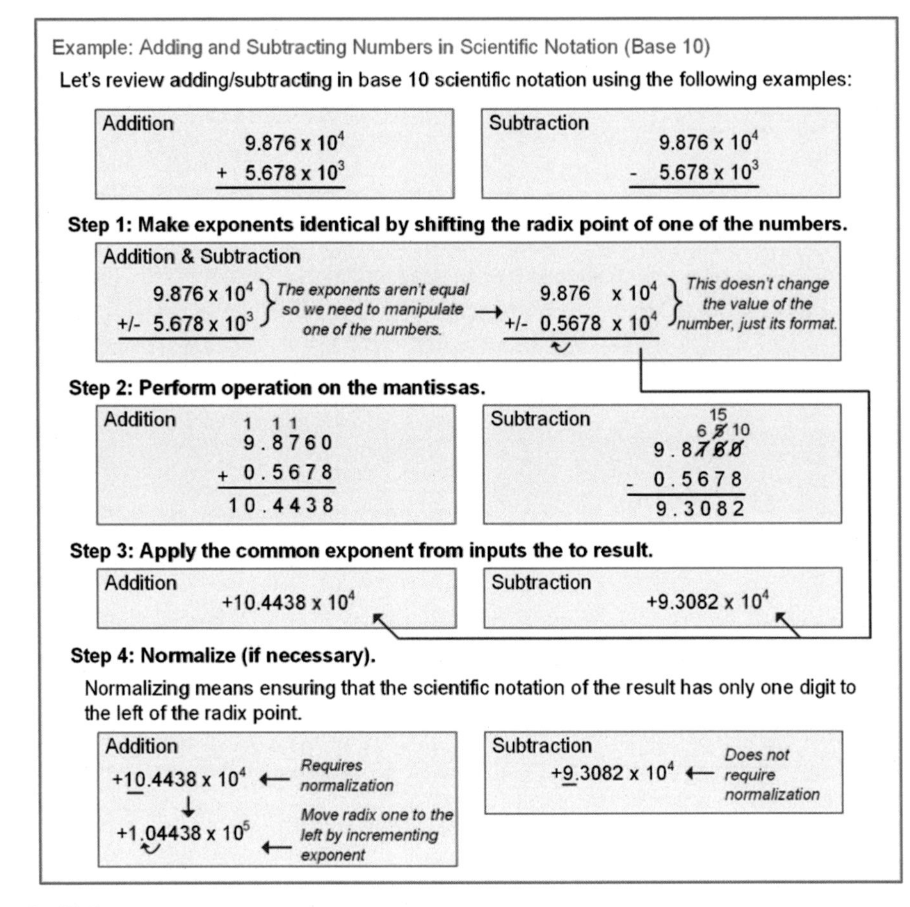

Example 13.5
Adding and subtracting numbers in scientific notation (base 10)

When doing addition/subtraction of IEEE 754 numbers, there are a few additional steps that need to be performed beyond the base 10 SN process. The first step is to convert the input arguments into normalized binary SN form. This step involves taking the original IEEE 754 encoded word and breaking it into its three distinct fields (sign, mantissa, and exponent). This step is called *unpacking*. During this step the implied 1 is applied to the mantissa field extending its size by 1 bit, and the bias is removed from the exponent.

The second step is to modify the input arguments to have the same exponents. This is accomplished by moving the radix point of one of the input arguments until it has an exponent that matches the other argument. Each time the radix point is moved, the exponent must be incremented or decremented accordingly. To minimize the loss of significant bits, this step is always performed on the input argument with the smaller exponent. A logical shift right is performed on this input argument, and the exponent is incremented accordingly until the exponents match. The logical shift right brings in 0s in the most significant position of the mantissa and can potentially shift the least significant positions out of the word. The bits lost in the lower significant position leads to *truncation error*. However, truncation error is less severe, in terms of precision, than losing bits in the most significant position as would be the case if the input argument with the larger exponent was altered using a logical shift left.

The third step involves addressing any negative inputs. Note that IEEE 754 uses a signed magnitude encoding approach for negative numbers. This means we either need to build adder/subtractor circuitry that can handle signed magnitude arithmetic or convert the signed magnitude numbers into two's complement form and reuse arithmetic circuitry that already exists for two's complement integer operations. The most typical approach is the latter (i.e., convert the signed magnitude values into two's complement representations). To convert a signed magnitude number into two's complement representation, a 0 sign bit is first applied to the mantissa. This does not alter the value being held in the mantissa but does make the word one bit larger. If the input argument is positive as indicated by the value in the IEEE 754 sign bit, then no modifications are performed on the argument as it is already in two's complement form for a positive number. If the input argument is negative as indicated by the value in the IEEE 754 sign bit, then two's complement negation is performed to convert the input into its negative two's complement representation.

The fourth step is to perform addition/subtraction on the mantissas with the radix points aligned. Note that the radix points will already be aligned based on the logical shift rights performed in step 2. Remember that the addition/subtraction is being performed on two's complement numbers, which means if there is a resulting carry out, it is discarded.

The fifth step is to address the sign bit of the result. If the result is positive, store a 0 in the IEEE 754 sign field for the result, and then remove the MSB of the result. Recall that the MSB of the result is an additional sign bit that was added in step 3 to convert the numbers from signed magnitude into two's complement representation. If the result is negative, store a 1 in the IEEE 754 sign field for the result, perform two's complement negation to convert the result into a magnitude, and then remove the MSB of the result. After the temporary sign bit is removed from the result, it will have the same number of bits as the original inputs' mantissas.

The sixth step is to apply the input arguments' exponent to the result. Recall that when we made the input arguments' exponents the same in step 2, we altered the smaller exponent to match the larger exponent. This means the larger number's exponent was not altered and can be used directly as the exponent for the result.

The seventh step is to normalize the result (if necessary). Recall that normalization of an IEEE 754 number means shifting the result so that there is a single one to the left of the radix point. Any shifts that are applied to the result to normalize it will be reflected by incrementing/decrementing the exponent. Note that this incrementing/decrementing can be performed on the exponent even if the bias has already been applied.

The final step is to convert the final binary SN into the binary32 encoding. This involves dropping the leading 1 of the mantissa and only storing the fractional part of the number into the 23 bits of the mantissa field. This also involves adding back in the bias to the exponent and storing it into the 8 bits of the exponent field. This final step is also called *packing*.

The steps in performing IEEE 754 addition are summarized below:

1) Convert input arguments from IEEE 754 into their normalized binary SN (also referred to as *unpacking* the original 32-bit word into its three individual parts).
2) Modify the arguments to have the same exponents.
3) Apply a leading 0 sign bit to the arguments, and convert any negative numbers into two's complement representation.
4) Perform addition on the mantissas with the radix points aligned.
5) Address the sign bit of the result.
6) Apply the input arguments' exponent to the result.
7) Normalize the result so that there is a leading 1 on the mantissa.
8) Put all three components of the result back into binary32 format (i.e., pack the results into a 32-bit single-precision word).

Example 13.6 shows steps 1 through 4 of adding binary32 numbers. This example shows adding two positive binary32 numbers that result in a positive sum (i.e., POS + POS = POS).

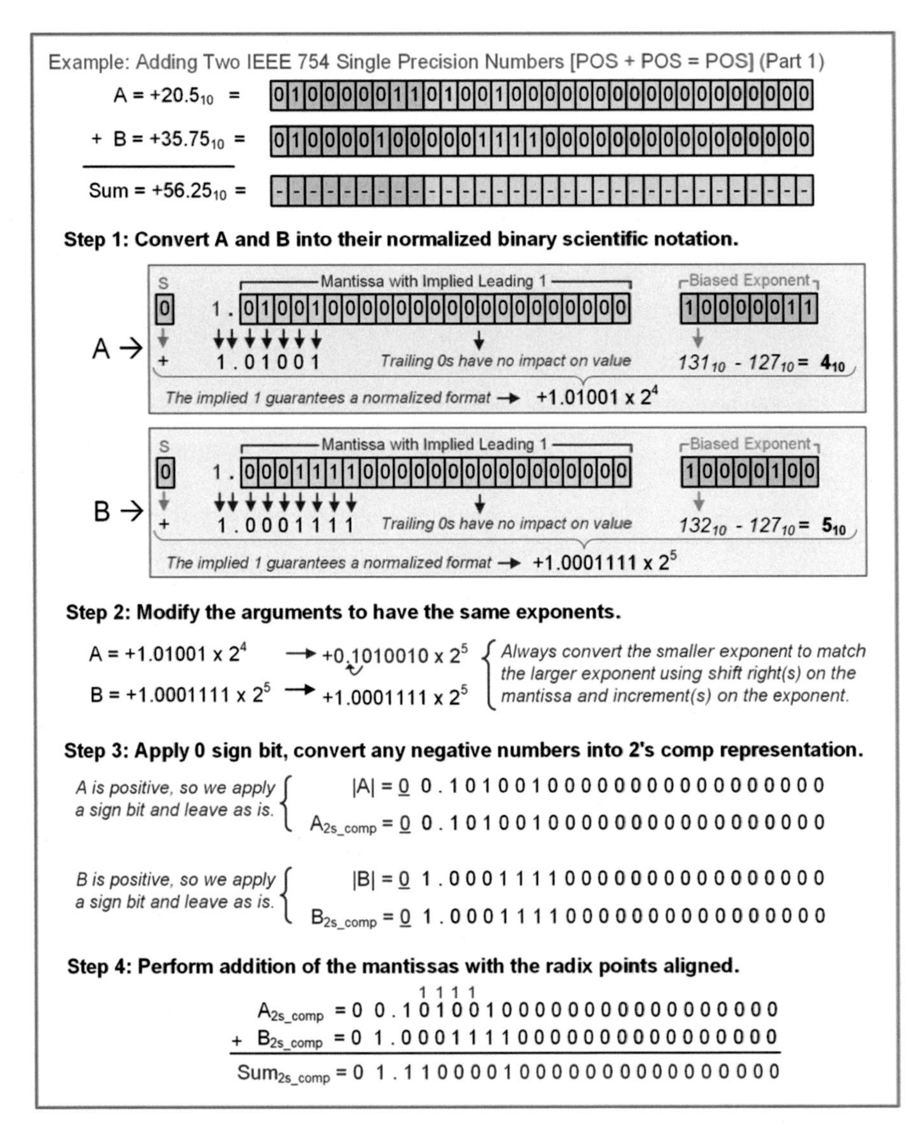

Example 13.6
Adding two IEEE 754 single-precision numbers [POS + POS = POS] (Part 1)

Example 13.7 shows steps 5 through 8 of adding binary32 numbers (POS + POS = POS).

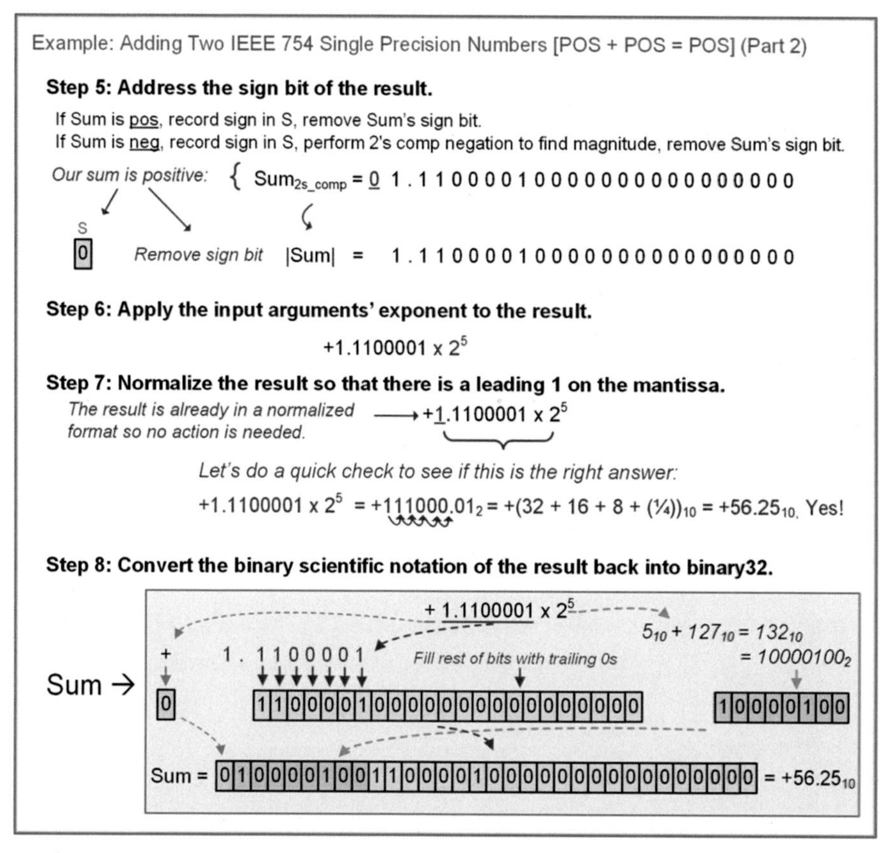

Example 13.7
Adding two IEEE 754 single-precision numbers [POS + POS = POS] (Part 2)

Example 13.8 shows steps 1 through 4 of adding binary32 numbers. This example shows adding a negative number to a positive number that results in a negative sum (i.e., NEG + POS = NEG).

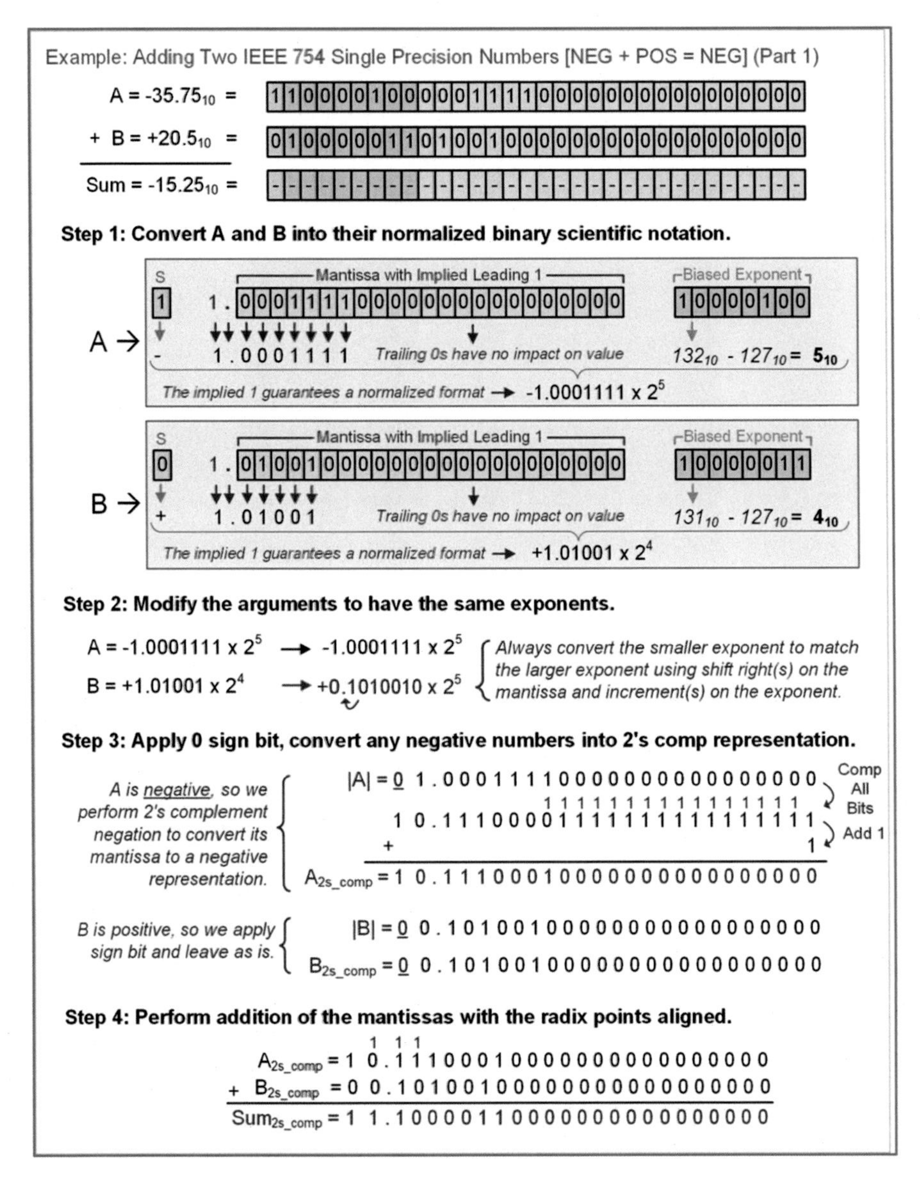

Example 13.8
Adding two IEEE 754 single-precision numbers [NEG + POS = NEG] (Part 1)

Example 13.9 shows steps 5 through 8 of adding binary32 numbers (NEG + POS = NEG).

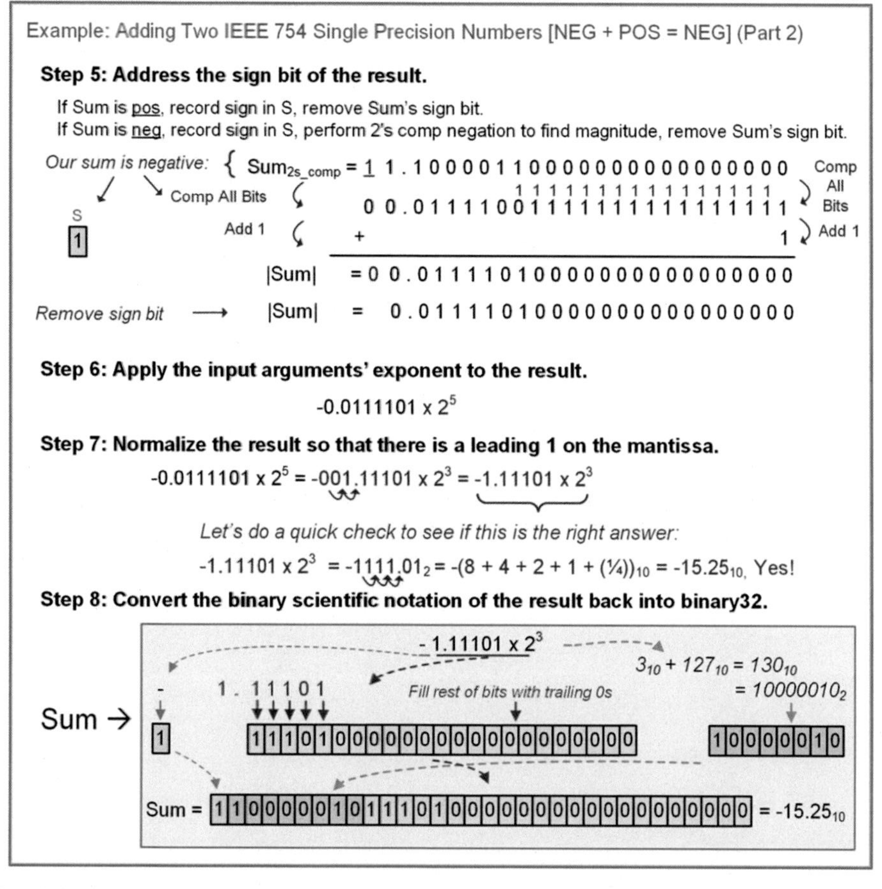

Example 13.9
Adding two IEEE 754 single-precision numbers [NEG + POS = NEG] (Part 2)

Subtraction of IEEE 754 numbers can take advantage of two's complement negation of the subtrahend in order to reuse existing addition circuits within the system. This inserts an additional step into the prior addition process as follows:

1) Convert input arguments from IEEE 754 into their normalized binary SN (i.e., unpack the inputs).
2) Modify the arguments to have the same exponents.
3) Apply a leading 0 sign bit to the arguments, and convert any negative numbers into two's complement representation.
4) Perform two's complement negation on the subtrahend so that we can use addition. *This is the extra step to perform subtraction that is included to the addition process described earlier.*
5) Perform addition on the mantissas with the radix points aligned.
6) Address the sign bit of the result.
7) Apply the input arguments' exponent to the result.
8) Normalize the result so that there is a leading 1 on the mantissa.
9) Put all three components of the result back into binary32 format (i.e., pack the results into a 32-bit single-precision word).

Example 13.10 shows steps 1 through 5 of subtracting binary32 numbers. This example shows subtracting a positive number from a positive number that results in a positive difference (i.e., POS − POS = POS). This example illustrates how to use an adder and two's complement negation of the subtrahend to accomplish subtraction.

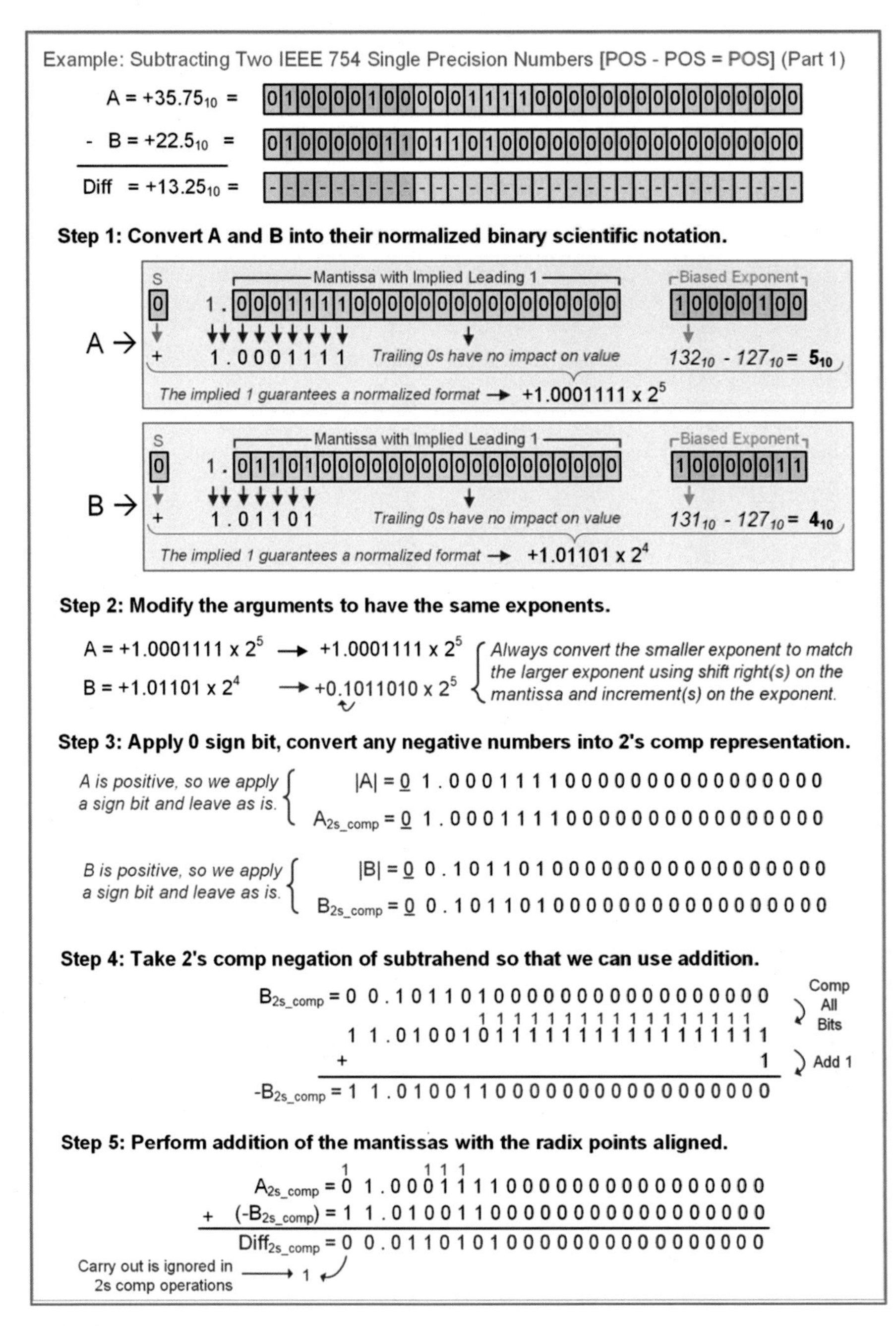

Example 13.10
Subtracting two IEEE 754 single-precision numbers [POS − POS = POS] (Part 1)

Example 13.11 shows steps 6 through 9 of subtracting binary32 numbers.

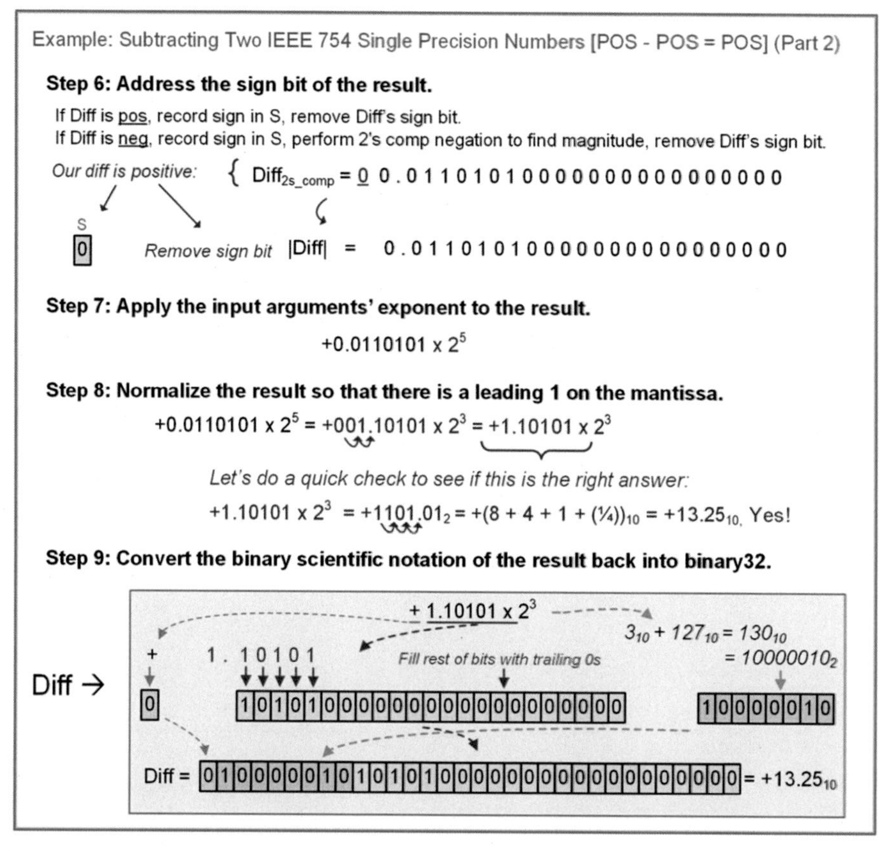

Example 13.11
Subtracting two IEEE 754 single-precision numbers [POS − POS = POS] (Part 2)

13.3.2 Multiplication and Division of IEEE 754 Numbers

The process of multiplying and dividing IEEE 754 numbers follows a similar algorithm as when using base 10 SN. For multiplication, the multiplication is performed on the mantissas, and then the exponents are added. The resulting exponent is applied to the product of the mantissa multiplication to get the final result. In division, the division is performed on the mantissas, and then the exponents are subtracted (dividend exponent – divisor exponent). The resulting exponent is applied to the quotient of the mantissa division for the final result. The final step is to normalize the result if necessary. These steps are summarized below:

1) Perform multiplication/division on the mantissas.
2) Add/subtract the exponents.
3) Apply the new exponent from step 2 to the result.
4) Normalize the result (if necessary).
5) Apply sign.

Example 13.12 shows an example of performing multiplication/division numbers in base 10 SN to illustrate these steps.

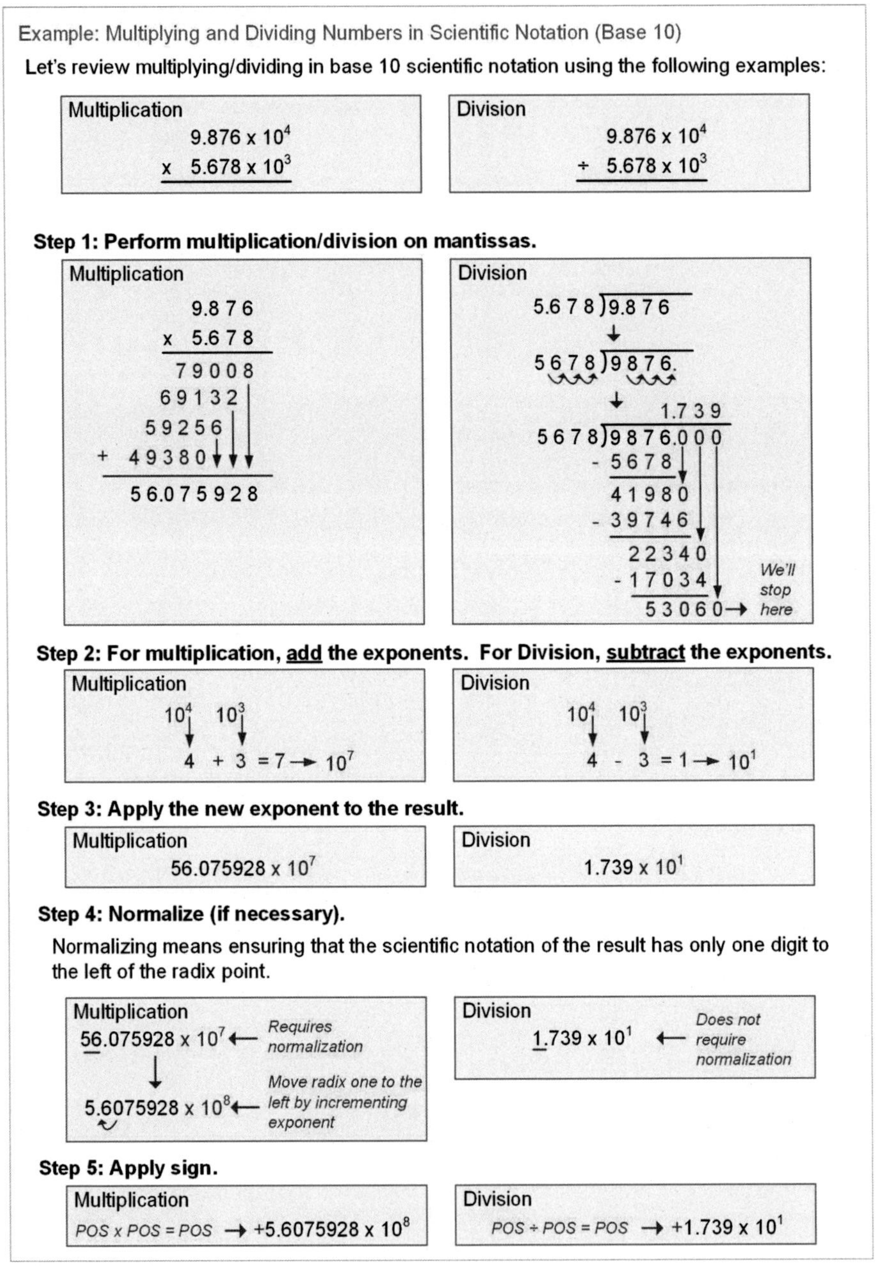

Example: Multiplying and Dividing Numbers in Scientific Notation (Base 10)

Let's review multiplying/dividing in base 10 scientific notation using the following examples:

Multiplication	Division
9.876×10^4 $\times\ 5.678 \times 10^3$	9.876×10^4 $\div\ 5.678 \times 10^3$

Step 1: Perform multiplication/division on mantissas.

Multiplication

```
          9.876
      x   5.678
      ---------
          79008
         69132
        59256
    +  49380
    -----------
     56.075928
```

Division

```
5.678)9.876
      ↓
5678)9876.
      ↓
         1.739
5678)9876.000
    - 5678
      41980
    - 39746
      22340
    - 17034
       53060 → We'll stop here
```

Step 2: For multiplication, <u>add</u> the exponents. For Division, <u>subtract</u> the exponents.

Multiplication	Division
10^4 10^3: $4 + 3 = 7 \rightarrow 10^7$	10^4 10^3: $4 - 3 = 1 \rightarrow 10^1$

Step 3: Apply the new exponent to the result.

Multiplication	Division
56.075928×10^7	1.739×10^1

Step 4: Normalize (if necessary).

Normalizing means ensuring that the scientific notation of the result has only one digit to the left of the radix point.

Multiplication	Division
56.075928×10^7 ← *Requires normalization* ↓ 5.6075928×10^8 ← *Move radix one to the left by incrementing exponent*	1.739×10^1 ← *Does not require normalization*

Step 5: Apply sign.

Multiplication	Division
POS x POS = POS → $+5.6075928 \times 10^8$	*POS ÷ POS = POS* → $+1.739 \times 10^1$

Example 13.12
Multiplying and dividing numbers in scientific notation (base 10)

When performing multiplication/division on IEEE 754 numbers, the sign of the result can be found using an exclusive-OR operation on the sign bits of the inputs. This is independent of the operations on the mantissa and exponent due to the signed magnitude encoding approach used in IEEE 754. The multiplication/division operations can then be performed directly on the mantissa without any consideration of the sign. The steps to perform multiplication/division on IEEE 754 numbers are as below:

1) Convert input arguments from IEEE 754 into their normalized binary SN (i.e., unpack).
2) Perform multiplication/division on the mantissas.
3) Add/subtract the exponents.
4) Apply the new exponent from step 2 to the result.
5) Normalize the result (if necessary).
6) Compute the sign of the result.
7) Put all three components of the result back into binary32 format (i.e., pack the results into a 32-bit single-precision word).

Example 13.13 shows steps 1 through 2 of multiplying binary32 numbers.

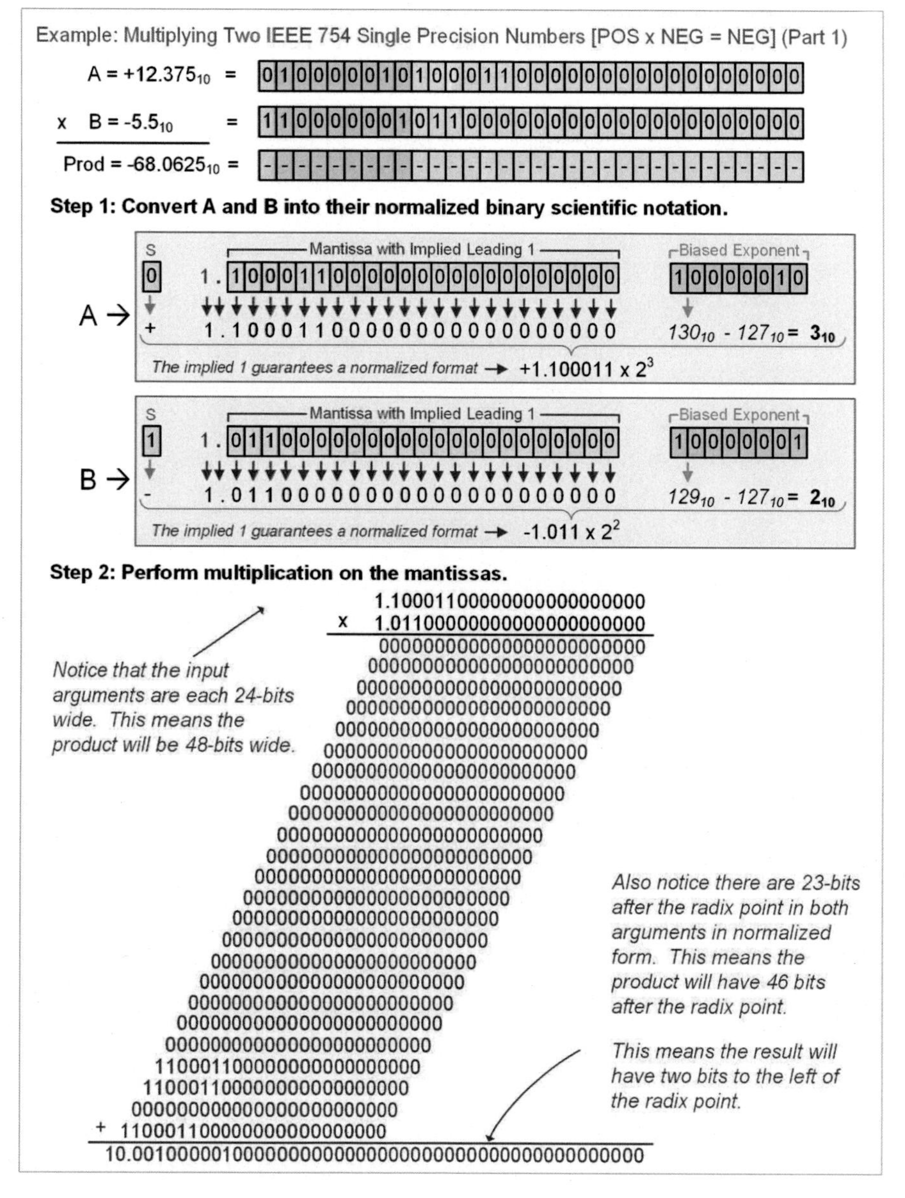

Example 13.13

Multiplying two IEEE 754 single-precision numbers [POS × NEG = NEG] (Part 1)

Example 13.14 shows steps 3 through 6 of multiplying binary32 numbers.

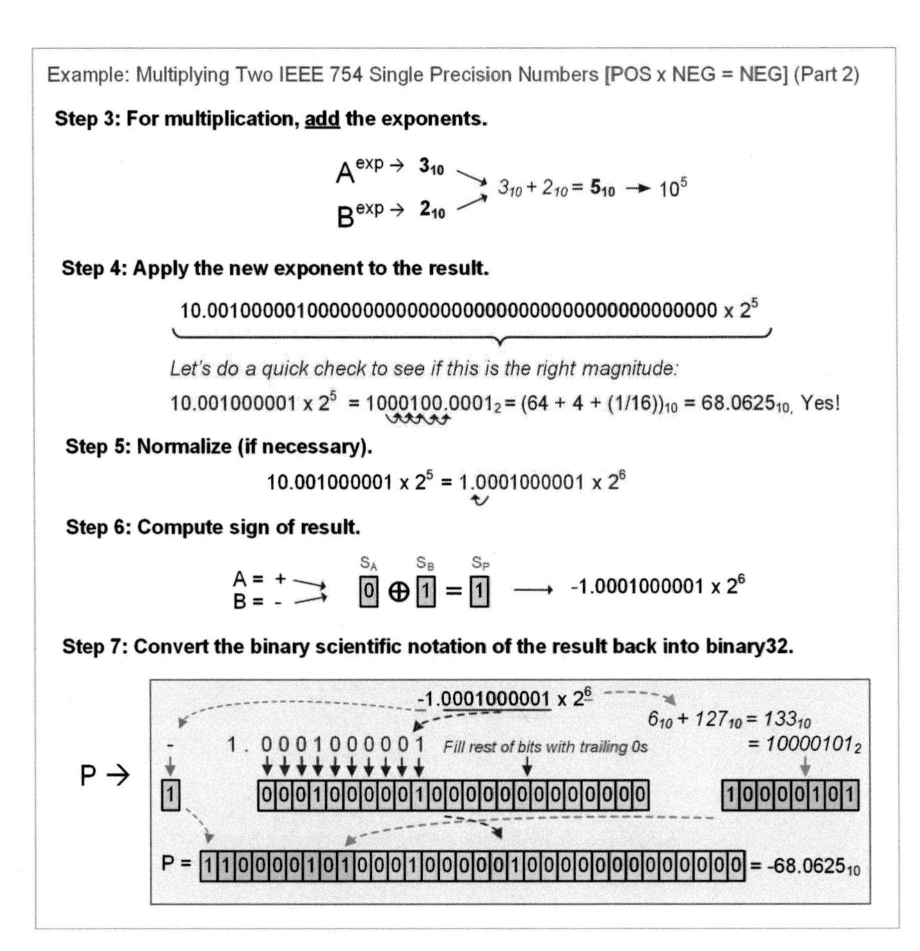

Example 13.14
Multiplying two IEEE 754 single-precision numbers [POS × NEG = NEG] (Part 2)

Example 13.15 shows steps 1 through 2 of dividing binary32 numbers.

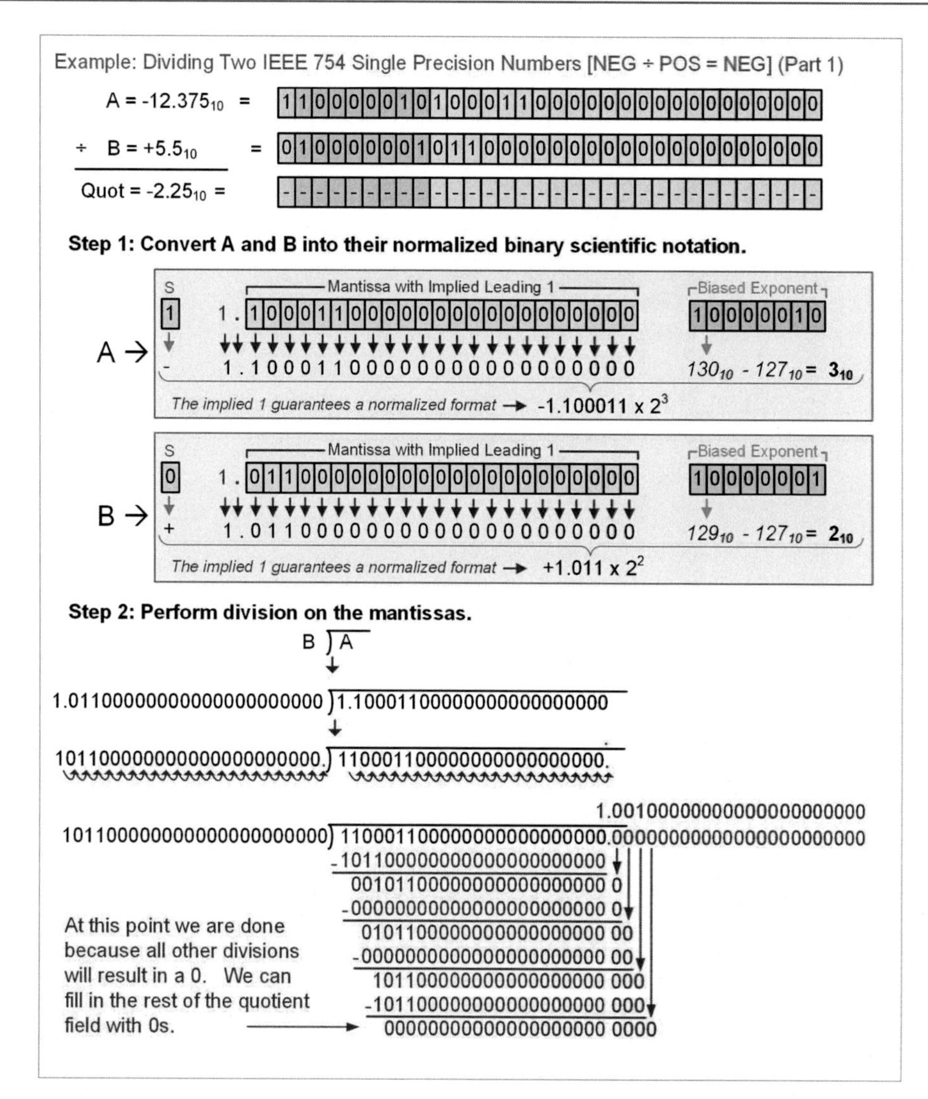

Example 13.15
Dividing two IEEE 754 single-precision numbers [NEG ÷ POS = NEG] (Part 1)

Example 13.16 shows steps 3 through 6 of dividing binary32 numbers.

Example: Dividing Two IEEE 754 Single Precision Numbers [NEG ÷ POS = NEG] (Part 2)

Step 3: For division, <u>subtract</u> the exponents.

$A^{exp} \rightarrow \mathbf{3_{10}}$

$B^{exp} \rightarrow \mathbf{2_{10}}$

$3_{10} - 2_{10} = \mathbf{1_{10}} \rightarrow 10^1$

Step 4: Apply the new exponent to the result.

$1.00100000000000000000000 \times 2^1$

Let's do a quick check to see if this is the right magnitude:

$1.001 \times 2^1 = 10.01_2 = (2 + (1/4))_{10} = 2.25_{10}$, Yes!

Step 5: Normalize (if necessary).

1.001×2^1 ← *The result is already in a normalized format so no action is needed.*

Step 6: Compute sign of result.

A = -, B = +

S_A [1] ⊕ S_B [0] = S_P [1] → -1.001×2^1

Step 7: Convert the binary scientific notation of the result back into binary32.

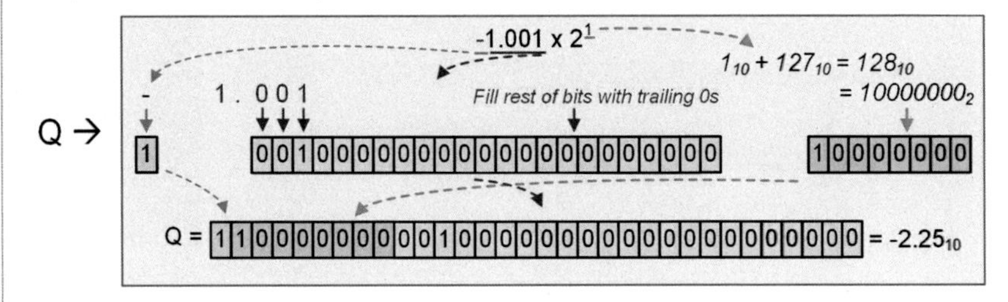

Example 13.16
Dividing two IEEE 754 single-precision numbers [NEG ÷ POS = NEG] (Part 2)

CONCEPT CHECK

CC13.3 The process for performing floating-point arithmetic seems much more complicated than for integer and fixed-point numbers. Is this why not all computers implement floating-point operations in hardware?

A) Yes

13.4 Floating-Point Modeling in VHDL

13.4.1 Floating-Point Packages in the IEEE Library

VHDL contains a set of packages that support floating-point data types, functions, and procedures. These packages leverage other IEEE packages that have been covered, such as IEEE.*std_logic_vector* and IEEE.*numeric_std*. The first package of interest for floating-point modeling is called *IEEE.float_pkg*. This package instantiates a floating-point framework by calling the *IEEE.float_generic_pkg*, which defines the generic structure of a floating-point number along with parameters that can be used when performing arithmetic. The parameters include rounding; support for subnormal numbers; checking for 0s, NaN, and infinities; and the use of guard bits when performing operations to improve rounding precision.

The IEEE.float_generic_pkg begins by defining a general-purpose subtype called *float* that can be parameterized to any number of bits along with how many are allocated for the exponent and mantissa. It then defines three additional subtypes to hold encoding techniques defined in the IEEE 754 standard, including **float32** (to hold IEEE 754 binary32 values), **float64** (to hold IEEE binary64 values), and **float128** (to hold IEEE binary128 values). These types are inherently of type *std_ulogic* but reassign the indices to make operating on specific fields within the vector more intuitive. Floating-point vectors are defined to have positive indices for the sign bit and exponent and negative indices for the mantissa. As an example, a float32 will have a vector definition of (8 downto −23); a float64 will have a vector definition of (11 downto −52); and a float128 will have a vector definition of (15 downto −112).

The IEEE.float_generic_pkg has conversion functions that can convert other VHDL types such as integer, real, signed, unsigned, and std_logic_vector into floating-point vector format. These functions are called **to_float()**, **to_float32()**, **to_float64()**, and **to_float128()**. Converting std_logic_vector and integer numbers into floating-point are mostly synthesizable operations and are used for designing circuitry. Converting real numbers into floating-point is useful for test benches where it is desired to drive the device under test with known real values without having to determine the stimulus vectors' floating-point binary codes. There are also conversion functions that take floating-point inputs and convert to other VHDL types. These include **to_std_logic_vector()**, **to_signed()**, **to_unsigned()**, **to_integer()**, and **to_real().** These functions can take in types of float, float32, float64, or float128.

The IEEE.float_generic_pkg also defines functions to support the IEEE 754 standard along with a variety of error-checking routines. The following arithmetic functions support floating-point arguments without any parameters. Note that the return type of *float* can be replaced with float32, float64, or float128. The return type will match the type of the input arguments.

Function name	Return type	Description
abs	float	Absolute value
-	float	Unary minus

The following functions will use the default parameters of the package implicitly. These defaults are defined when the floating-point framework is created using IEEE.float_generic_pkg. These defaults are round to the nearest; support for subnormal numbers; error-check for NaN, zeros, and infinity; use three guard bits; and turn on warnings.

Function name	Return type	Description
+	float	Addition
-	float	Subtraction
*****	float	Multiplication
/	float	Division
rem	float	Remainder
mod	float	Modulo

The following functions provide the ability to explicitly alter the default parameters. The parameter for the rounding approach to use (*round_style*) can take on enumerated values of *round_nearest* (default), *round_inf*, *round_neginf*, and *round_zero* corresponding to the techniques described in Sect. 13.1.7. The subtype names are defined in another package that must be included if specifying a non-default rounding type called *IEEE.fixed_float_types*. The parameter to dictate whether there is support for subnormal numbers (*denormalize*) takes a BOOLEAN value of either *TRUE* (default) or *FALSE*. The parameter that dictates whether to perform error-checking on special values (*check_error*) takes a BOOLEAN value of either *TRUE* (default) or *FALSE*. The parameter for the number of guard bits to use (*guard*) can take on natural values with a default of 3. Finally, the parameter that suppresses compilation and simulation warnings (*no_warning*) takes a BOOLEAN value of either *TRUE* or *FALSE* (default).

Function name	Return type	Description
add(l, r, <params>,...)	float	Addition with optional parameters
subtract(l, r, <params>,...)	float	Subtraction with optional parameters
multiply(l, r, <params>,...)	float	Multiplication with optional parameters
divide(l, r, <params>,...)	float	Division with optional parameters
remainder(l, r, <params>,...)	float	Remainder with optional parameters
modulo(l, r, <params>,...)	float	Modulo with optional parameters
reciprocal(arg, <params>,...)	float	Reciprocal with optional parameters
dividebyp2(l, r, <params>,...)	float	Divide by power of two with optional parameters
mac(l, r, c, <params>,...)	float	Multiply and accumulate (l*r + c) with optional parameters
sqrt(arg, <params>,...)	float	Square root with optional parameters

The IEEE.float_generic_pkg supports the following compare functions for floating-point numbers. There are compare functions that do not use parameters, while others support the parameters for defining *check_error* and *denormalize*:

Function name	Return type	Description
=	BOOLEAN	Equality
/=	BOOLEAN	Inequality
<	BOOLEAN	Less than
>	BOOLEAN	Greater than
<=	BOOLEAN	Less than or equal
>=	BOOLEAN	Greater than or equal
eq(l, r, <params>,...)	BOOLEAN	Equality with optional parameters
ne(l, r, <params>,...)	BOOLEAN	Inequality with optional parameters
lt(l, r, <params>,...)	BOOLEAN	Less than with optional parameters
gt(l, r, <params>,...)	BOOLEAN	Greater than with optional parameters
le(l, r, <params>,...)	BOOLEAN	Less than or equal with optional parameters
ge(l, r, <params>,...)	BOOLEAN	Greater than or equal with optional parameters

The IEEE.float_generic_pkg supports the following logic functions for floating-point numbers with no parameters:

Function name	Return type	Description
not	float	Complement
and	float	AND
or	float	OR
nand	float	NAND
nor	float	NOR
xor	float	XOR
xnor	float	XNOR

Two more useful functions in the IEEE.float_generic_pkg are **valid_ftpstate()** and **normalize()**. The valid_fpstate() will check to see if the vector contains a special value or a valid floating-point number. The normalize() function takes the exponent and mantissa, converts them into a floating-point number, performs the needed shifting and rounding to put into a normalized binary SN, and then converts back into a floating-point binary code.

The following examples show the VHDL modeling of a floating-point arithmetic circuit that performs addition, subtraction, multiplication, and division using the IEEE.float_generic_pkg functions +, −, *, and /. Note that these functions do not support user-provided parameters but instead use the default parameters defined during the IEEE.float_generic_pkg creation. Example 13.17 shows the simulation

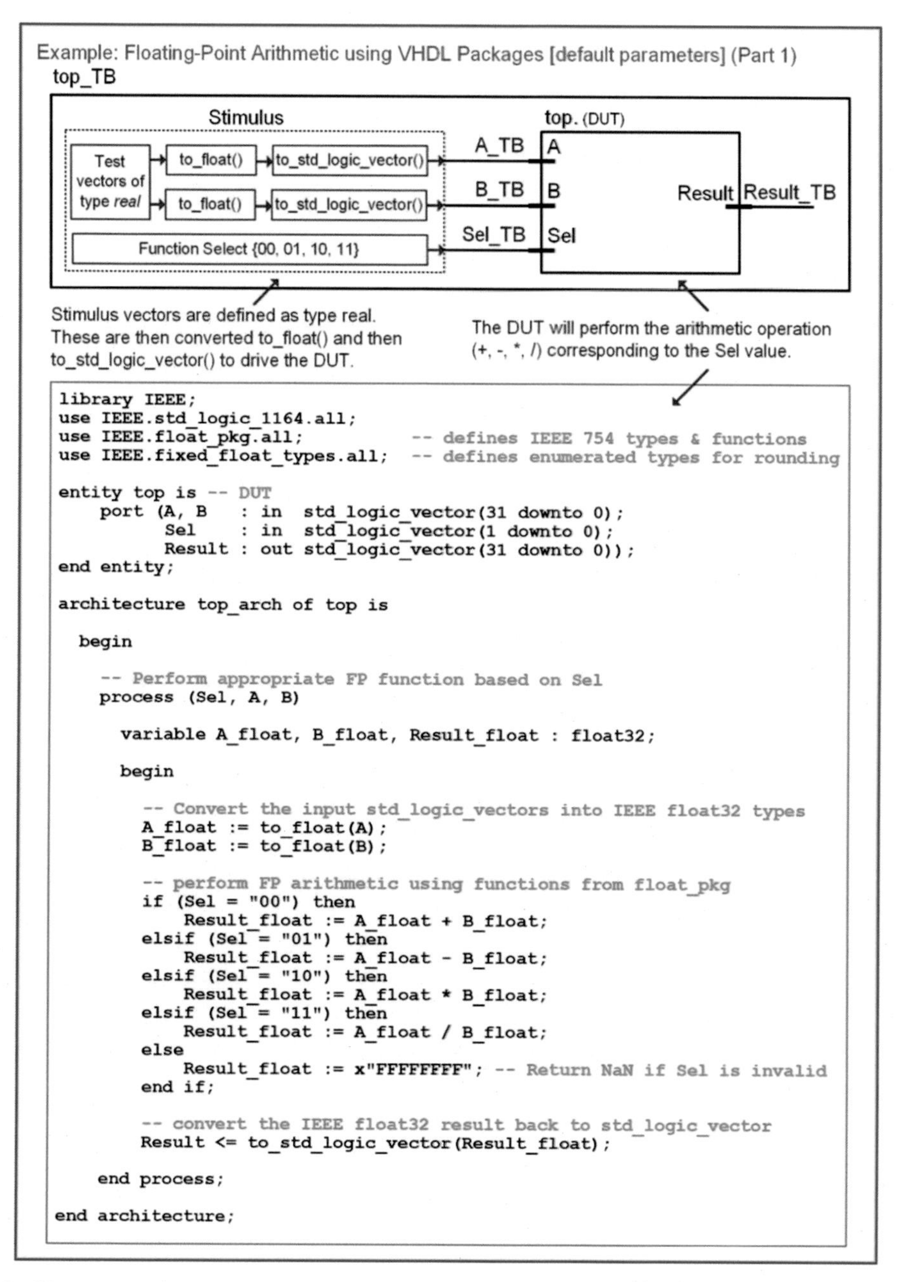


```
library IEEE;
use IEEE.std_logic_1164.all;
use IEEE.float_pkg.all;           -- defines IEEE 754 types & functions
use IEEE.fixed_float_types.all;   -- defines enumerated types for rounding

entity top is -- DUT
    port (A, B   : in  std_logic_vector(31 downto 0);
          Sel    : in  std_logic_vector(1 downto 0);
          Result : out std_logic_vector(31 downto 0));
end entity;

architecture top_arch of top is

  begin

    -- Perform appropriate FP function based on Sel
    process (Sel, A, B)

      variable A_float, B_float, Result_float : float32;

      begin

        -- Convert the input std_logic_vectors into IEEE float32 types
        A_float := to_float(A);
        B_float := to_float(B);

        -- perform FP arithmetic using functions from float_pkg
        if (Sel = "00") then
            Result_float := A_float + B_float;
        elsif (Sel = "01") then
            Result_float := A_float - B_float;
        elsif (Sel = "10") then
            Result_float := A_float * B_float;
        elsif (Sel = "11") then
            Result_float := A_float / B_float;
        else
            Result_float := x"FFFFFFFF"; -- Return NaN if Sel is invalid
        end if;

        -- convert the IEEE float32 result back to std_logic_vector
        Result <= to_std_logic_vector(Result_float);

    end process;

end architecture;
```

Example 13.17
Floating-point arithmetic using VHDL packages [default parameters] (Part 1)

setup and the VHDL for the circuit that will perform the arithmetic operations. This circuit takes in two, 32-bit vectors of type std_logic_vector (A and B) and produces a 32-bit output vector (Result) also of type std_logic_vector. The circuit treats the 32-bit I/O vectors as IEEE 754 binary32 numbers. The circuit takes in a 2-bit select vector (Sel) that dictates which arithmetic operation to perform (+, −, *, or /). The arithmetic is performed within a process. The input std_logic_vectors A and B are first converted to float32 using a variable assignment. The arithmetic operation is then performed on the float32 variables. At the end, the result is converted from float32 back to std_logic_vector.

Example 13.18 gives the test bench for the floating-point arithmetic example started in Example 13.17. This test bench stimulates the DUT with a variety of input values. The input values are inserted into the test bench as real numbers (i.e., +20.5, +35.75, −35.75, etc.). In order to convert these real

Example: Floating-Point Arithmetic using VHDL Packages [default parameters] (Part 2)

```
library IEEE;
use IEEE.std_logic_1164.all;
use IEEE.float_pkg.all;          -- defines IEEE 754 types & functions

entity top_TB is -- Test Bench
end entity;

architecture top_TB_arch of top_TB is

  signal A_TB, B_TB : std_logic_vector(31 downto 0);
  signal Sel_TB     : std_logic_vector(1 downto 0);
  signal Result_TB  : std_logic_vector(31 downto 0);

  component top is
    port (A, B   : in  std_logic_vector(31 downto 0);
          Sel    : in  std_logic_vector(1 downto 0);
          Result : out std_logic_vector(31 downto 0));
  end component;

  begin

     DUT : top port map (A_TB, B_TB, Sel_TB, Result_TB);

     STIM : process
       begin

         -- Add Testing
         Sel_TB <= "00"; -- ADD
         A_TB   <= to_std_logic_vector(to_float(+20.5));
         B_TB   <= to_std_logic_vector(to_float(+35.75));
         wait for 50 ps;

         A_TB   <= to_std_logic_vector(to_float(-35.75));
         B_TB   <= to_std_logic_vector(to_float(+20.5));
         wait for 50 ps;

         -- Sub Testing
         Sel_TB <= "01"; -- SUB
         A_TB   <= to_std_logic_vector(to_float(+35.75));
         B_TB   <= to_std_logic_vector(to_float(+22.5));
         wait for 50 ps;

         -- Mul Testing
         Sel_TB <= "10"; -- MUL
         A_TB   <= to_std_logic_vector(to_float(+12.375));
         B_TB   <= to_std_logic_vector(to_float(-5.5));
         wait for 50 ps;

         -- DIV Testing
         Sel_TB <= "11"; -- DIV
         A_TB   <= to_std_logic_vector(to_float(-12.375));
         B_TB   <= to_std_logic_vector(to_float(+5.5));
         wait for 50 ps;

     end process;

end architecture;
```

Example 13.18
Floating-point arithmetic using VHDL packages [default parameters] (Part 2)

values into std_logic_vector to drive the DUT, they are first encoded in IEEE 754 single-precision format using the to_float() function. The resulting float32 value is then converted into a 32-bit std_logic_vector using the to_std_logic_vector() function.

Example 13.19 gives the simulation waveforms for the floating-point arithmetic example started in Example 13.19. The simulation waveform gives the ability to show the inputs and outputs in different formats. The input and output vectors are shown in the *float32* radix to show their values in decimal. Copies of select portions of the inputs and output vectors are also shown in the waveforms in binary format. These portions are the sign, exponent, and mantissa. These binary formats allow verification that the output does indeed get calculated correctly using the IEEE 754 arithmetic algorithms.

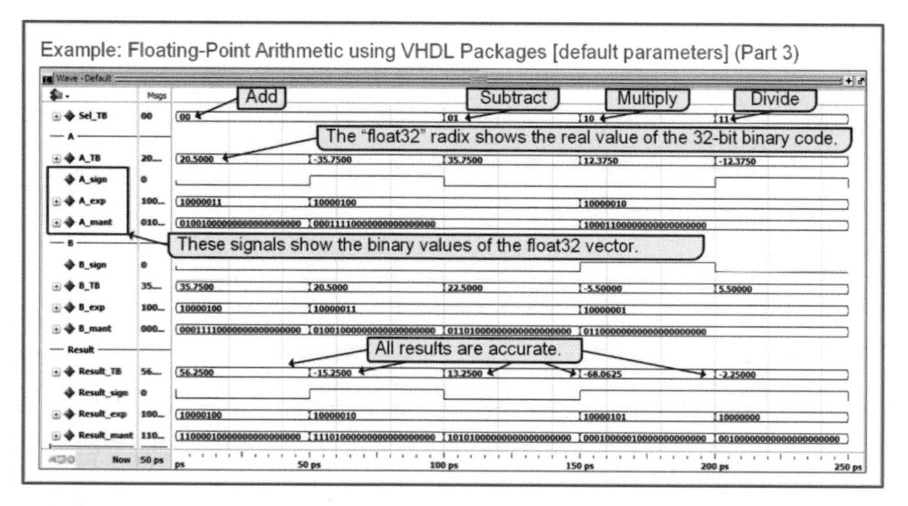

Example 13.19
Floating-point arithmetic using VHDL packages [default parameters] (Part 3)

Example 13.20 shows an example of performing the same arithmetic operations as shown in Example 13.17 but using the add(), subtract(), multiply(), and divide() functions from the IEEE. float_generic_pkg package. These functions allow the user to explicitly define parameters for the operation. In this example, the rounding technique chosen is *round to zero*, which takes the least amount of hardware when implemented. These functions are also configured to not use guard bits and to not support subnormal numbers. These choices further reduce the complexity of the hardware implementation if synthesized. The function is configured to support special value error-checking.

Example: Floating-Point Arithmetic using VHDL Packages [with parameters] (Part 1)

```
library IEEE;
use IEEE.std_logic_1164.all;
use IEEE.float_pkg.all;              -- defines IEEE 754 types & functions
use IEEE.fixed_float_types.all;      -- defines enumerated types for rounding

entity top is -- DUT
    port (A, B   : in  std_logic_vector(31 downto 0);
          Sel    : in  std_logic_vector(1 downto 0);
          Result : out std_logic_vector(31 downto 0));
end entity;

architecture top_arch of top is

  begin

    -- Perform appropriate FP function based on Sel
    process (Sel, A, B)

      variable A_float, B_float, Result_float : float32;

      begin

        -- Convert the input std_logic_vectors into IEEE float32 types
        A_float := to_float(A);
        B_float := to_float(B);

        if (Sel = "00") then
          Result_float := add (l => A_float, r => B_float,
                               round_style => round_zero,
                               guard       => 0,
                               check_error => true,
                               denormalize => false);
        elsif (Sel = "01") then
          Result_float := subtract (l => A_float, r => B_float,
                               round_style => round_zero,
                               guard       => 0,
                               check_error => true,
                               denormalize => false);
        elsif (Sel = "10") then
          Result_float := multiply (l => A_float, r => B_float,
                               round_style => round_zero,
                               guard       => 0,
                               check_error => true,
                               denormalize => false);
        elsif (Sel = "11") then
          Result_float := divide (l => A_float, r => B_float,
                               round_style => round_zero,
                               guard       => 0,
                               check_error => true,
                               denormalize => false);
        else
          Result_float := x"FFFFFFFF"; -- NaN

        end if;

        -- convert the IEEE float32 result back to std_logic_vector
        Result <= to_std_logic_vector(Result_float);

    end process;

end architecture;
```

Example 13.20
Floating-point arithmetic using VHDL packages [with parameters] (Part 1)

Example 13.21 shows the simulation waveforms for the VHDL model given in Example 13.20. The same test bench from Example 13.18 is used to stimulate the DUT. These waveforms prove that the parameterized functions yield the same results as when using the default parameter functions in Example 13.17.

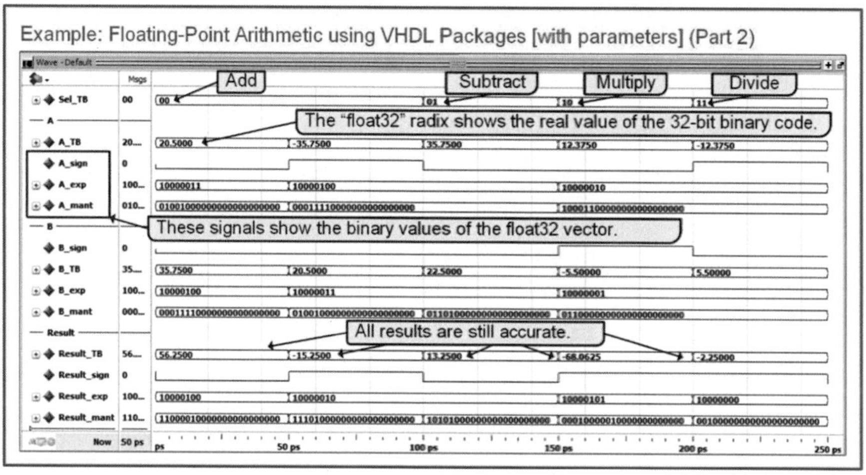

Example: Floating-Point Arithmetic using VHDL Packages [with parameters] (Part 2)

Example 13.21
Floating-point arithmetic using VHDL packages [with parameters] (Part 2)

13.4.2 The IEEE_Proposed Library

The VHDL models covered in Sect. 13.4.1 that use the IEEE.float_pkg package are widely supported in functional simulation tools but are often unsynthesizable. As a result, designers are left to create custom floating-point circuitry to match the results of the functional simulations given in Examples 13.19 and 13.21. In 2008, an *IEEE_Proposed* library was preliminarily released that provided floating-point packages with a far greater level of implementation details for the floating-point arithmetic functions. This greater level of detail, along with the use of only synthesizable HDL constructs, led to a broader level of synthesis support for floating-point models. The IEEE_Proposed library needs to be installed as part of the synthesis tools. If it is, then the library and package declarations for the VHDL shown in Examples 13.17 and 13.20 can be changed to the following to create a fully synthesizable floating-point model:

```
library IEEE;
use IEEE.std_logic_1164.all;

library IEEE_Proposed;
use IEEE_Proposed.fixed_float_types.all;
use IEEE_Proposed.fixed_pkg.all;
use IEEE_Proposed.float_pkg.all;
```

The VHDL source code for the floating-point arithmetic system given in Examples 13.17 and 13.20 was altered to have the above library and package declaration and then synthesized for an AMD *Artix-7 100T* FPGA using the *Vivado 2023.1* integrated design environment. The utilization reports for each modeling approach are shown in Example 13.22. This illustrates the hardware resource impact of reducing the complexity of the floating-point functions through user-provided parameterization. This also verifies that the floating-point VHDL models shown are capable of being implemented in hardware using the IEEE_Proposed library.

Example: Resource Utilization for Floating-Point Arithmetic Synthesis (default vs parametrized)

The following are the hardware resource utilizations for implementing floating-point addition, subtraction, multiplication, and division operations on an AMD Artix-7 100T FPGA. Different implementation parameters are used within the *IEEE_proposed.float_pkg* to view the impact on resource usage.

	VHDL Implementation	
FPGA Resources (XC7A100T)	**VHDL Syntax: +, -, *, /** (Rounding = round_nearest, Guard Bits = 3, Error Checking = TRUE, Subnormal Numbers = TRUE)	**VHDL Syntax: add(), subtract(), multiply(), divide()** (Rounding = round_zero, Guard Bits = 0, Error Checking = TRUE, Subnormal Numbers = FALSE)
Slices	998 (6.3%)[1]	632 (4%)[1]
LUT as Logic	3553 (5.6%)[1]	2220 (3.5%)[1]
DSP Block	2 (0.83%)[1]	2 (0.83%)[1]

[1] The % is the amount of total FPGA resources used on the AMD Artix-7 100T device.

Example 13.22
Floating-point arithmetic synthesis resource utilization (default vs. user-provided parameters)

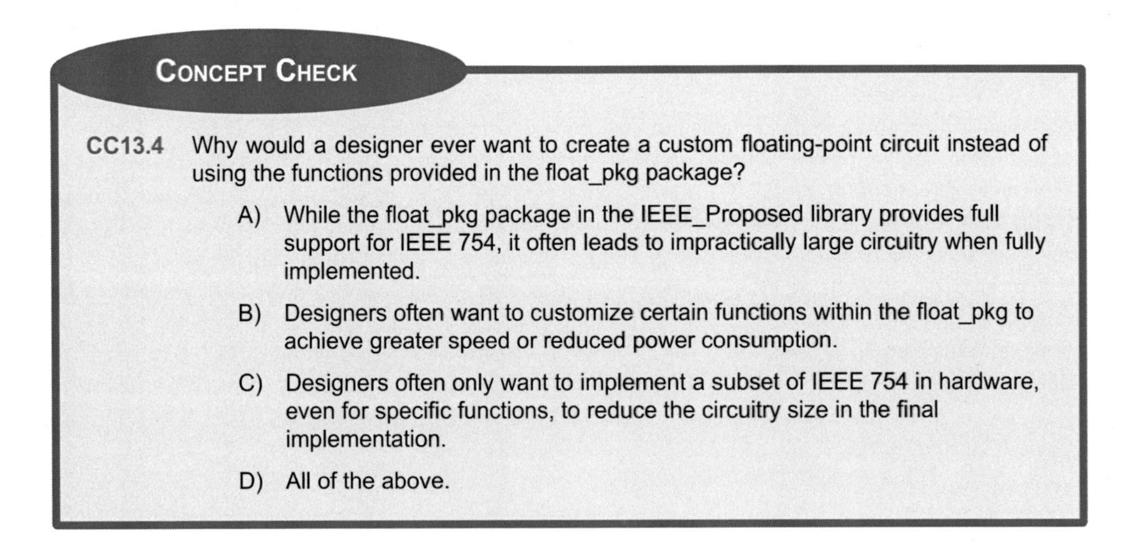

CONCEPT CHECK

CC13.4 Why would a designer ever want to create a custom floating-point circuit instead of using the functions provided in the float_pkg package?

A) While the float_pkg package in the IEEE_Proposed library provides full support for IEEE 754, it often leads to impractically large circuitry when fully implemented.

B) Designers often want to customize certain functions within the float_pkg to achieve greater speed or reduced power consumption.

C) Designers often only want to implement a subset of IEEE 754 in hardware, even for specific functions, to reduce the circuitry size in the final implementation.

D) All of the above.

Summary

- Floating-point number encoding provides a way to represent very large and very small numbers by expressing the value in binary scientific notation with the ability to move the radix point.
- IEEE 754 is a standard that defines the encoding for specific floating-point types, rounding techniques, and functions that should be supported.
- An IEEE 754 floating-point number consists of a sign bit, an exponent field, and a mantissa field.
- In IEEE 754, a sign bit of 0 indicates a positive number, and a sign bit of 1 indicates a negative number. This is a signed magnitude approach to represent the polarity of the number.

- IEEE 754 uses a biased exponent, which shifts the exponent's center point from 0 to a predetermined bias. This allows the exponent to always be positive instead of having both positive and negative values.
- IEEE 754 uses an implied 1 when storing the mantissa. Since in normalized binary scientific form there is always a single 1 to the left of the radix point, this 1 is not stored in the mantissa field but is rather *implied* in the number. This gives one more bit for precision.
- The exponent of an IEEE floating-point number dictates its range.
- The mantissa of an IEEE floating-point number dictates its precision.
- A single-precision IEEE 754 number uses 32 bits that consist of 1 sign bit, 8 bits of exponent, and 23 bits of mantissa.
- A single-precision IEEE 754 number has a range of $+/-3.4028235 \times 10^{38}{}_{10}$ and a precision of ~7 significant digits.
- A double-precision IEEE 754 number uses 64 bits that consist of 1 sign bit, 11 bits of exponent, and 52 bits of mantissa.
- A double-precision IEEE 754 number has a range of $+/-1.7976931348623158 \times 10^{308}{}_{10}$ and a precision of ~16 significant digits.
- The steps to convert from a decimal number into IEEE 754 floating-point format include the following: (1) convert the decimal number into a fixed-point binary representation; (2) convert the fixed-point number into normalized binary scientific notation; (3) from the binary SN, determine the sign bit; (4) from the binary SN, determine the biased exponent; (5) from the binary SN, determine the mantissa with implied leading 1; and (6) combine the three fields from steps 3–5 into the final 32-bit binary number.
- The steps to convert from an IEEE 754 floating-point format into decimal include the following: (1) reassemble the original mantissa by adding back in the implied 1; (2) determine the decimal value of the original exponent from the biased exponent; (3) determine whether the number is positive or negative from sign bit; (4) assemble the extracted information from steps 1–3 into binary scientific notation; (5) shift the radix point in the binary SN per the exponent value to get back into fixed-point binary; and (6) convert the fixed-point binary number to decimal.
- The steps to perform addition and subtraction on IEEE 754 numbers follow a similar algorithm as when doing these operations on decimal scientific notation. The numbers are reformatted until the exponents are the same. Then the operations are performed on the mantissas. Then the common exponent is applied to the result, and the final value is normalized.
- The steps to perform multiplication and division on IEEE 754 numbers follow a similar algorithm as when doing these operations on decimal scientific notation. The operations are performed on the mantissas. For multiplication, the exponents are added. For division, the exponents are subtracted. The new exponent is then applied to the resulting mantissa, and the final value is normalized.
- The standard VHDL package supports operations on the type *real*, but these are not widely synthesizable.
- The IEEE.float_pkg package provides support for IEEE 754 numbers including types (float, float32, float64, and float128), conversions, and arithmetic functions.
- The IEEE.float_pkg provides algorithms for rounding; the ability to selectively support subnormal numbers; the ability to use guard bits on certain operations to maintain precision; the ability to selectively check for special values; and the ability to report floating-point specific warnings.
- The released version of IEEE.float_pkg is mostly unsynthesizable by modern design tools. As a result, the IEEE_Proposed library was created.
- The IEEE_Proposed.float_pkg provides greater detail for floating-point arithmetic functions along with using only synthesizable constructs. This allows the IEEE_Proposed.float_pkg functions to be more readily synthesizable.
- When synthesizing floating-point arithmetic circuits, the parameters for rounding, error-checking, guard bits, and subnormal number support greatly impact the hardware resource utilization needed for the implementation. The less features used in the VHDL model results in a lower resource utilization.

Exercise Problems

Section 13.1: Overview of Floating-Point Numbers

13.1.1 How many bits are used to encode an IEEE 754 single-precision number?

13.1.2 How many bits does an IEEE 754 single-precision number use to hold the exponent?

13.1.3 How many bits does an IEEE 754 single-precision number use to hold the mantissa?

13.1.4 What is the smallest normal, positive number that an IEEE 754 single-precision number can hold?

13.1.5 What is the largest normal, positive number that an IEEE 754 single-precision number can hold?

13.1.6 How many bits are used to encode an IEEE 754 double-precision number?

13.1.7 How many bits does an IEEE 754 double-precision number use to hold the exponent?

13.1.8 How many bits does an IEEE 754 double-precision number use to hold the mantissa?

13.1.9 What is the smallest normal, positive number that an IEEE 754 double-precision number can hold?

13.1.10 What is the largest normal, positive number that an IEEE 754 double-precision number can hold?

Section 13.2: IEEE 754 Base Conversions

13.2.1 Convert 10.5_{10} into its IEEE 754 single-precision binary representation.

13.2.2 Convert -55.25_{10} into its IEEE 754 single-precision binary representation.

13.2.3 Convert $7.75 \times 10^{-5}{}_{10}$ into its IEEE 754 single-precision binary representation.

13.2.4 Convert 10.5_{10} into its IEEE 754 double-precision binary representation.

13.2.5 Convert -55.25_{10} into its IEEE 754 double-precision binary representation.

13.2.6 Convert the IEEE 754 single-precision code $42B1C000_{16}$ into decimal.

13.2.7 Convert the IEEE 754 single-precision code $44E90000_{16}$ into decimal.

13.2.8 Convert the IEEE 754 single-precision code $44ECA000_{16}$ into decimal.

13.2.9 Convert the IEEE 754 single-precision code $4382A168_{16}$ into decimal.

13.2.10 Convert the IEEE 754 single-precision code $43B7AEF9_{16}$ into decimal.

Section 13.3: Floating-Point Arithmetic

13.3.1 Calculate the sum of $101_{10} + 261_{10}$ by hand using IEEE 754 single-precision encoding. Show your work.

13.3.2 Calculate the sum of $261_{10} + 367_{10}$ by hand using IEEE 754 single-precision encoding. Show your work.

13.3.3 Calculate the sum of $(-367_{10}) + 101_{10}$ by hand using IEEE 754 single-precision encoding. Show your work.

13.3.4 Calculate the difference of $367_{10} - 261_{10}$ by hand using IEEE 754 single-precision encoding. Show your work.

13.3.5 Calculate the difference of $261_{10} - 101_{10}$ by hand using IEEE 754 single-precision encoding. Show your work.

13.3.6 Calculate the difference of $(-367_{10}) - 261_{10}$ by hand using IEEE 754 single-precision encoding. Show your work.

13.3.7 Calculate the product of $101_{10} \times 101_{10}$ by hand using IEEE 754 single-precision encoding. Show your work.

13.3.8 Calculate the product of $(-261_{10}) \times 367_{10}$ by hand using IEEE 754 single-precision encoding. Show your work.

13.3.9 Calculate the quotient of $367_{10} \div 261_{10}$ by hand using IEEE 754 single-precision encoding. Show your work.

13.3.10 Calculate the quotient of $(-367_{10}) \div 101_{10}$ by hand using IEEE 754 single-precision encoding. Show your work.

Section 13.4: Floating-Point Modeling in VHDL

13.4.1 Design and simulate a VHDL model for an IEEE 754 single-precision *remainder* operation with the default parameters defined in the IEEE.float_pkg package. Use the **rem** function.

13.4.2 Design and simulate a VHDL model for an IEEE 754 single-precision *remainder* operation with default parameters defined in the IEEE.float_pkg package for all settings except rounding. Set the rounding type to *round*_zero. Use the **remainder()** function.

13.4.3 Design and simulate a VHDL model for an IEEE 754 single-precision *modulo* operation with the default parameters defined in the IEEE.float_pkg package. Use the **mod** function.

13.4.4 Design and simulate a VHDL model for an IEEE 754 single-precision *modulo* operation with default parameters defined in the IEEE.float_pkg package for all settings except rounding. Set the rounding type to *round*_zero. Use the **modulo()** function.

13.4.5 Design and simulate a VHDL model for an IEEE 754 single-precision *square root* operation with default parameters defined in the IEEE.float_pkg package for all settings except error-checking. Set the error-checking to FALSE. Use the **sqrt()** function.

Appendix A: List of Worked Examples

B. J. LaMeres, *Quick Start Guide to VHDL*, https://doi.org/10.1007/978-3-031-42543-1

Index

B. J. LaMeres, *Quick Start Guide to VHDL*, https://doi.org/10.1007/978-3-031-42543-1